Student Solutions Manual
to accompany

Advanced Engineering Mathematics

Eighth Edition

Herbert Kreyszig

Erwin Kreyszig
Professor of Mathematics
Ohio State University
Columbus, Ohio

John Wiley & Sons, Inc.
New York • Chichester • Weinheim • Brisbane • Singapore • Toronto

Further Material:

E. Kreyszig and E. J. Norminton,
MAPLE Computer Manual

E. Kreyszig and E.J. Norminton,
MATHEMATICA Computer Manual

Please contact your local Wiley representative,
or write to:

John Wiley & Sons, Inc.
Attn: Mathematics Marketing Manager, College Division
605 Third Avenue
New York, NY 10158-0012
FAX us at (212) 850-6118
or Email us at math@wiley.com

ISBN 0-471-33375-1

Printed in the United States of America

10 9 8 7 6 5 4 3 2 1

Printed and bound by Bradford & Bigelow, Inc.

PREFACE

ADVANCED ENGINEERING MATHEMATICS (AEM), 8th Edition (New York: J. Wiley and Sons, Inc., 1999) introduces students of engineering, physics, mathematics, and computer science to those areas of mathematics which from a modern viewpoint are most important in connection with practical problems. The book consists of the following independent parts;

A Ordinary Differential Equations (Chaps. 1-5)
B Linear Algebra, Vector Calculus (Chaps. 6-9)
C Fourier Analysis, Partial Differential Equations (Chaps. 10, 11)
D Complex Analysis (Chaps. 12-16)
E Numerical Methods (Chaps. 17-19)
F Optimization, Graphs (Chaps. 20, 21)
G Probability, Statistics (Chaps. 22, 23)

AEM includes a problem set after each section of every chapter. These problem sets consist of about 4500 problems and team projects.

This **Student Solutions Manual** will enhance the effectiveness of AEM. The manual contains worked-out solutions and helpful suggestions to carefully selected odd-numbered problems (to which Appendix 2 of AEM gives only the final answers). The solutions discussed in this Manual illustrate the basic ideas, techniques, and applications of the material in the text. This will provide a representative cross-section of the main topics in the book as a learning environment for the student.

The Manual may give help in working assignments, in clarifying conceptual and technical difficulties, and in developing skills. It can also be used in reviewing the course material, in studying for exams, and for self-study.

The following two facts are important to fully understanding the role and character of this Manual;

1. Whereas the book AEM itself gives only answers to (odd-numbered) problems (in Appendix 2), *this Manual provides complete solutions* with all the conceptual and technical details, considered from a practical point of view.

2. In addition to those problem sets, AEM contains numerous examples in the text. It is true that these examples include worked-out solutions, but we wish to emphasize very distinctly that *the presentation in this Manual is much more detailed and leisurely and of a substantially lower level than the level of those examples in AEM*. Thus, worked-out examples in the book and worked-out problems in this Manual belong to two different categories of objects designed to enhance understanding, intuition, and skill in two different ways and on two different levels.

For best results in using this Manual follow these rules.

RULE 1. First try to solve your problem without help. A solution obtained completely by yourself will be much more valuable to you than one obtained with outside help. The more you have to struggle, the greater will be your gain in knowledge, skill, and self-confidence. It is of minor importance whether your way of solution is shortest or most elegant; essential is that you obtain a solution at all.

RULE 2. Look at worked-out examples in the text. Find out whether one of them is similar to your problem or, at least, one or another idea in an example might be useful in solving your problem.

RULE 3. Use this Manual stepwise. That is, if you reach an impasse, find out from the Manual the next step of the solution, without looking at the further steps. This act of self-discipline is part of a process of maturing, involving great rewards for you.

RULE 4. Analyze your work critically. Find reasons for the difficulties in solving the problem, which you wish to overcome, such as lack of understanding of word problems in general, difficulties in grasping the meaning of a concept or a theorem needed, insufficient insight into the general idea on which a solution method is based, deficiencies in differentiation or integration, lack of technical skill in algebraic manipulations, and so on.

RULE 5. Find a better method. Try to design a solution method that is shorter than that given in this Manual, more elegant, logically more appealing to you, or more adequate from the viewpoint of applications.

Organize your worksheets as clearly and legibly as you can. Use letter size paper, don't use little scraps of paper. You will be surprised to see how improvement in form will entail improvement in content.

Best wishes for success and fun in using this Manual.

Acknowledgement: The authors wish to thank Professor E. J. Norminton for various valuable suggestions as well as for his great help in the design of this book.

Herbert Kreyszig
and
Erwin Kreyszig

CONTENTS

PART A.   ORDINARY DIFFERENTIAL EQUATIONS

CHAPTER 1.   First-Order Differential Equations

Sec. 1.1   Basic Concepts and Ideas

Comment on (5). $y = cx - c^2$, hence $y' = c$, and $y'^2 - xy' + y = c^2 - xc + (cx - c^2) = 0$.

Problem Set 1.1. Page 8

1. **Calculus.** This is a problem of calculus, namely, to integrate x^2, giving $\frac{1}{3}x^3 + c$, where the constant of integration c is arbitrary. This is essential. It means that the differential equation $y' = x^2$ has infinitely many solutions, each of these cubical parabolas corresponding to a certain value of c. Sketch some of them.

13. **Initial value problem.** $y' = -2ce^{-2x}$ by differentiation. Hence the left side becomes
$$y' + 2y = -2ce^{-2x} + 2(ce^{-2x} + 1.4) = 2.8.$$
This verifies the given solution $y = ce^{-2x} + 1.4$. For $x = 0$ you have $e^0 = 1$ and thus $y(0) = c + 1.4$, which is required to be equal to 1.0. Hence $1.0 = c + 1.4, c = -0.4$, and the answer is $y = -0.4 e^{-2x} + 1.4$.

23. **Falling body.** $s = gt^2/2 = 100$ [m]. Here $g = 9.80$ m/sec^2 since s is measured in meters. Using $s = 100$ and solving for t gives
$$t = \sqrt{\frac{100}{g/2}} = 10\sqrt{\frac{1}{4.9}} = 4.52 \text{ [sec]}.$$
The second result, 6.389 sec, is less than twice the first because the motion is accelerated, the velocity increases.

Sec. 1.2   Geometrical Meaning of $y' = f(x,y)$. Direction Fields

Problem Set 1.2. Page 12

1. **Calculus.** Note that the solution curves are *not* congruent because c is a factor, not an additive constant (as, for instance, in Prob. 5).

5. **Verification of solution.** Geometrically, the solution curves are obtained from each other by translations in the y-direction; they are congruent because c is an additive constant.

7. **Verification of solution.** At each point (x, y) the tangent direction of the solution is $-x/y$, hence perpendicular to the slope y/x of the ray from $(0, 0)$ to (x, y), suggesting that the solutions are concentric circles about the origin. You can prove this by calculus, as follows. Multiply the equation by y, obtaining $yy' = -x$. Then integrate on both sides with respect to x. This gives
$$\frac{1}{2}y^2 = -\frac{1}{2}x^2 + c \quad \text{or} \quad y^2 + x^2 = 2c.$$

15. **Initial value problem.** The idea from calculus just applied in Prob. 7 here gives $(9/2)y^2 + 2x^2 = c$ or $4x^2 + 9y^2 = 2c$; these are the ellipses $x^2/9 + y^2/4 = c/18$.

17. Initial value problem. In this section the usual notation is (2), that is, $y' = f(x, y)$, and the direction field lies in the xy-plane. In Prob. 17 the equation is $v' = f(t, v) = g - bv^2/m$. Hence the direction field lies in the tv-plane. With $m = 1$ and $b = 1$ the equation becomes $v' = g - v^2$. Then $v = 3.13$ gives $g - v^2 = 9.80 - 3.13^2 = 0$, approximately. The differential equation now shows that v' must be identically zero. Conclude that $v = 3.13$ must be a solution. For $v < 3.13$ you have $v' > 0$ (increasing curves) and for $v > 3.13$ you have $v' < 0$ (decreasing curves). Note that the isoclines are the horizontal parallel straight lines $g - v^2 = const$, thus $v = const$.

Sec. 1.3   Separable Differential Equations

Problem Set 1.3. Page 18

3. General solution by separation. Dividing by the right side gives

$$\frac{y'}{1 + 0.01\,y^2} = 1 \quad \text{or} \quad \frac{dy}{1 + 0.01\,y^2} = dx. \tag{A}$$

Now integrate. This is one of the more important integrals; set $v = 0.1y$ to get $y = 10\,v$, $dy = 10\,dv$, and from (A),

$$10\,dv/(1 + v^2) = dx, \qquad \text{integrated} \qquad 10 \arctan v = x + C.$$

Recalling that $v = 0.1\,y$ gives $10 \arctan 0.1y = x + C$. This implies

$$y = 10 \tan (0.1(x + C)) \; = 10 \tan (0.1x + c), \qquad c = 0.1C.$$

15. Initial value problem. Separate variables and integrate on both sides (by parts on the right) to get

$$dy/y^2 = 2(x + 1)e^{-x}\,dx, \qquad -1/y = (-2x - 4)e^{-x} + c.$$

Multiply by -1 and take the reciprocal,

$$y = 1/[(2x + 4)e^{-x} - c].$$

From the initial condition $y(0) = 1/6$ obtain by setting $x = 0$

$$1/6 = y(0) = 1/(4 - c), \qquad \text{hence} \qquad 6 = 4 - c, \qquad c = -2.$$

Inserting this into y gives the answer.

23. Initial value problem. Dividing the given equation by x^2 and setting $y/x = u$, hence $y = xu$ and $y' = u + xu'$, gives

$$(y/x)y' = u(u + xu') = 2u^2 + 4.$$

Subtracting u^2 on both sides gives $xuu' = u^2 + 4$. Separate variables, then multiply both sides by 2, and integrate with respect to x on both sides,

$$2u\,du/(u^2 + 4) = 2\,dx/x, \qquad \ln (u^2 + 4) = \ln (x^2) + C, \qquad u^2 + 4 = cx^2.$$

Solving for u^2 and taking roots gives $y/x = u = \sqrt{cx^2 - 4}$, so that

$$y = ux = \sqrt{cx^4 - 4x^2}.$$

From this and the initial condition,

$$y(2) = 4 = \sqrt{16c - 16} = 4\sqrt{c - 1}, \qquad c - 1 = 1, \qquad c = 2.$$

This gives the answer in Appendix 2.

26. Team Project. (b) In finding a differential equation you always have to get rid of the arbitrary constant c. For $xy = c$ this is very simple because this equation is solved for c (differentiate this equation implicitly with respect to x); in other cases it is usually best to first solve algebraically for c.
(d) This orthogonality condition is usually considered in calculus. You will need it again in Sec. 1.8.

27. CAS Project. This integral (the error function, except for a constant factor; see (35) in Appendix A3.1) is important in heat conduction (see Sec. 11.6). A similar integral is basic in statistics (see Sec. 22.8).

Sec. 1.4　Modeling: Separable Equations

Problem Set 1.4. Page 23

1. Exponential growth. Let $y(0) = y_0$ be the initial amount at $t = 0$. The model equation $y' = ky$ has the solution $y = ce^{kt}$. For the given initial amount y_0 this becomes $y = y_0 e^{kt}$. For $t = 1$ (1 day) this gives $y(1) = y_0 e^k$. By assumption this is twice the initial amount (doubling in 1 day). Hence $y_0 e^k = 2y_0$. Divide this by y_0 to get $e^k = 2$. After 3 days you have $y(3) = y_0 e^{3k} = y_0 \cdot 2^3$, where we used $e^{ab} = (e^a)^b$. Similarly for 1 week ($t = 7$).

11. Sugar inversion. $y' = ky$, $y(t) = 0.01e^{kt}$ from the first condition and $y(4) = 0.01e^{4k} = 1/300 = 0.01/3$ from the second. Hence $e^{4k} = 1/3$, $k = 1/4 \ln (1/3) = -0.275$.

15. Curves (ellipses) From calculus you know that the slope of the tangent of a curve $y = y(x)$ is the derivative $y'(x)$. From the given data you thus obtain immediately the differential equation $y' = -4x/y$. Solve it by separation of variables (multiply by y),
$$y\,dy = -4\,x\,dx, \qquad y^2/2 = -2x^2 + c, \qquad y^2/4 + x^2 = c/2.$$
For instance, $c = 2$ gives the ellipse with semi-axes 1 (in the x-direction) and 2 (in the y-direction). Sketch this ellipse and some of the others.

Sec. 1.5　Exact Differential Equations. Integrating Factors

Example 3. A nonexact equation. You can write the given equation as $y' = y/x$. Separate variables, obtaining $dy/y = dx/x$, $\ln y = \ln x + \tilde{c}$, $y = cx$.

Problem Set 1.5. Page 31

17. Test for exactness. Initial value problem. Exactness is seen from
$$\frac{\partial}{\partial y} M = \frac{\partial}{\partial y} ((x+1)e^x - e^y) = -e^y,$$
$$\frac{\partial}{\partial x} N = \frac{\partial}{\partial x} (-xe^y) = -e^{-y},$$
where the minus sign in the second line results from taking the dy-term to the left in order to have the standard form of the equation. You see that the equation is exact. Integrating M with respect to x gives $u = xe^x - xe^y + k(y)$ with arbitrary $k(y)$. Differentiating this with respect to y and equating the result to N gives $-xe^y + k'(y) = -xe^y$, hence $k'(y) = 0$ and $k = const$. This shows that a general solution is $u = xe^x - xe^y = c$. Because of the initial condition set $x = 1$ and $y = 0$, obtaining $u = e - 1$. This gives the answer $u = xe^x - xe^y = e - 1$.

23. Several integrating factors. From this problem you can learn that if an equation has an integrating factor, it has many such factors, giving essentially the same (implicit) general solution. Taking $F = y$, you obtain the equation $y^2\,dx + 2xy\,dy = 0$. To check exactness, calculate $\frac{\partial}{\partial y}(y^2) = 2y$ and $\frac{\partial}{\partial x}(2xy) = 2y$, which proves exactness. Integrating y^2 with respect to x gives $xy^2 + k(y)$. Differentiating this with respect to y and equating the result to $2xy$, you obtain for $k(y)$ the condition $2xy + k'(y) = 2xy$, $k'(y) = 0$, $k(y) = const$. The solution is $xy^2 = const$.
　Choosing $F = xy^3$ as an integrating factor gives the exact equation $xy^4\,dx + 2x^2y^3\,dy = 0$. Proceeding

as before, you obtain
$$u = (1/2)x^2 y^4 + k(y), \qquad 2x^2 y^3 + k'(y) = 2x^2 y^3, \qquad u = (1/2)x^2 y^4 = C,$$
which implies $xy^2 = c$, as before.

25. Integrating factor. $P\,dx + Q\,dy = 0$ in (12) is the nonexact equation. $FP\,dx + FQ\,dy = 0$ is the exact equation obtained by multiplying with an integrating factor F. Hence $FP = M$ and $FQ = N$ play the role of M and N in an exact equation. Accordingly, the exactness condition is $\partial(FP)/\partial y = \partial(FQ)/\partial x$. In the present problem,

$$\frac{\partial}{\partial y}(FP) = \frac{\partial}{\partial y}(e^x \sin y) = e^x \cos y,$$

$$\frac{\partial}{\partial x}(FQ) = \frac{\partial}{\partial x}(e^x \cos y) = e^x \cos y,$$

which shows exactness. Integrating FP with respect to x gives $u = e^x \sin y + k(y)$. To determine $k(y)$, differentiate u with respect to y and equate the result to FQ (which now plays the role of N). This gives

$$e^x \cos y + k'(y) = e^x \cos y, \qquad k'(y) = 0, \qquad k(y) = const.$$

Hence the answer is

$$u = e^x \sin y = c = const.$$

Note that in the present case you can solve this for y; this gives

$$y = \arcsin (ce^{-x}).$$

Sec. 1.6   Linear Differential Equations. Bernoulli Equation

Example 2. The integral can be solved by integration by parts or more simply by "undetermined coefficients", that is, by setting

$$\int e^{0.05t} \cos t\, dt = e^{0.05t}(A \cos t + B \sin t)$$

and differentiating on both sides. This gives

$$e^{0.05t} \cos t = e^{0.05t}[0.05\,(A \cos t + B \sin t) - A \sin t + B \cos t].$$

Now equate the coefficients of $\sin t$ and $\cos t$ on both sides. The sine terms give $0 = 0.05\,B - A$, hence $A = 0.05B$. The cosine terms give

$$1 = 0.05A + B = 0.05^2 B + B,$$

hence $B = 1/1.0025 = 0.997506$ and $A = 0.05B = 0.049875$. Multiplying A and B by 50 (the factor that we did not carry along) gives a and b in Example 2. The integrals in Example 3 can be handled similarly.

Problem Set 1.6. Page 38.

7. General solution. Multiplying the given equation by e^{kx}, you obtain
$$(y' + ky)\,e^{kx} = (ye^{kx})' = e^{kx}e^{-kx} = 1$$
and by integration, $ye^{kx} = x + c$. Division by e^{kx} gives the solution $y = (x + c)e^{-kx}$. Note that in (4) you have the integral of $e^{kx}e^{-kx} = 1$, which has the value $x + c$, so that the use of (4) is very simple, too.

17. Initial value problem. In any case the first task is to write the equation in the form (1). In the present problem,
$$y' - 2\,y \tanh 2x = -2 \tanh 2x.$$
In (4) you thus have $p = -2 \tanh 2x = -(\ln \cosh 2x)'$. Hence the integral h of p is $h = -\ln (\cosh 2x)$. In (4) you need $e^{-h} = \cosh 2x$ and under the integral sign $e^h = 1/(\cosh 2x)$. Since $r = -2 \tanh 2x$, the integrand is

$$-2\tanh 2x/\cosh 2x = -2\sinh 2x/(\cosh 2x)^2 = (1/\cosh 2x)'.$$

Hence the integral equals $1/(\cosh 2x) + c$. Multiplying this by $e^{-h} = \cosh 2x$ gives the general solution $y = 1 + c\cosh 2x$. From this and the initial condition, $y(0) = 1 + c = 4$, $c = 3$. *Answer:* $y = 1 + 3\cosh 2x$.

33. Bernoulli equation. This is a Bernoulli equation with $a = 4$. Hence you have to set $u = 1/y^3$. By differentiation (chain rule!) $u' = -3y^{-4}y'$. This suggests multiplying the given equation by $-3y^{-4}$, obtaining

$$-3y^{-4}y' - y^{-3} = -1 + 2x.$$

The first term is u' and the second is $-u$; thus $u' - u = 2x - 1$. Formula (4) with u instead of y gives the general solution $u = ce^x - 2x - 1$. Hence the answer is

$$y = u^{-1/3} = (ce^x - 2x - 1)^{-1/3}.$$

Sec. 1.7 Modeling: Electric Circuits

Example 1 Step 5. For the idea of evaluating the integral by undetermined coefficients, see this Manual, Sec. 1.6.

Problem Set 1.7. Page 47

7. Choice of L. This is a problem on the exponential approach to the limit, as it also occurs in various other applications. For constant $E = E_0$ the model of the circuit is $I' + (R/L)I = E_0/L$. The initial condition is $I(0) = 0$ since the current is supposed to start from zero. The general solution and the particular solution are

$$I = ce^{-Rt/L} + \frac{E_0}{R}, \qquad I = \frac{E_0}{R}(1 - e^{-Rt/L}).$$

25% of the final value of I is reached if the exponential term has the value 0.75, that is, $\exp(-Rt/L) = 0.75$. With $R = 1000$, $t = 1/10000$ by taking logarithms you obtain $0.1/L = \ln(1/0.75) = 0.2877$, so that $L = 0.1/0.2877 = 0.3476$.

9. RL-circuit. The two cases can first be handled jointly; the difference will appear in evaluating the integral. The model is $I' + RI/L = e^{-t}/L$. You can solve it by (4) in Sec. 1.6. Since $p = R/L$, integration gives $h = Rt/L$. Hence $e^{-h} = e^{-Rt/L}$ and $e^h = e^{Rt/L}$. This yields the integrand $(1/L)\exp(Rt/L)\exp(-t) = (1/L)\exp[(R/L - 1)t]$. If $R/L - 1 = 0$, the integrand is $1/L$, and the integral is $t/L + c$. This is Case (b), the solution being

$$I = (t/L + c)e^{-t}.$$

If $R/L - 1$ is not zero, you have to integrate an exponential function, obtaining $\exp[(R/L - 1)t]/(R - L)$. This is Case (a), the solution being

$$I = \frac{e^{-t}}{R - L} + ce^{-Rt/L},$$

where the first term became simple because $\exp(-h)\exp h = 1$. The figure shows the two solutions for $I_0 = 0$, $L = 1$ and (a) $R = 3$, (b) $R = 1$. Find out which curve corresponds to (a) and which to (b). Sketch the solutions when $L = 1$, $R = 3$, and $I_0 = 1$, and compare.

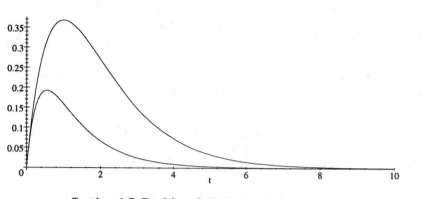

Section 1.7. Problem 9. Solutions in both cases

19. Periodic electromotive forces are particularly important in practice. The simplest way of obtaining steady-state solutions is by substituting an expression of the form of the electromotive force with undetermined coefficients and determining the latter by equating corresponding coefficients on both sides of the equation. In the problem, the model equation, divided by a common factor 25, is

$$2Q' + Q = 4 \cos 2t + \sin 2t + 8 \cos 4t + \sin 4t.$$

The right side suggests setting

$$Q = a \cos 2t + b \sin 2t + c \cos 4t + k \sin 4t$$

By differentiation and multiplication by 2,

$$2Q' = -4a \sin 2t + 4b \cos 2t - 8c \sin 4t + 8k \cos 4t.$$

Hence you must have $a + 4b = 4$ (from cos $2t$), $-4a + b = 1$ (from sin $2t$). The solution is $a = 0, b = 1$. Similarly, $c + 8k = 8$ (from cos $4t$), $-8c + k = 1$ (from sin $4t$). The solution is $c = 0, k = 1$. Hence there are no cosine terms. The answer is $Q = \sin 2t + \sin 4t$. This "method of undetermined coefficients" will be very important in connection with vibrations in the next chapter.

Sec. 1.8   Orthogonal Trajectories of Curves.   *Optional*

Problem Set 1.8. Page 51

3. Family of curves. cosh $(x - c)$ is a translate of cosh x through the distance c to the right ($x - c = 0$ or $x = c$ corresponds to the lowest point of the curve, which is now at $x = c, y = 1$). Adding $-c$ moves the translated curve down. Thus, $y = \cosh (x - c) - c$. If $x = c$, then $y = -c + 1$; this is the lowest point of the corresponding curve. Make a sketch.

9. Differential equation of a family of curves. The differential equation to be derived must not contain c. This is quite essential. You accomplish this as follows. Solve the given equation algebraically for c^2,

$$c^2 (x^2 - 1) + y^2 = 0, \qquad - c^2 = y^2/(x^2 - 1).$$

Differentiation with respect to x gives (chain rule!)

$$0 = \frac{2yy'}{x^2 - 1} - \frac{y^2}{(x^2 - 1)^2} \cdot 2x.$$

Dividing by $2y$ and solving algebraically for y' yields the answer shown in Appendix 2 of the text.

21. Orthogonal trajectories derive their importance from applications in electrostatics, fluid flow, heat flow, and so on. The given curves $xy = c$ are the familiar hyperbolas with the coordinate axes as asymptotes (the solid curves in Fig. 30 of the text). Differentiation with respect to x gives their differential equation

$y + xy' = 0$ or $y' = -y/x$. Formula (2) in Sec. 1.8 gives the differential equation of the trajectories $y' = +x/y$ or $yy' = x$. By integration on both sides you obtain $y^2/2 = x^2/2 + C$ or $x^2 - y^2 = c^*$, the dashed hyperbolas in Fig. 30, whose asymptotes are $y = x$ and $y = -x$ (the latter in the quadrants not shown in the figure).

Sec. 1.9   Existence and Uniqueness of Solutions. Picard Iteration

Problem Set 1.9. Page 58

1. No solution. Obtain the general solution by separating variables.

3. Vertical strip. α is the smaller of the numbers a and b/K. Since K is constant and you can now choose b as large as you please (there is no restriction in the y-direction), the smaller number is a, as claimed.

7. Linear differential equation. $y' = f(x,y) = r - p(x)\,y$ shows that the continuity of r and p makes both f and $\partial f/\partial y = -p(x)$ continuous.

11. Picard iteration. Proof by induction. You have to show that $y_n = 1 + x + \dots + x^n/n!$. This is true for $n = 0$ because $y_0 = y(0) = 1$; see (6) in Sec. 1.9. Since $y' = f(x,y) = y$, the integrand in (6) is $y_{n-1}(t)$. Make the induction hypothesis that this equals $1 + t + \dots + t^{n-1}/(n-1)!$ According to (6) you have to integrate this expression from 0 to x, obtaining $x + x^2/2 + \dots + x^n/n!$ (because $(n-1)!\,n = n!$), and to add $y_0 = 1$. This gives y_n, the next partial sum of the Maclaurin series of e^x, and completes the proof.

13. Picard iteration. $y' = x + y$, $y_0 = -1$.

$$y_n = -1 + \int_0^x (t + y_{n-1}(t))\,dt = -1 + \int_0^x y_{n-1}(t)\,dt + \frac{x^2}{2},$$

thus

$$y_1 = -1 - x + \frac{x^2}{2},$$

$$y_2 = -1 - x - \frac{x^2}{2} + \frac{x^3}{6} + \frac{x^2}{2} = -1 - x + \frac{x^3}{6}, \quad \text{etc.}$$

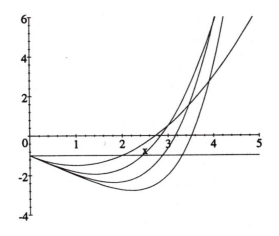

Section 1.9. Problem 13. Picard approximations of the solution $y = -1 - x$

CHAPTER 2. Linear Differential Equations of Second and Higher Order

Sec. 2.1   Homogeneous Linear Equations of Second Order

Problem Set 2.1. Page 71

7. **Reduction to first order.** $y'' + e^y y'^3 = 0$ is of the form $F(y, y', y'') = 0$, so that you can set $z = y'$ and $y'' = (dz/dy) z$ (see Prob. 2). Substitution of this and division by z gives $dz/dy + e^y z^2 = 0$. By separation of variables, $dz/z^2 = -e^y \, dy$. Integration on both sides and multiplication by -1 gives $1/z = e^y + c_1$. Now by calculus, $z = dy/dx$ implies $dx/dy = 1/z$. Hence you can separate again and then integrate,

$$dx = (e^y + c_1) \, dy$$
$$x = e^y + c_1 y + c_2.$$

13. **Motion.** Expressing the given data in formulas gives $y'y'' = 1$, $y(0) = 2$, $y'(0) = 2$. By integration, $y'^2/2 = t + C$, hence $y' = \sqrt{2t + c_1}$, where $c_1 = 2C$. If you wish, you can now use the second initial condition to get $y'(0) = \sqrt{c_1} = 2$, hence $c_1 = 4$, so that $y' = \sqrt{2t + 4}$. By another integration and the use of the first initial condition you obtain

$$y = \frac{1}{3}(2t + 4)^{3/2} + c_2, \quad y(0) = \frac{1}{3} 4^{3/2} + c_2 = \frac{8}{3} + c_2 = 2, \quad c_2 = -\frac{2}{3}.$$

This gives the answer

$$y = \frac{1}{3}(2t + 4)^{3/2} - \frac{2}{3}.$$

Sec. 2.2   Second-Order Homogeneous Equations with Constant Coefficients

Problem Set 2.2. Page 75

7. **General solution.** Problems 1-9 amount to solving a quadratic equation (3), the characteristic equation. Observe that the solutions (4) refer to the case that y'' has the coefficient 1. For the present equation you can write $y'' - (30/9) y' + (25/9) y = 0$. Then the radicand in (4) is $225/81 - 25/9 = 0$, so that you have a double root $15/9 = 5/3$. The corresponding general solution is $y = (c_1 + c_2 x) \exp(5x/3)$.

15. **Initial value problem.** To solve an initial value problem, first determine a general solution by solving the characteristic equation $\lambda^2 + 2.2\lambda + 1.17 = 0$. The roots (4) are -1.3 and -0.9. The corresponding general solution is

$$y = c_1 e^{-1.3x} + c_2 e^{-0.9x}. \tag{a}$$

Because of the second initial condition you also need the derivative

$$y' = -1.3 c_1 e^{-1.3x} - 0.9 c_2 e^{-0.9x}. \tag{b}$$

In (a) and (b) you now put $x = 0$ and equate the result to 2 and -2.6, respectively (the given initial values), that is,

$$c_1 + c_2 = 2, \quad -1.3 c_1 - 0.9 c_2 = -2.6.$$

The solution is $c_1 = 2$, $c_2 = 0$, so that you get the answer $y = 2 e^{-1.3x}$. Note that, in general, both solutions of a basis of solutions would appear; in that sense our present initial conditions are special.

21. **Linear independence and dependence.** This problem is typical of cases where one must use functional relations to prove linear dependence. Namely, $\ln x$ and $\ln(x^4) = 4 \ln x$ are linearly dependent on any

interval of the positive semi-axis. Graphs may help when the functions are very complicated and transformations are not so obvious as in this problem; then you may find out whether the curves of the functions look "proportional".

Sec. 2.3   Case of Complex Roots. Complex Exponential Function

Problem Set 2.3. Page 80

5. General solution. $y'' + 1.6\,y' + 0.64\,y = 0$ (the given equation divided by 2.5) has the characteristic equation $\lambda^2 + 1.6\,\lambda + 0.64 = (\lambda + 0.8)^2 = 0$ with the double root -0.8. This is Case II, with the general solution as given in Appendix 2.

7. General solution. Division by 16 gives $y'' - 0.5y' + 0.3125y = 0$. From (3) you thus obtain the roots

$$\lambda_1 = 0.25 + 0.5\,\sqrt{0.25 - 1.25} = 0.25 + 0.5\,i \quad \text{and} \quad \lambda_2 = 0.25 - 0.5\,i.$$

Note that if an equation (with real coefficients) has a complex root, the conjugate of the root must also be a root. The real part is 0.25 and gives the exponential function $\exp(0.25x)$. The imaginary parts are 0.5 and -0.5 and give the cosine and sine terms. Together,

$$y = e^{0.25x}\,(A\,\cos 0.5\,x + B\,\sin 0.5\,x),$$

which is oscillating with an increasing maximum amplitude.

21. Boundary value problems will be less important to us than initial value problems. The determination of a particular solution by using given boundary conditions is similar to that for an initial value problem. In the present problem the characteristic equation is $\lambda^2 + 2\,\lambda + 2 = 0$. Its roots are

$$\lambda_1 = -1 + \sqrt{1 - 2} = -1 + i \quad \text{and} \quad \lambda_2 = -1 - i.$$

This gives the real general solution

$$y = e^{-x}\,(A\,\cos x + B\,\sin x).$$

On the left boundary, $y(0) = A = 1$. On the right boundary, $y(\pi/2) = B\,\exp(-\pi/2) = 0$, hence $B = 0$. Hence the answer is $y = e^{-x}\,\cos x$.

Sec. 2.4   Differential Operators.   *Optional*

Problem Set 2.4. Page 83

3. Differential operators. $(D - 2)(D + 1)\,e^{2x} = 0$ because
$$(D - 2)\,e^{2x} = 2\,e^{2x} - 2\,e^{2x} = 0.$$

For the second of the four given functions you first have
$$(D - 2)x\,e^{2x} = e^{2x} + 2x\,e^{2x} - 2x\,e^{2x} = e^{2x}$$

and then

$$(D + 1)\,e^{2x} = 2\,e^{2x} + e^{2x} = 3\,e^{2x}.$$

Similarly for the other functions.

13. General solution. The optional Sec. 2.4 introduces to the operator notation and shows how it can be applied to linear differential equations with constant coefficients. The facts considered are essentially as before, merely the notation changes. The given equation, divided by 10, is

$$(D^2 + 1.2D + 0.36)\,y = (D + 0.6)^2\,y = 0.$$

It shows that the characteristic equation has the double root -0.6, so that the corresponding general solution is

$$y = (c_1 + c_2 x) e^{-0.6x}.$$

Sec. 2.5   Modeling: Free Oscillations (Mass-Spring Systems)

Problem Set 2.5. Page 90

1. **Harmonic oscillations.** Formula (4*) gives a better impression than a sum of cosine and sine terms because the maximum amplitude C and phase shift δ readily characterize the harmonic oscillation. The result follows by direct calculation, starting from the general solution

$$y = A \cos \omega_0 t + B \sin \omega_0 t$$

and using the initial conditions, first $y(0) = A = y_0$ and then

$$y' = \text{a sine term} + \omega_0 B \cos \omega_0 t, \qquad y'(0) = \omega_0 B = v_0,$$

where v suggests 'velocity'. This gives the particular solution

$$y = y_0 \cos \omega_0 t + (v_0/\omega_0) \sin \omega_0 t.$$

Accordingly, in (4*),

$$C = \sqrt{y_0{}^2 + \left(\frac{v_0}{\omega_0}\right)^2}, \qquad \tan \delta = \frac{v_0/\omega_0}{y_0}.$$

The derivation of (4*) suggested in the text begins with

$$y(t) = C \cos(\omega_0 t - \delta) = C(\cos \omega_0 t \cos \delta + \sin \omega_0 t \sin \delta)$$

$$= C \cos \delta \cos \omega_0 t + C \sin \delta \sin \omega_0 t = A \cos \omega_0 t + B \sin \omega_0 t.$$

By comparing you see that

$$A^2 + B^2 = C^2 \cos^2 \delta + C^2 \sin^2 \delta = C^2$$

and

$$\tan \delta = \frac{\sin \delta}{\cos \delta} = \frac{C \sin \delta}{C \cos \delta} = \frac{B}{A}.$$

7. **Determination of frequencies.** $\omega_0 = \sqrt{k/m}$; see (4). Hence the frequencies are

$$\frac{\omega_0}{2\pi} = \frac{1}{2\pi}\sqrt{\frac{k_1}{m}} \quad \text{and} \quad \frac{1}{2\pi}\sqrt{\frac{k_2}{m}},$$

respectively. To prove $k = k_1 + k_2$, fix $s = s_0$ (for instance, $s_0 = 1$), choose $W_1 = k_1 s_0$ and $W_2 = k_2 s_0$, and add (couple the two systems), where k is the spring constant of the two systems

$$W = W_1 + W_2 = (k_1 + k_2) s_0 = k s_0$$

combined.

15. **Underdamping.** Equate the derivative to zero.

Sec. 2.6   Euler-Cauchy Equation

Problem Set 2.6. Page 96

3. **General solution.** Problems 2-13 are solved as explained in the text by determining the roots of the auxiliary equation (3). This is similar to the method for constant-coefficient equations in Secs. 2.2 and 2.3, but note well that the linear term in (3) is $(a - 1) m$, not am. Thus in Prob. 3 you have

$$m(m - 1) - 20 = m^2 - m - 20 = 0.$$

The roots are −4 and 5. Hence a general solution is $y = c_1 x^{-4} + c_2 x^5$. The value $x = 0$ is excluded.

Similarly, the case of a double root of (3) gives a logarithmic term [see (7) in Sec. 2.6] and $x = 0$ and all negative x must be excluded.

7. Pure imaginary roots. The auxiliary equation is $m^2 + 1 = 0$. It has the roots $i = \sqrt{-1}$ and $-i$. Hence in (8) of Sec, 2.6 you have $\mu = 0$ (the real part of the roots is zero) and $v = 1$, so that (8) becomes simply $y = A\cos(\ln x) + B\sin(\ln x)$.

15. Initial value problems for Euler-Cauchy equations are solved as for constant-coefficient equations by first determining a general solution. The initial values must not be given at 0, where the coefficients of (1), written in standard form

$$y'' + \frac{a}{x}\,y' + \frac{b}{x^2}\,y = 0,$$

become infinite, but must refer to some other point, for instance, to $x = 1$. In Prob. 15 the auxiliary equation is

$$4m(m-1) + 24m + 25 = 0 \quad\text{or}\quad m^2 + 5m + 6.25 = 0.$$

It has the double root -2.5. The corresponding general solution (7), Sec. 2.6, is

$$y = (c_1 + c_2 \ln x)x^{-2.5}.$$

The first initial condition gives $y(1) = c_1 = 2$. For the second initial condition $y'(1) = -6$ you need the derivative. With $c_1 = 2$ the latter is

$$y' = \frac{c_2}{x}x^{-2.5} - 2.5(2 + c_2 \ln x)x^{-3.5}.$$

Setting $x = 1$, you thus obtain (since $\ln 1 = 0$)

$$y'(1) = c_2 - 5 = -6, \quad\text{hence}\quad c_2 = -1.$$

The figure shows the particular solution obtained, $y = (2 - \ln x)x^{-2.5}$. For $x > 7.4$ the logarithm is greater than 2, so that for these x the solution becomes negative, but this can hardly be seen from the figure because the x-factor is very small in absolute value when x is large.

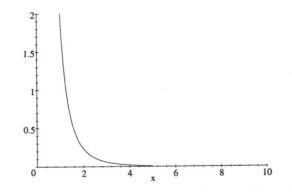

Section 2.6. Problem 15. Particular solution satisfying $y(1) = 2$, $y'(1) = -6$

Sec. 2.7   Existence and Uniqueness Theory. Wronskian

The **Wronskian** $W(y_1, y_2)$ of two solutions y_1 and y_2 of a differential equation is defined by (5), Sec. 2.7. It is conveniently written as a second-order determinant (but this is not essential for using it; you need not be familiar with determinants here). It serves for checking linear independence or dependence, which is important in obtaining bases of solutions. The latter are needed, for instance, in connection with initial value problems, where a single solution will generally not be sufficient for satisfying two given initial conditions. Of course, two functions are linearly independent if and only if their quotient is not constant. To check this, you would not need Wronskians, but we discuss them here in the simple case of second-order differential equations as a preparation for Secs. 2.13-2.15 on higher order equations, where Wronskians will show their power and are extremely useful.

Problem Set 2.7. Page 100

3. Basis, Wronskian. For $a > 0$ these solutions
$$y_1 = e^{-ax/2} \cos 3x \quad \text{and} \quad y_2 = e^{-ax/2} \sin 3x$$
represent damped vibrations, x being time. Their Wronskian is obtained by straightforward differentiation or by the following trick. From the quotient rule and (5), Sec. 2.7, it follows that
$$W = (y_2/y_1)' y_1^2, \tag{A}$$
where the prime denotes the derivative. In the present problem, $y_2/y_1 = \tan 3x$ has the derivative $3/\cos^2 3x$ (chain rule!). Furthermore, $y_1^2 = e^{-ax} \cos^2 3x$. The product of the two expressions is the Wronskian $W = 3e^{-ax}$.

5. Wronskian. Formula (A) in Prob. 3 gives $(x^4(\ln x)/x^4)' x^8 = (\ln x)' x^8 = x^7$.

7. Wronskian. Formula (A) in Prob. 3 contains
$$(\tan(2\ln x))' = [1/\cos^2(2\ln x)] \cdot \frac{2}{x},$$
the last factor resulting from the chain rule. Now $y_1^2 = x^{2\mu} \cos^2(2\ln x)$, and the product is $W = 2x^{2\mu-1}$.

11. Equation for a given basis. Problems 9-17 survey the most important types of equations discussed so far. The form in Prob. 11 suggests an Euler-Cauchy equation with a double root (because of the logarithmic term). Now
$$(m-2)^2 = m(m-1) - 3m + 4 \quad \text{shows that} \quad x^2 y'' - 3xy' + 4y = 0.$$
From (5) you obtain the Wronskian
$$W = x^2(2x \ln x + x) - (x^2 \ln x)2x = x^3.$$
Check this by (A) in Prob. 3, obtaining $(\ln x)' x^4 = x^3$.

Sec. 2.8   Nonhomogeneous Equations

Verification of solutions proceeds for nonhomogeneous equations as it does for homogeneous equations, namely, by the calculation of y' and y'' and substitution of y, y', and y''. It is interesting that in Probs. 1-8 most solutions to some extent resemble the form of the functions on the right side of the equation. This observation gives the idea of a method for determining particular solutions to be discussed in the next section.

Problem Set 2.8. Page 103

7. General solution. $y_p = \ln \pi x = \ln x + \ln \pi$ has the derivatives $1/x$ and $-1/x^2$. Substitution gives $y'' + y = -1/x^2 + \ln \pi x$. A general solution of the homogeneous equation is $A \cos x + B \sin x$. Hence the answer (a general solution of the nonhomogeneous equation) is $y = A \cos x + B \sin x + \ln \pi x$.

11. Initial value problem. To solve an initial value problem, you must first determine a general solution of the nonhomogeneous equation (because if you first determine a particular solution of the homogeneous equation satisfying the initial conditions, the addition of a solution y_p will generally change the value of the entire solution and its derivative at the point at which the initial conditions are given). Now a general solution of the homogeneous equation $y'' - y = 0$ is $c_1 e^x + c_2 e^{-x}$. The particular solution $y_p = xe^x$ may come as a surprise because $2e^x$ on the right might have suggested $y_p = ke^x$ with a suitable constant k, but if you substitute this, you get $0 = 2e^x$. Choosing $y_p = xe^x$ gives $y_p' = (x + 1)e^x$ and $y_p'' = (x + 2)e^x$, so that substitution yields $(x + 2)e^x - xe^x = 2e^x$ and verifies that y_p is indeed a solution. Hence a general solution of the given equation is

$$y = c_1 e^x + c_2 e^{-x} + xe^x.$$

From the first initial condition, $y(0) = c_1 + c_2 = -1$. From the derivative and the second initial condition,

$$y' = c_1 e^x - c_2 e^{-x} + (x + 1)e^x, \quad y'(0) = c_1 - c_2 + 1 = 0.$$

The solution of this system of two equations is $c_1 = -1$, $c_2 = 0$. This gives the answer $y = -e^x + xe^x = (x - 1)e^x$ shown in the figure.

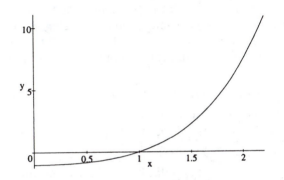

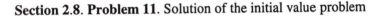

Section 2.8. Problem 11. Solution of the initial value problem

Sec. 2.9   Solution by Undetermined Coefficients

New in this section and problem set is the determination of a particular solution y_p by the method of undetermined coefficients. Because of the Modification Rule it is necessary to first determine a general solution of the homogeneous equation since the form of y_p differs depending on whether or not the function (or a term of it) on the right side of the differential equation is a solution of the homogeneous equation. If you forget to take this into account, you will not be able to determine the coefficients; in this sense the method will warn you that you made a mistake.

Problem Set 2.9. Page 107

  1. **General solution.** A general solution of the homogeneous equation (1) $y'' + 4y = 0$ is $y_h = A\cos 2x + B\sin 2x$. The function $\sin 3x$ on the right is not a solution of (1). Hence the Modification Rule does not apply. Table 2.1 requires that you start from $y_p = K\cos 3x + M\sin 3x$. Two differentiations give $y_p'' = -9K\cos 3x - 9M\sin 3x$. Substituting this and y_p into the given equation yields

$$-9K\cos 3x - 9M\sin 3x + 4(K\cos 3x + M\sin 3x) = \sin 3x.$$

  Since there is no cosine term on the right, this implies $-9K + 4K = 0$, hence $K = 0$. For the sine terms, $-9M + 4M = 1$, hence $M = -0.2$. This gives the answer $y = A\cos 2x + B\sin 2x - 0.2\sin 3x$.

11. **Modification rule.** The characteristic equation of the homogeneous equation is $\lambda^2 + 10\lambda + 25 = (\lambda + 5)^2 = 0$. Hence it has the double root -5, so that a general solution of the homogeneous equation is $y_h = (c_1 + c_2 x)e^{-5x}$. This shows that e^{-5x} is a solution of the homogeneous equation. Hence you must apply the Modification Rule. More precisely, since you are dealing with a *double root*, you must multiply the usual choice e^{-5x} by x^2, (In the case of a simple root you would have to multiply by x.) Accordingly, choose $y_p = kx^2 e^{-5x}$. By differentiation,

$$y_p' = k(2x - 5x^2)e^{-5x}, \qquad y_p'' = k(2 - 10x - 10x + 25x^2)e^{-5x}.$$

Substitution of these expressions into the differential equation $y'' + 10y' + 25y = e^{-5x}$ and omission of the common factor e^{-5x} on both sides of the equation gives

$$k(2 - 20x + 25x^2) + 10k(2x - 5x^2) + 25kx^2 = 1.$$

In this equation, x^2 has the coefficient $25k + 10k(-5) + 25k = 0$. Similarly, x has the coefficient $-20k + 10k \cdot 2 = 0$. Finally, the constant terms give $2k = 1$, $k = 0.5$. Hence the answer (a general solution of the given nonhomogeneous equation) is

$$y = (c_1 + c_2 x)e^{-5x} + 0.5x^2 e^{-5x}.$$

17. Initial value problem. A general solution of the homogeneous equation $y'' - 4y = 0$ is $y_h = c_1 e^{2x} + c_2 e^{-2x}$. The right side $e^{-2x} - 2x$ has two terms. The first is a solution of the homogeneous equation, the corresponding root of the characteristic equation being simple. Hence the Modification rule calls for kxe^{-2x} instead of the usual ke^{-2x}. By Table 2.1 in Sec. 2.9 the second term $-2x$ calls for the choice $ax + b$ (line 2 of the table, with a more convenient notation). Together, $y_p = kxe^{-2x} + ax + b$. By differentiation,

$$y_p' = k(1 - 2x)e^{-2x} + a, \qquad y_p'' = k(-2 - 2 + 4x)e^{-2x}.$$

Substitution into the nonhomogeneous equation gives

$$k(-4 + 4x)e^{-2x} - 4kxe^{-2x} - 4(ax + b) = e^{-2x} - 2x.$$

The terms in xe^{-2x} drop out. The e^{-2x}-terms give $-4k = 1$, $k = -1/4$. The x-terms give $-4a = -2$, $a = 1/2$. The constant terms give $b = 0$. Hence a general solution of the given equation is

$$y = c_1 e^{2x} + c_2 e^{-2x} + 0.5x - 0.25xe^{-2x}.$$

$y(0) = 0$ gives $y(0) = c_1 + c_2 = 0$. By differentiation of y,

$$y' = 2c_1 e^{2x} - 2c_2 e^{-2x} + 0.5 - 0.25(1 - 2x)e^{-2x}.$$

$y'(0) = 0$ thus gives $y'(0) = 2c_1 - 2c_2 + 0.5 - 0.25 = 0$. The solution of these two equations is $c_1 = -1/16, c_2 = 1/16$. Hence the answer is

$$y = -\frac{1}{16}(e^{2x} - e^{-2x}) + \frac{1}{2}x - \frac{1}{4}xe^{-2x}.$$

The exponential terms combine into $-(\sinh 2x)/8$, as given in Appendix 2.

Sec. 2.10   Solution by Variation of Parameters

Problem Set 2.10. Page 111

1. General solution. The right side e^{2x}/x does not permit the method of undetermined coefficients (which would be simpler than the present method). The homogeneous equation $y'' - 4y' + 4y = 0$ has the characteristic equation $\lambda^2 - 4\lambda + 4 = (\lambda - 2)^2 = 0$. It has the double root 2. Hence a basis of solutions is $y_1 = e^{2x}$ and $y_2 = xe^{2x}$. Now determine a particular solution of the given equation by (2), Sec. 2.10. In (2) you need the Wronskian

$$W = e^{2x}(xe^{2x})' - xe^{2x}(e^{2x})' = e^{4x}(1 + 2x) - e^{4x}2x = e^{4x}.$$

The integrands of the integrals in (2) are

$$xe^{2x}(e^{2x}/x)/e^{4x} = 1 \quad \text{and} \quad e^{2x}(e^{2x}/x)/e^{4x} = 1/x.$$

Integration gives x and $\ln |x|$, respectively. From (2) you thus obtain the particular solution

$$y_p = -xe^{2x} + xe^{2x}\ln |x|.$$

Hence the corresponding general solution of the given nonhomogeneous equation is

$$y = (c_1 + c_2 x)e^{2x} + (-x + x\ln |x|)e^{2x}.$$

3. General solution. This equation can also be solved by undetermined coefficients, starting from $y_p = e^{-x}(K \cos x + M \sin x)$. Try it.

11. Euler-Cauchy equation. The homogeneous equation $x^2 y'' - 4x y' + 6y = 0$ has the auxiliary equation $m^2 - 5m + 6 = 0$. The roots are $m_1 = 2$ and $m_2 = 3$. Hence a general solution of the homogeneous equation is $y_h = c_1 x^2 + c_2 x^3$. Try to find a particular solution of the nonhomogeneous equation by undetermined coefficients, setting $y_p = C x^{-4}$. Then $y_p' = -4 C x^{-5}$, $y_p'' = 20 C x^{-6}$, and substitution into the given equation yields

$$20 C x^{-4} + 16 C x^{-4} + 6 C x^{-4} = 21 x^{-4}, \qquad 42 C = 21, \qquad C = 1/2.$$

Hence a general solution of the given nonhomogeneous equation is

$$y = c_1 x^2 + c_2 x^3 + \frac{1}{2} x^{-4}.$$

15. Euler-Cauchy equation. Determine y_p by (2) in Sec. 2.10. It is quite important that you first write the given equation in standard form

$$y'' - 2y'/x + 2y/x^2 = x \cos x. \qquad \text{Hence } r = x \cos x \qquad (\text{not } x^3 \cos x\,!).$$

The auxiliary equation of the homogeneous differential equation is $m^2 - 3m + 2 = 0$ and has the roots $m_1 = 1$, $m_2 = 2$. This gives the basis of solutions $y_1 = x$, $y_2 = x^2$. In (2) you need the Wronskian $W = x(2x) - 1 \cdot x^2 = x^2$. Hence the first integral in (2) has the integrand $x^2(x \cos x)/x^2 = x \cos x$. Integration by parts gives $x \sin x$ minus the integral of $\sin x$, which is $+\cos x$. Together, $x \sin x + \cos x$. The second integral in (2) has the integrand $x(x \cos x)/x^2 = \cos x$. Integration gives $\sin x$. From this and (2) you obtain

$$y_p = -x(x \sin x + \cos x) + x^2 \sin x = -x \cos x.$$

The answer (a general solution of the nonhomogeneous equation) is

$$y = c_1 x + c_2 x^2 - x \cos x.$$

Sec. 2.11   Modeling: Forced Oscillations. Resonance

In the solution a, b of (4) (the formula after (4) without number) the denominator is the coefficient determinant. The numerator of a is the determinant

$$\begin{vmatrix} F_0 & \omega c \\ 0 & k - m\omega^2 \end{vmatrix} = F_0(k - m\omega^2).$$

Similarly for b.

Problem Set 2.11. Page 117

Problems 1-17 involve driving forces such that the method of undetermined coefficients (Sec. 2.9) can be applied.

3. Steady-state solution. Because of the function $\sin 0.2t$ on the right you have to choose $y_p = K \cos 0.2t + M \sin 0.2t$. By differentiation,

$$y_p' = -0.2K \sin 0.2t + 0.2M \cos 0.2t,$$

$$y_p'' = -0.04K \cos 0.2t - 0.04M \sin 0.2t.$$

Substitute this into the equation $y'' + 2y' + 4y = \sin 0.2t$. To get a simple formula, use the abbreviations $C = \cos 0.2t$ and $S = \sin 0.2t$. Then

$$-0.04KC - 0.04MS + 2(-0.2KS + 0.2MC) + 4(KC + MS) = S.$$

Now collect the C-terms and the S-terms on the left and equate their sums to 0 (there is no C-term on the right) and 1, respectively,

$$-0.04K + 0.4M + 4K = 3.96K + 0.4M = 0$$

$$-0.04M - 0.4K + 4M = -0.4K + 3.96M = 1.$$

Elimination or Cramer's rule (Sec. 6.6) gives the solution $K = -0.02525, M = 0.2500$ (more exactly, $K = -0.025\,249\,975, M = 0.249\,974\,750$). Hence the steady-state solution is

$$y = -0.02525 \cos 0.2t + 0.2500 \sin 0.2t.$$

15. **Initial value problem.** Divide by 4 to have the standard form $y'' + 2y' + 0.75y = 106.25 \sin 2t$. (This is convenient, although not absolutely necessary.). The characteristic equation of the homogeneous equation is $\lambda^2 + 2\lambda + 0.75 = 0$. The roots are $-1/2$ and $-3/2$. Hence a general solution of the homogeneous equation is $y_h = c_1 e^{-0.5t} + c_2 e^{-1.5t}$. Now determine a particular solution y_p. The right side calls for the choice $y_p = K \cos 2t + M \sin 2t$. Both terms will be needed because the equation has a damping term, which causes a phase shift (in contrast to Prob. 13, where there is no damping and y_p is a sine term). By differentiation,

$$y_p' = -2K \sin 2t + 2M \cos 2t, \qquad y_p'' = -4K \cos 2t - 4M \sin 2t.$$

Using the abbreviations $C = \cos 2t, S = \sin 2t$ and substituting y_p and its derivatives into the given equation yields

$$-4KC - 4MS + 2(-2KS + 2MC) + 0.75(KC + MS) = 106.25\,S.$$

The sum of the cosine terms on the left must equal 0 since there is no cosine term on the right. Similarly, the sum of the sine terms on the left must equal 106.25. This gives the linear system of two equations

$$-4K + 4M + 0.75K = -3.25K + 4.00M = 0$$

$$-4M - 4K + 0.75M = -4.00K - 3.25M = 106.25.$$

By elimination or by Cramer's rule (Sec. 6.6) you obtain the solution $K = -16, M = -13$. This gives the general solution

$$y(t) = c_1 e^{-0.5t} + c_2 e^{-1.5t} - 16 \cos 2t - 13 \sin 2t.$$

From this and the first initial condition, $y(0) = c_1 + c_2 - 16 = -16$. The derivative is

$$y'(t) = -0.5c_1 e^{-0.5t} - 1.5c_2 e^{-1.5t} + 32 \sin 2t - 26 \cos 2t,$$

and the second initial condition gives $y'(0) = -0.5c_1 - 1.5c_2 - 26 = -26$. By inspection or by elimination, $c_1 = 0, c_2 = 0$. This gives the answer $y = -16 \cos 2t - 13 \sin 2t$. This is a harmonic oscillation of period π and maximum amplitude $\sqrt{16^2 + 13^2} = 20.62$ (see the figure). It is interesting that because of the initial conditions the solution of the homogeneous equation does not contribute to the answer, so that there is no transition period.

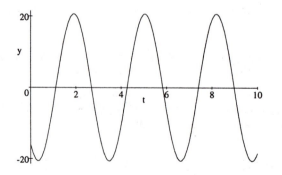

Section 2.11. Problem 15. Solution without transition period

Sec. 2.12   Modeling of Electric Circuits

Example 1. The linear system of equations near the end of the example consists of (7) and an unnumbered equation, namely,

$$c_1 + c_2 = 0.484 \qquad \text{(from } I(0) = 0\text{)}$$
$$-10c_1 - 990c_2 = -1.380 \cdot 377 = -520.26 \qquad \text{(from } I'(0) = 0\text{)}.$$

It can be solved by elimination, namely, $c_2 = 0.484 - c_1$, hence

$$-10c_1 - 990(0.484 - c_1) = -520.26$$

and from this,

$$c_1 = \frac{1}{980}(479.16 - 520.26) = -0.041\,939,$$

so that $c_2 = 0.484000 + 0.041939 = 0.525939$.

Problem Set 2.12. Page 122

7. **Transient current.** In the model (1) the right side is the derivative of the elctromotive force $E = 25\cos 100t$, that is, $E' = -2500\sin 100t$. Hence (1), divided by $L = 0.5$, is

$$I'' + 80I' + 1500I = -5000\sin 100t. \qquad (a)$$

The characteristic equation $\lambda^2 + 80\lambda + 1500 = 0$ has the roots -30 and -50. Hence a general solution of the homogeneous equation is

$$I_h = c_1 e^{-30t} + c_2 e^{-50t}.$$

This solution approaches zero as t goes to infinity, regardless of initial conditions. A particular solution I_p of the nonhomogeneous equation is obtained by substituting $I_p = K\cos 100t + M\sin 100t$ and its derivatives

$$I_p' = -100K\sin 100t + 100M\cos 100t,$$
$$I_p'' = -10000K\cos 100t - 10000M\sin 100t$$

into (a). Writing $C = \cos 100t$, $S = \sin 100t$, you obtain

$$-1000KC - 10000MS + 80(-100KS + 100MC) + 1500(KC + MS) = -5000S.$$

The sum of the cosine coefficients must be zero since there is no cosine term on the right side of (a). Similarly, the sum of the sine coefficients must equal -5000. This gives a system for determining K and M, namely,

$$-10000K + 8000M + 1500K = -8500K + 8000M = 0$$
$$-10000M - 8000K + 1500M = -8000K - 8500M = -5000.$$

Solving the first equation for M gives $M = 1.0625K$. Substituting this into the second equation, you find

$$-8000K - 8500M = -17031.25K = -5000, \qquad K = 0.293578.$$

From this, $M = 1.0625K = 0.311927$. Hence the answer is

$$I = c_1 e^{-30t} + c_2 e^{-50t} + 0.293578\cos 100t + 0.311927\sin 100t.$$

The exponential terms go to zero and the steady-state solution is a harmonic oscillation whose frequency equals that of the electromotive force. (The decimal fractions are approximations of the exact coefficients $32/109$ and $34/109$ given in the answer in Appendix 2.)

15. **LC-circuit.** Differentiating $E = 220\sin 4t$ gives $E' = 880\cos 4t$. Hence the model of the circuit is $2I'' + 200I = 880\cos 4t$. Division by 2 gives

$$I'' + 100I = 440\cos 4t.$$

A general solution of the homogeneous equation is $I_h = A\cos 10t + B\sin 10t$. You can find a particular solution of the form $I_p = K\cos 4t$. It is not necessary to add a term $M\sin 4t$ because there is no term in I'

(physically: no damping, no phase shift). Substitution gives $K(-16 + 100) \cos 4t = 440 \cos 4t$. Hence $K = 110/21 = 5.238$. Consequently, a general solution of the model is

$$I(t) = A \cos 10t + B \sin 10t + 5.238 \cos 4t. \tag{b}$$

This is a superposition of two harmonic oscillations.

Now use the initial conditions. For the first condition this is simple:

$$I(0) = A + 5.238 = 0, \qquad \text{hence} \quad A = -5.238.$$

The second initial condition is $Q(0) = 0$, meaning that at $t = 0$ the capacitor is uncharged. To use this condition, proceed as in Example 1 on p. 121. From $(1')$ on p. 119 with $R = 0$ (an LC-circuit has $R = 0$!) and $\int I\,dt = Q$ you have

$$LI' + Q/C = 220 \sin 4t.$$

For $t = 0$ and $Q(0) = 0$ this gives $LI'(0) + 0 = 0$, hence $I'(0) = 0$. Now by differentiating (b) you obtain two sine terms, which are 0 when $t = 0$, and the cosine term $10B \cos 10t$, which equals $10B$ when $t = 0$. From this and $I'(0) = 0$ you have $B = 0$. You thus obtain the answer

$$I = 5.238 (\cos 4t - \cos 10t).$$

Note that the term in $\cos 4t$ appears regardless of the initial conditions, whereas the other term is present because of these conditions. The figure shows that the oscillation is periodic (what is the shortest period?) and looks rather complicated. Additional insight into its character is obtained from (12) in Appendix A3.1, which gives

$$\cos 4t - \cos 10t = 2 \sin 7t \sin 3t.$$

This is similar to Fig. 59 in Sec. 2.11, but less distinct because $3t$ is not small enough compared to $7t$.

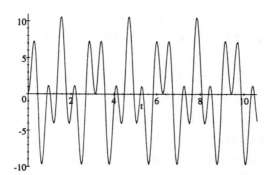

Section 2.12. Problem 15. Superposition of two harmonic oscillations in an LC-circuit

Sec. 2.13   Higher Order Linear Differential Equations

Example 2. The determinant of this homogeneous system is not zero (add Row 1 to Row 2, then develop by Row 2); hence the system has only the trivial solution (all unknowns zero). Or, calling the equations (a), (b), (c), you obtain $k_2 = k_1 + k_3$ from (a), then $k_2 = 0$ from this and (b), then $k_3 = -k_1$ from (a), then $-6k_1 = 0$ from (c), then $k_3 = 0$ from (b).

Problem Set 2.13. Page 131

3. **Wronskian. Initial value problem.** Calculate the Wronskian W. Since $f = e^{-3x}$ has the derivatives $f' = -3f$ and $f'' = 9f$, you obtain a factor f in each column, so that you can factor out $f^3 = e^{-9x}$ from the determinant, the remaining determinant being

$$\begin{vmatrix} 1 & x & x^2 \\ -3 & 1-3x & 2x-3x^2 \\ 9 & -6+9x & 2-12x+9x^2 \end{vmatrix}.$$

To simplify this, add 3 times Row 1 to Row 2 and subtract 9 times Row 1 from Row 3. The result is a determinant that can readily be developed by the first column, giving the value 2, that is,

$$\begin{vmatrix} 1 & x & x^2 \\ 0 & 1 & 2x \\ 0 & -6 & 2-12x \end{vmatrix} = 1(2-12x)+12x = 2.$$

Hence the Wronskian is $W = 2e^{-9x}$. The corresponding differential equation is obtained by first noting that because of the form of these solutions the characteristic equation must have the triple root -3, that is, it must be of the form (use the binomial formula)

$$(\lambda+3)^3 = \lambda^3 + 9\lambda^2 + 27\lambda + 27 = 0.$$

Hence these functions are solutions of the differential equation

$$y''' + 9y'' + 27y' + 27y = 0,$$

as claimed, and they are linearly independent because their Wronskian is not zero. Now write down the corresponding general solution and its first and second derivatives for general x as well as for $x = 0$, and use the initial conditions to determine the three arbitrary constants. This looks as follows.

$$y(x) = (c_1 + c_2 x + c_3 x^2)e^{-3x},$$

$$y(0) = c_1 = 4$$

$$y'(x) = (c_2 + 2c_3 x - 3c_1 - 3c_2 x - 3c_3 x^2)e^{-3x}$$

$$y'(0) = c_2 - 3\cdot 4 = -13, \qquad c_2 = -1$$

$$y''(x) = (2c_3 - 3c_2 - 3(c_2 - 3c_1)) + \textit{further terms})\, e^{-3x}$$

$$y''(0) = 2c_3 - 3(-1) - 3(-1 - 3\cdot 4) = 2c_3 + 42 = 46, \quad c_3 = 2$$

where "*further terms*" are those that give zero when $x = 0$, so you do not need to write them down. This gives the answer

$$y = (4 - x + 2x^2)e^{-3x}.$$

11. Linear dependence. Use $\cos^2 x + \sin^2 x = 1$.

15. Linear dependence. Consider the difference of the first two functions.

Sec. 2.14   Higher Order Homogeneous Equations with Constant Coefficients

Example 1. In the Wronskian, pull out a factor e^{-x} from the first column, e^x from the second, and e^{2x} from the third. Hence the Wronskian equals e^{2x} times the determinant

$$\begin{vmatrix} 1 & 1 & 1 \\ -1 & 1 & 2 \\ 1 & 1 & 4 \end{vmatrix}.$$

The latter is not zero (subtract Column 1 from Column 2 and develop by Column 2, to get $2(4-1) = 6$).

Problem Set 2.14. Page 137

3. General solution. Use that the characteristic equation is a quadratic equation in λ^2. Thus,
$$\lambda^4 - 2\lambda^2 + 1 = (\lambda^2 - 1)^2 = ((\lambda + 1)(\lambda - 1))^2 = (\lambda + 1)^2(\lambda - 1)^2 = 0.$$
In some of the other problems in this set it may be necessary to overcome the practical difficulty of determining the roots by using a numerical method, such as Newton's method (Sec. 17.2) (although we have chosen values such that one root, λ_1, may often be found by inspection and the remaining roots then by dividing the characteristic equation by $\lambda - \lambda_1$).

13. Initial value problem. $\lambda_1 = 1$ by inspection. Dividing by $\lambda - 1$ now gives
$$(\lambda^3 - \lambda^2 - \lambda + 1) \div (\lambda - 1) = \lambda^2 - 1 = (\lambda + 1)(\lambda - 1).$$
From this and the given initial values you get the corresponding general solution, its derivatives, the values at zero, and conditions (a)-(c) for the arbitrary constants in the general solution, as follows.

$$y(x) = (c_1 + c_2 x)e^x + c_3 e^{-x}$$

$$y(0) = c_1 + c_3 = 2 \quad \text{(from the first initial condition)} \tag{a}$$

$$y'(x) = (c_2 + c_1 + c_2 x)e^x - c_3 e^{-x}$$

$$y'(0) = c_2 + c_1 - c_3 = 1 \quad \text{(from the second)} \tag{b}$$

$$y''(x) = (c_2 + c_2 + c_1 + c_2 x)e^x + c_3 e^{-x}$$

$$y''(0) = c_1 + 2c_2 + c_3 = 0 \quad \text{(from the third).} \tag{c}$$

Write the system (a)-(c) more orderly,

$$c_1 \qquad + c_3 = 2 \tag{a}$$

$$c_1 + \ c_2 - c_3 = 1 \tag{b}$$

$$c_1 + 2c_2 + c_3 = 0. \tag{c}$$

To solve this, apply the Gauss elimination or Cramer's rule or simply form (c) minus (a) to get $2c_2 = -2$, hence $c_2 = -1$, then form (a) plus (b), obtaining $2c_1 + c_2 = 2c_1 - 1 = 3$, $c_1 = 2$, and finally use (a), obtaining $c_3 = 2 - c_1 = 0$. Together, this gives the answer $y = (2 - x)e^x$.

The figure shows that y has a maximum at $x = 1$; this can be confirmed by using the derivative. From the change of the tangent direction with increasing x conclude that for positive x the second derivative must always be negative. Indeed, the previous formula for $y''(x)$ with the constants as just determined shows that $y'' = -xe^x$.

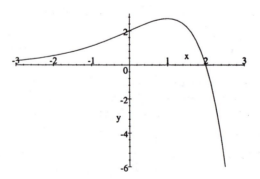

Section 2.14. Problem 13. Solution of the initial value problem

19. CAS Project. **(c)** Without a computer, the equation can be solved as follows. The auxiliary equation is
$$m(m-1)(m-2) + m(m-1) - 2m + 2 = 0.$$
The sum of the last two terms is $-2(m-1)$. Hence you now have a common factor $m-1$ and can write the auxiliary equation as
$$(m-1)(m(m-2)+m-2) = (m-1)(m^2-m-2) = (m-1)(m-2)(m+1).$$
The corresponding general solution is $y = c_1 x + c_2 x^2 + c_3/x$.

Sec. 2.15   Higher Order Nonhomogeneous Equations

Problem Set 2.15. Page 141

1. General solution. The characteristic equation of the homogeneous equation is
$$\lambda^3 + 3\lambda^2 + 3\lambda + 1 = (\lambda+1)^3.$$
It has the triple root -1, so that a basis of solutions is
$$y_1 = e^{-x}, \quad y_2 = xe^{-x}, \quad y_3 = x^2 e^{-x}$$
and the corresponding general solution of the homogeneous equation is
$$y_h = (c_1 + c_2 x + c_3 x^2) e^{-x}.$$
The right side is such that you can use the method of undetermined coefficients. The Modification Rule is not needed since none of the terms on the right is a solution of the homogeneous equation. Start from (see Table 2.1 in Sec, 2.9 if necessary)
$$y_p = Ce^x + K_1 x + K_0.$$
Substitution of this and the derivatives
$$y_p' = Ce^x + K_1, \quad y_p'' = Ce^x, \quad y_p''' = Ce^x$$
into the given equation
$$y''' + 3y'' + 3y' + y = 8e^x + x + 3$$
gives
$$C(1+3+3+1)e^x + 3K_1 + K_1 x + K_0 = 8e^x + x + 3.$$
From this you see that $C = 1$, $K_1 = 1$, $3K_1 + K_0 = 3$, $K_0 = 0$, and the answer is
$$y = y_h + y_p = (c_1 + c_2 x + c_3 x^2) e^{-x} + e^x + x.$$

11. Initial value problem. The auxiliary equation of this Euler-Cauchy equation is
$$m(m-1)(m-2) - 3m(m-1) + 6m - 6 = 0. \tag{A}$$
Ordering terms gives
$$m^3 - 6m^2 + 11m - 6 = 0. \tag{B}$$
$m = 1$ is a root. This can be seen from (B) by inspection or from (A) by noting that $6m - 6 = 6(m-1)$, so that (A) has a common factor $m-1$. Division of (B) by $m-1$ gives
$$(m^3 - 6m^2 + 11m - 6) \div (m-1) = m^2 - 5m + 6 = (m-2)(m-3).$$
Hence 2 and 3 are roots, and a general solution of the homogeneous equation is
$$y_h = c_1 x + c_2 x^2 + c_3 x^3.$$
Now determine a particular solution of the nonhomogeneous equation. Try the method of undetermined coefficients, setting
$$y_p = Kx^5. \quad \text{Then} \quad y_p' = 5Kx^4, \quad y_p'' = 20Kx^3, \quad y_p''' = 60Kx^2.$$
Differentiation has reduced the exponent, but this will be compensated by the increasing power in

successive coefficients, so that you obtain a common factor x^5, namely, since the equation is

$$x^3 y''' - 3x^2 y'' + 6xy' - 6y = 24x^5,$$

you obtain

$$(60 - 3 \cdot 20 + 6 \cdot 5 - 6)Kx^5 = 24Kx^5 = 24x^5, \quad \text{hence} \quad K = 1.$$

This gives the general solution of the nonhomogeneous equation, its derivatives, their values at $x = 1$, and three equations (a), (b), (c) for determining the arbitrary constants by using the initial conditions, as follows.

$$y(x) = c_1 x + c_2 x^2 + c_3 x^3 + x^5$$

$$y(1) = c_1 + c_2 + c_3 + 1 = 1, \qquad c_1 + c_2 + c_3 = 0 \tag{a}$$

$$y'(x) = c_1 + 2c_2 x + 3c_3 x^2 + 5x^4$$

$$y'(1) = c_1 + 2c_2 + 3c_3 + 5 = 3, \qquad c_1 + 2c_2 + 3c_3 = -2 \tag{b}$$

$$y''(x) = 2c_2 + 6c_3 x + 20x^3$$

$$y''(1) = 2c_2 + 6c_3 + 20 = 14, \qquad 2c_2 + 6c_3 = -6 \ . \tag{c}$$

You can solve (a), (b), (c) by elimination. (b) minus (a) gives

$$c_2 + 2c_3 = -2. \tag{d}$$

(c) minus 2(d) gives $2c_3 = -2$, hence $c_3 = -1$. From this and (d) there follows $c_2 = -2 - 2c_3 = 0$. From this and (a) you finally have $c_1 = -c_2 - c_3 = 1$. This gives the answer $y = x - x^3 + x^5$.

CHAPTER 3.   Systems of Differential Equations, Phase Plane, Qualitative Methods

Sec. 3.1   Introductory Examples

Example 2. Spend time on Fig. 76 until you feel that you fully understand the difference between (b) (the usual representation in calculus) and (c), because trajectories will play an important role throughout this chapter. Try to understand the reasons for the following. The trajectory starts at the origin. It reaches its highest point where y_2 has a maximum (before $t = 1$). It has a vertical tangent where I_1 has a maximum, short after $t = 1$. As t increases from there to $t = 5$, the trajectory goes downward until it almost reaches the I_1-axis at 3; this point is a limit as $t \to \infty$. In terms of t the trajectory goes up faster than it comes down.

Problem Set 3.1. Page 158

5. **Electrical network.** The problem amounts to the determination of the two arbitrary constants in a general solution of a system of two differential equations in two unknown functions I_1 and I_2, representing the currents in an electrical network shown in Fig. 76 in Sec. 3.1. You will see that this is quite similar to the corresponding task for a single second-order differential equation. That solution is given by (6), in components

$$I_1(t) = 2c_1 e^{-2t} + c_2 e^{-0.8t} + 3, \qquad I_2(t) = c_1 e^{-2t} + 0.8 c_2 e^{-0.8t}.$$

Setting $t = 0$ and using the given initial conditions $I_1(0) = 9$, $I_2(0) = 0$ gives

$$I_1(0) = 2c_1 + c_2 + 3 = 9 \tag{a}$$

$$I_2(0) = c_1 + 0.8 c_2 = 0. \tag{b}$$

From (b) you have $c_1 = -0.8 c_2$. Substituting this into (a) and simplifying gives $2(-0.8 c_2) + c_2 = 6$, hence $-0.6 c_2 = 6$ or $c_2 = -10$, and $c_1 = 8$. The answer is (note that $2c_1 = 16$)

$$I_1(t) = 16 e^{-2t} - 10 e^{-0.8t} + 3$$

$$I_2(t) = 8e^{-2t} - 8 e^{-0.8t}.$$

These currents are shown in the figure. $I_1(t)$ has the limit 3, as expected. $I_2(t)$ comes out negative; this means it is directed opposite to the arrows shown in Fig. 76(a), which had been assumed arbitrarily at the beginning of the process of modeling; this had to be done because at the beginning, one does not know in what directions the currents will actually flow.

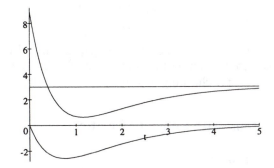

Section 3.1. Problem 5. Currents I_1 (upper curve) and its limit 3 and I_2

9. **Conversion of single differential equations to a system** is an important process, which always follows the pattern shown in formulas (9) and (10) of Sec. 3.1. The present equation $y'' - 9y = 0$ can be readily

solved. A general solution is $y = c_1 e^{3t} + c_2 e^{-3t}$. The point of the problem is not to explain a (complicated) solution method for a simple problem, but to explain the relation between systems and single equations and their solutions. In the present case the formulas (9) and (10) give $y_1 = y$, $y_2 = y'$ and

$$y_1' = y_2$$
$$y_2' = 9y_1$$

(because the given equation can be written $y'' = 9y$, hence $y_1'' = y_1$, but $y_1'' = y_2'$). In matrix form (as in Example 3 of the text) this is

$$\mathbf{y}' = \mathbf{Ay} = \begin{bmatrix} 0 & 1 \\ 9 & 0 \end{bmatrix} \mathbf{y}.$$

The characteristic equation is

$$\det(\mathbf{A} - \lambda\mathbf{I}) = \begin{vmatrix} -\lambda & 1 \\ 9 & -\lambda \end{vmatrix} = \lambda^2 - 9 = 0.$$

The eigenvalues are $\lambda_1 = 3$ and $\lambda_2 = -3$. For λ_1 you obtain an eigenvector from (13) in Sec. 3.0 with $\lambda = \lambda_1$, that is,

$$(\mathbf{A} - \lambda_1\mathbf{I})\mathbf{x} = \begin{bmatrix} -3 & 1 \\ 9 & -3 \end{bmatrix} \begin{bmatrix} x_1 \\ x_2 \end{bmatrix} = \begin{bmatrix} -3x_1 + x_2 \\ 9x_1 - 3x_2 \end{bmatrix} = \mathbf{0}.$$

From the first equation $-3x_1 + x_2 = 0$ you have $x_2 = 3x_1$. An eigenvector is determined only up to a nonzero constant. Hence, in the present case, a convenient choice is $x_1 = 1$, $x_2 = 3$. The second equation gives the same result and is not needed. For the second eigenvalue, $\lambda_2 = -3$, the procedure is the same, namely,

$$(\mathbf{A} - \lambda_2\mathbf{I})\mathbf{x} = \begin{bmatrix} 3 & 1 \\ 9 & 3 \end{bmatrix} \begin{bmatrix} x_1 \\ x_2 \end{bmatrix} = \begin{bmatrix} 3x_1 + x_2 \\ 9x_1 + 3x_2 \end{bmatrix} = \mathbf{0}.$$

You now have $3x_1 + x_2 = 0$, hence $x_2 = -3x_1$, and can choose $x_1 = 1$, $x_2 = -3$. The eigenvectors obtained are

$$\mathbf{x}^{(1)} = [1 \quad 3]^T \quad \text{and} \quad \mathbf{x}^{(2)} = [1 \quad -3]^T.$$

Multiplying these by e^{3t} and e^{-3t}, respectively, and taking a linear combination involving two arbitrary constants c_1 and c_2 gives a general solution of the present system in the form

$$\mathbf{y} = c_1[1 \quad 3]^T e^{3t} + c_2[1 \quad -3]^T e^{-3t}.$$

In components, this is

$$y_1 = c_1 e^{3t} + c_2 e^{-3t}$$
$$y_2 = 3c_1 e^{3t} - 3c_2 e^{-3t}.$$

Here you see that $y_1 = y$ is a general solution of the given equation, and $y_2 = y_1' = y'$ is the derivative of this solution, as had to be expected because of the definition of y_2 at the beginning of the process. Incidentally, you can use $y_2 = y_1'$ for checking your result.

Sec. 3.3   Homogeneous Systems with Constant Coefficients. Phase Plane, Critical Points

Example 2. The characteristic equation is

$$\det(\mathbf{A} - \lambda\mathbf{I}) = \begin{vmatrix} 1-\lambda & 0 \\ 0 & 1-\lambda \end{vmatrix} = (\lambda - 1)^2 = 0.$$

Thus $\lambda = 1$ is an eigenvalue. Any nonzero vector with two components is an eigenvector because $\mathbf{Ax} = \mathbf{x}$ for any $\mathbf{x}$; indeed, $\mathbf{A}$ is the 2×2 unit matrix! Hence you can take $\mathbf{x}^{(1)} = [1 \quad 0]^T$ and $\mathbf{x}^{(2)} = [0 \quad 1]^T$ or any other two linearly independent vectors with two components. This gives the solution on p. 165.

Example 3. $(1 - \lambda)(-1 - \lambda) = (\lambda - 1)(\lambda + 1) = 0$, and so on.

Problem Set 3.3. Page 169

3. General solution. The matrix of the system is

$$A = \begin{bmatrix} 1 & 1 \\ 3 & -1 \end{bmatrix}.$$

The characteristic equation is

$$\det(A - \lambda I) = \begin{vmatrix} 1 - \lambda & 1 \\ 3 & -1 - \lambda \end{vmatrix} = (1 - \lambda)(-1 - \lambda) - 1 \cdot 3 = \lambda^2 - 4 = 0.$$

Hence the eigenvalues are $\lambda_1 = 2$ and $\lambda_2 = -2$. An eigenvector corresponding to λ_1 is obtained by solving $(A - \lambda_1 I)x = (A - 2I)x = 0$, in components,

$$(1 - 2)x_1 + x_2 = 0$$
$$3x_1 + (-1 - 2)x_2 = 0.$$

Each of the equations gives $x_1 = x_2$ (and you need only one of them). Hence you can take $x^{(1)} = [1 \ 1]^T$ as an eigenvector corresponding to $\lambda_1 = 2$. For $\lambda_2 = -2$ those component equations are $(1 + 2)x_1 + x_2 = 0$ (both of the same form) and you can take $x_1 = 1, x_2 = -3$, so that an eigenvector corresponding to $\lambda_2 = -2$ is $x^{(2)} = [1 \ -3]^T$. This gives as a general solution of the system

$$y = c_1 \begin{bmatrix} 1 \\ 1 \end{bmatrix} e^{2t} + c_2 \begin{bmatrix} 1 \\ -3 \end{bmatrix} e^{-2t},$$

in components,

$$y_1 = c_1 e^{2t} + c_2 e^{-2t}$$
$$y_2 = c_1 e^{2t} - 3c_2 e^{-2t}.$$

This agrees with the answer in Appendix 2, with c_1 and c_2 interchanged. (Of course, the notation for arbitrary constants is up to us.)

5. General solution. In this problem you will see the typical calculations in the case of complex eigenvalues. The matrix of the system is

$$A = \begin{bmatrix} 1 & -1 \\ 1 & 1 \end{bmatrix}.$$

This gives the characteristic equation

$$\det(A - \lambda I) = \begin{vmatrix} 1 - \lambda & -1 \\ 1 & 1 - \lambda \end{vmatrix} = (1 - \lambda)^2 + 1 = \lambda^2 - 2\lambda + 2 = 0.$$

The eigenvalues are complex conjugates, $\lambda_1 = 1 + i$ and $\lambda_2 = 1 - i$. Eigenvectors x are obtained from $(A - \lambda I)x = 0$, as before. This is a vector equation, and you need only the first of the two corresponding scalar equations. For $\lambda = \lambda_1 = 1 + i$ it is

$$(1 - (1 + i))x_1 - x_2 = 0,$$

thus

$$-ix_1 = x_2; \quad \text{say,} \quad x_1 = 1 \quad \text{and} \quad x_2 = -i.$$

So the new aspect is that this eigenvector $x^{(1)} = [1 \quad -i]^T$ is no longer real but is complex. Similarly, for $\lambda = \lambda_2 = 1 - i$ you get an eigenvector from

$$(1 - (1 - i))x_1 - x_2 = 0,$$

thus

$$ix_1 = x_2; \quad \text{say,} \quad x_1 = 1 \quad \text{and} \quad x_2 = i.$$

This gives the eigenvector $\mathbf{x}^{(2)} = [1 \quad i]^T$. Using the Euler formula

$$e^{it} = \cos t + i \sin t, \quad e^{-it} = \cos t - i \sin t$$

(see Sec. 2.3) you can write a complex general solution

$$\mathbf{y} = c_1 \begin{bmatrix} 1 \\ -i \end{bmatrix} e^t(\cos t + i \sin t) + c_2 \begin{bmatrix} 1 \\ i \end{bmatrix} e^t(\cos t - i \sin t).$$

Writing this in terms of components and collecting cosine and sine terms, you obtain

$$y_1 = c_1 e^t(\cos t + i \sin t) + c_2 e^t(\cos t - i \sin t)$$
$$= e^t(A \cos t + B \sin t), \quad\quad A = c_1 + c_2, \quad B = ic_1 - ic_2$$
$$y_2 = -ic_1 e^t(\cos t + i \sin t) + ic_2 e^t(\cos t - i \sin t)$$
$$= e^t(C \cos t + D \sin t), \quad\quad C = -ic_1 + ic_2, \quad D = c_1 + c_2.$$

You see that $C = -B$ and $D = A$. You can now write a real general solution in vector form, namely,

$$\mathbf{y} = e^t \left(\begin{bmatrix} A \\ -B \end{bmatrix} \cos t + \begin{bmatrix} B \\ A \end{bmatrix} \sin t \right).$$

15. Initial value problem. The matrix of the given system is

$$\mathbf{A} = \begin{bmatrix} -14 & 10 \\ -5 & 1 \end{bmatrix}.$$

From $\mathbf{A}$ you obtain the characteristic equation

$$\det (\mathbf{A} - \lambda \mathbf{I}) = \begin{vmatrix} -14 - \lambda & 10 \\ -5 & 1 - \lambda \end{vmatrix} = (-14 - \lambda)(1 - \lambda) + 50 = 0.$$

Simplification gives $\lambda^2 + 13\lambda + 36 = 0$. Hence the eigenvalues are $\lambda_1 = -9$ and $\lambda_2 = -4$. Corresponding eigenvectors are obtained from

$$[-14 - (-9)]x_1 + 10x_2 = 0, \quad \text{thus} \quad -5x_1 + 10x_2 = 0, \quad \text{say,} \quad x_1 = 2, \quad x_2 = 1$$

and

$$[-14 - (-4)]x_1 + 10x_2 = 0, \quad \text{thus} \quad -10x_1 + 10x_2 = 0, \quad \text{say,} \quad x_1 = 1, \quad x_2 = 1.$$

Hence the corresponding general solution is

$$\mathbf{y} = c_1 \begin{bmatrix} 2 \\ 1 \end{bmatrix} e^{-9t} + c_2 \begin{bmatrix} 1 \\ 1 \end{bmatrix} e^{-4t}.$$

From this and the initial conditions $y_1(0) = -1$, $y_2(0) = 1$, written in vector form, you obtain

$$\mathbf{y}(0) = c_1 \begin{bmatrix} 2 \\ 1 \end{bmatrix} + c_2 \begin{bmatrix} 1 \\ 1 \end{bmatrix} = \begin{bmatrix} -1 \\ 1 \end{bmatrix}, \quad \text{thus} \quad \begin{matrix} 2c_1 + c_2 = -1 \\ c_1 + c_2 = 1 \end{matrix}.$$

The solution is $c_1 = -2$, $c_2 = 3$, so that you obtain the particular solution

$$\mathbf{y} = -2 \begin{bmatrix} 2 \\ 1 \end{bmatrix} e^{-9t} + 3 \begin{bmatrix} 1 \\ 1 \end{bmatrix} e^{-4t}, \quad \text{thus} \quad \begin{matrix} y_1 = -4e^{-9t} + 3e^{-4t} \\ y_2 = -2e^{-9t} + 3e^{-4t} \end{matrix}.$$

The figure shows that both y_1 (the lower curve) and y_2 have a maximum and then approach zero in a monotone fashion.

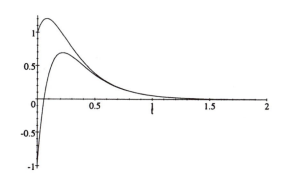

Section 3.3. Problem 15. Particular solutions y_1 (lower curve) and y_2

Sec. 3.4   Criteria for Critical Points. Stability

Problem Set 3.4. Page 174

3. Saddle. The type of a critical point is determined by quantities closely related to the eigenvalues of the matrix of a system, namely, the trace p, which is the sum of the eigenvalues, the determinant q, which is the product of the eigenvalues, and the discriminant Δ, which equals $p^2 - 4q$; see (9) in Sec. 3.4. In Prob. 3 the matrix is

$$ \mathbf{A} = \begin{bmatrix} 1 & 2 \\ 2 & 1 \end{bmatrix}. $$

Hence $p = 1 + 1 = 2$, $q = 1 - 4 = -3$, and $\Delta = 4 - 4\cdot(-3) = 16$. Since $q < 0$, the system has a saddle point at 0, which is always unstable, as follows from (10c) in Sec. 3.4, and is plausible from Fig. 80 in Sec. 3.3. To solve the system, you need the eigenvalues, which you obtain as solutions of the characteristic equation

$$ \det (\mathbf{A} - \lambda\mathbf{I}) = \begin{vmatrix} 1-\lambda & 2 \\ 2 & 1-\lambda \end{vmatrix} = \lambda^2 - 2\lambda - 3 = (\lambda - 3)(\lambda + 1) = 0. $$

Hence the eigenvalues are -1 and 3. Their signs differ, which makes q negative and causes a saddle point. An eigenvector $\mathbf{x}^{(1)}$ for -1 is obtained from

$$ (1 - (-1))x_1 + 2x_2 = 0, \quad \text{thus} \quad x_2 = -x_1, \quad \text{say,} \quad x_1 = 1, \quad x_2 = -1. $$

Hence $\mathbf{x}^{(1)} = [1 \quad -1]^T$. Similarly, for the eigenvalue 3 you obtain an eigenvector from

$$ (1 - 3)x_1 + 2x_2 = 0, \quad \text{thus} \quad x_2 = x_1, \quad \text{say,} \quad x_1 = 1, \quad x_2 = 1. $$

This gives the eigenvector $\mathbf{x}^{(2)} = [1 \quad 1]^T$. Hence a general solution is

$$ y = c_1 \begin{bmatrix} 1 \\ -1 \end{bmatrix} e^{-t} + c_2 \begin{bmatrix} 1 \\ 1 \end{bmatrix} e^{3t}. $$

17. Perturbation of center. Example 4 in Sec. 3.3, to which the problem refers, shows two methods of solution, a systematic method and the shortcut. The first of them is similar to the procedure explained in this Manual in Prob. 5 of Problem Set 3.3 and can be completed following that method. The purpose of this Prob. 17 is to become aware of the fact that inaccuracies in the coefficients of a system (errors caused by rounding or in the process of physical measurements, etc.) can change the type of a critical point. In this problem it is suggested to go from $\mathbf{A}$ to $\mathbf{B} = \mathbf{A} + 0.1\mathbf{I}$, but it will be obvious from the analysis that smaller deviations would have a similar effect. Given

$$\mathbf{A} = \begin{bmatrix} 0 & 1 \\ -4 & 0 \end{bmatrix}, \quad \text{hence} \quad \mathbf{B} = \mathbf{A} + 0.1\mathbf{I} = \begin{bmatrix} 0.1 & 1 \\ -4 & 0.1 \end{bmatrix}.$$

For $\mathbf{A}$ you have $p = 0 + 0 = 0$, $q = -1 \cdot (-4) = 4$. This confirms that the system in Example 4 has a center as its critical point; see (9c) in Sec. 3.4. For $\mathbf{B}$ you have $p = 0.1 + 0.1 = 0.2 \neq 0$, $q = 0.01 + 4 = 4.01$, and $\Delta = p^2 - 4q = 0.04 - 16.04 = -16 < 0$, which gives a spiral point by (9d) in Sec. 3.4. The eigenvalues of $\mathbf{A}$ are pure imaginary, $2i$ and $-2i$ (see Example 4 in Sec. 3.3), and it is interesting that the eigenvalues of $\mathbf{B}$ are $0.1 + 2i$ and $0.1 - 2i$, that is, they were changed by the same amount by which the main diagonal entries were changed (this reflects a general "shifting property"). Indeed, the characteristic equation of $\mathbf{B}$ is

$$\det(\mathbf{B} - \lambda\mathbf{I}) = \begin{vmatrix} 0.1 - \lambda & 1 \\ -4 & 0.1 - \lambda \end{vmatrix} = (0.1 - \lambda)^2 + 4 = \lambda^2 - 0.2\lambda + 4.01 = 0.$$

The roots (the eigenvalues of $\mathbf{B}$) are $0.1 + 2i$ and $0.1 - 2i$.

Sec. 3.5   Qualitative Methods for Nonlinear Systems

Example 1. The critical point at $(0, 0)$ turns out to be a center. This follows from the general criteria in Sec. 3.4. This is the first result. The next result follows from this and the periodicity of $\sin\theta = \sin y_1$ with 2π. Namely, the points $\pm 2\pi$, $\pm 4\pi$, ... must also be centers. (Keep in mind that y_1 is just another notation for θ, introduced to fit the notation of our general discussions in this chapter.) The third result concerns the critical point $(\pi, 0)$ at $\theta = \pi$ of the θ-axis. The trick now is to move the origin to this point because our criteria were derived under the assumption that the critical point to be discussed is at the origin. This is the idea of the transformation (a translation)

$$\theta - \pi = y_1, \quad \text{thus} \quad \theta = \pi + y_1. \tag{A}$$

You see that $\theta = \pi$ now corresponds to our new $y_1 = 0$; we are at the new origin. Think about this before going on. From (A), $\sin\pi = 0$, and $\cos\pi = -1$ it follows that

$$\sin\theta = \sin(\pi + y_1) = \sin\pi\cos y_1 + \cos\pi\sin y_1 = -\sin y_1 = -y_1 + \frac{y_1^3}{6} - + ...,$$

as indicated in the example.

Problem Set 3.5. Page 183

5. Linearization begins with the determination of the positions of the critical points. As a system the given equation $y'' - y + y^2 = 0$ becomes (see Sec. 3.1 for the general formula)

$$y_1' = y_2 \tag{a}$$
$$y_2' = y_1 - y_1^2.$$

The critical points are obtained from $y_2 = 0$ (then $y_1' = 0$), $y_1 - y_1^2 = y_1(1 - y_1) = 0$ (then $y_2' = 0$). Hence they are at 0 and 1 on the y_1-axis. Linearization is then done for each critical point separately. and in each case the point is first shifted to the origin by a suitable change of coordinates, as explained in somewhat more detail just above (in connection with Example 1). Accordingly, begin with the critical point $(0, 0)$. No transformation of y_1 or y_2 is necessary because the point already has the required position. From (a) you obtain the system linearized at the origin simply by dropping the quadratic term. This linearized system is

$$y_1' = y_2$$
$$y_2' = y_1.$$

Its matrix is

$$\mathbf{A}_1 = \begin{bmatrix} 0 & 1 \\ 1 & 0 \end{bmatrix}.$$

Calculate $p = 0 + 0 = 0$, $q = -1 \cdot 1 = -1$, and $\Delta = p^2 - 4q = 4$. Since $q < 0$, this is a saddle by (9b) in Sec. 3.4. (You do not need p and Δ.) The second critical point is at $y_1 = 1$, $y_2 = 0$. Hence make a shift by setting $y_1 = 1 + \tilde{y}_1$, $y_2 = \tilde{y}_2$. Then $y_1' = \tilde{y}_1'$,

$$y_1 - y_1{}^2 = y_1(1 - y_1) = (1 + \tilde{y}_1)(-\tilde{y}_1) = -\tilde{y}_1 - \tilde{y}_1^2$$

and (a) takes the form

$$\tilde{y}_1' = \tilde{y}_2$$
$$\tilde{y}_2' = -\tilde{y}_1 - \tilde{y}_1^2.$$

Linearize this by dropping the nonlinear term (the last term in the second differential equation). This gives the linearized system

$$\begin{array}{l} \tilde{y}_1' = \tilde{y}_2 \\ \tilde{y}_2' = -\tilde{y}_1 \end{array} \quad \text{whose matrix is} \quad \mathbf{A}_2 = \begin{bmatrix} 0 & 1 \\ -1 & 0 \end{bmatrix}.$$

Calculate $p = 0$, $q = 1$, and conclude from (9c) in Sec. 3.4 that the critical point at $(1, 0)$ is a center.

13. Trajectories. $yy'' + y'^2 = (yy')' = 0$. By integration, $yy' = const$ or $y_1 y_2 = const$. These are the familiar hyperbolas with the coordinate axes as asymptotes.

Sec. 3.6   Nonhomogeneous Linear Systems

Example 1. The solution of the homogeneous system (not shown in the text) proceeds as before. That is, the characteristic equation of the matrix $\mathbf{A}$ is

$$(2 - \lambda)(-3 - \lambda) - (-4) \cdot 1 = \lambda^2 + \lambda - 2 = (\lambda - 1)(\lambda + 2) = 0.$$

Hence the eigenvalues are 1 and -2. Eigenvectors are obtained for $\lambda = 1$ from

$$(2 - 1)x_1 - 4x_2 = 0, \quad \text{say}, \quad x_1 = 4, \quad x_2 = 1$$

and for $\lambda = -2$ from

$$(2 - (-2))x_1 - 4x_2 = 0, \quad \text{say}, \quad x_1 = 1, \quad x_2 = 1.$$

This gives the solution of the homogeneous equation shown in the answer on p. 185.

Problem Set 3.6. Page 189

3. General solution. e^{3t} and $-3e^{3t}$ are such that you can apply the method of undetermined coefficients for determining a particular solution of the nonhomogeneous system. For this purpose you must first determine a general solution of the homogeneous system. The matrix of the latter is

$$\mathbf{A} = \begin{bmatrix} 0 & 1 \\ 1 & 0 \end{bmatrix}.$$

It has the characteristic equation $\lambda^2 - 1 = 0$. Hence the eigenvalues of $\mathbf{A}$ are $\lambda_1 = -1$ and $\lambda_2 = 1$. Eigenvectors $\mathbf{x} = \mathbf{x}^{(1)}$ and $\mathbf{x}^{(2)}$ are obtained from $(\mathbf{A} - \lambda\mathbf{I})\mathbf{x} = \mathbf{0}$ with $\lambda = \lambda_1 = -1$ and $\lambda = \lambda_2 = 1$, respectively. For $\lambda_1 = -1$ you obtain

$$x_1 + x_2 = 0, \quad \text{thus} \quad x_2 = -x_1, \quad \text{say}, \quad x_1 = 1, \quad x_2 = -1.$$

Similarly, for $\lambda_2 = 1$ you obtain

$$-x_1 + x_2 = 0, \quad \text{thus} \quad x_2 = x_1, \quad \text{say}, \quad x_1 = 1, \quad x_2 = 1.$$

Hence eigenvectors are $\mathbf{x}^{(1)} = [1 \quad -1]^T$ and $\mathbf{x}^{(2)} = [1 \quad 1]^T$. This gives the general solution of the homogeneous system

$$\mathbf{y}^{(h)} = c_1 \begin{bmatrix} 1 \\ -1 \end{bmatrix} e^{-t} + c_2 \begin{bmatrix} 1 \\ 1 \end{bmatrix} e^t.$$

Now determine a particular solution of the nonhomogeneous system. Using the notation in the text (Sec. 3.6) you have on the right $\mathbf{g} = [1 \quad -3]^T e^{3t}$. This suggests the choice

$$\mathbf{y}^{(p)} = \mathbf{u} e^{3t} = [u_1 \quad u_2]^T e^{3t}. \tag{a}$$

Here $\mathbf{u}$ is a constant vector to be determined. The Modification Rule is not needed because 3 is not an eigenvalue of $\mathbf{A}$. Substitution of (a) into the given system $\mathbf{y}' = \mathbf{Ay} + \mathbf{g}$ yields

$$\mathbf{y}^{(p)\prime} = 3\mathbf{u} e^{3t} = \mathbf{Ay}^{(p)} + \mathbf{g} = \begin{bmatrix} 0 & 1 \\ 1 & 0 \end{bmatrix} \begin{bmatrix} u_1 \\ u_2 \end{bmatrix} e^{3t} + \begin{bmatrix} 1 \\ -3 \end{bmatrix} e^{3t}.$$

Omitting the common factor e^{3t}, you obtain in terms of components

$$3u_1 = u_2 + 1 \qquad \text{ordered} \qquad 3u_1 - u_2 = 1$$
$$3u_2 = u_1 - 3 \qquad\qquad\qquad -u_1 + 3u_2 = -3.$$

Solution by elimination or by Cramer's rule (Sec. 6.6) gives $u_1 = 0$ and $u_2 = -1$. Hence the answer is

$$\mathbf{y} = c_1 \begin{bmatrix} 1 \\ -1 \end{bmatrix} e^{-t} + c_2 \begin{bmatrix} 1 \\ 1 \end{bmatrix} e^t + \begin{bmatrix} 0 \\ -1 \end{bmatrix} e^{3t}.$$

17. Network. First derive the model. For the left loop of the electrical network you obtain from Kirchhoff's voltage law

$$LI_1' + R_1(I_1 - I_2) = E \tag{a}$$

because both currents flow through R_1, but in opposite directions, so that you have to take their difference. For the right loop you similarly obtain

$$R_1(I_2 - I_1) + R_2 I_2 + \frac{1}{C} \int I_2 \, dt = 0. \tag{b}$$

Insert the given numerical values in (a). Do the same in (b) and differentiate (b) in order to get rid of the integral. This gives

$$I_1' + 2(I_1 - I_2) = 200$$
$$2(I_2' - I_1') + 8I_2' + 2I_2 = 0.$$

Write the terms in the first of these two equations in the usual order, obtaining

$$I_1' = -2I_1 + 2I_2 + 200. \tag{a1}$$

Do the same in the second equation as follows. Collecting terms and then dividing by 10, you first have

$$10I_2' - 2I_1' + 2I_2 = 0 \quad \text{or} \quad I_2' - 0.2I_1' + 0.2I_2 = 0.$$

To obtain the usual form, you have to get rid of the term in I_1', which you replace by using (a1). This gives

$$I_2' - 0.2(-2I_1 + 2I_2 + 200) + 0.2I_2 = 0.$$

Collecting terms and ordering them as usual, you obtain

$$I_2' = -0.4I_1 + 0.2I_2 + 40. \tag{b1}$$

(a1) and (b1) are the two equations of the system that you use in your further work. The matrix of the corresponding homogeneous system is

$$\mathbf{A} = \begin{bmatrix} -2 & 2 \\ -0.4 & 0.2 \end{bmatrix}.$$

Its characteristic equation is ($\mathbf{I}$ is the unit matrix)

$$\det(\mathbf{A} - \lambda\mathbf{I}) = (-2 - \lambda)(0.2 - \lambda) - (-0.4) \cdot 2 = \lambda^2 + 1.8\lambda + 0.4 = 0.$$

This gives the eigenvalues

$$\lambda_1 = -0.9 + \sqrt{0.41} = -0.259688$$

and

$$\lambda_2 = -0.9 - \sqrt{0.41} = -1.540312.$$

Eigenvectors are obtained from $(\mathbf{A} - \lambda\mathbf{I})\mathbf{x} = \mathbf{0}$ with $\lambda = \lambda_1$ and $\lambda = \lambda_2$. For λ_1 this gives

$$(-2 - \lambda_1)x_1 + 2x_2 = 0, \quad \text{say,} \quad x_1 = 2 \quad \text{and} \quad x_2 = 2 + \lambda_1.$$

Similarly, for λ_2 you obtain

$$(-2 - \lambda_2)x_1 + 2x_2 = 0, \quad \text{say,} \quad x_1 = 2 \quad \text{and} \quad x_2 = 2 + \lambda_2.$$

The eigenvectors thus obtained are

$$\mathbf{x}^{(1)} = \begin{bmatrix} 2 \\ 2 + \lambda_1 \end{bmatrix} = \begin{bmatrix} 2 \\ 1.1 + \sqrt{0.41} \end{bmatrix}$$

and

$$\mathbf{x}^{(2)} = \begin{bmatrix} 2 \\ 2 + \lambda_2 \end{bmatrix} = \begin{bmatrix} 2 \\ 1.1 - \sqrt{0.41} \end{bmatrix}.$$

This gives as a general solution of the homogeneous system

$$\mathbf{I}^{(h)} = c_1 \mathbf{x}^{(1)} e^{\lambda_1 t} + c_2 \mathbf{x}^{(2)} e^{\lambda_2 t}.$$

You finally need a particular solution $\mathbf{I}^{(p)}$ of the given nonhomogeneous system $\mathbf{J}' = \mathbf{A}\mathbf{J} + \mathbf{g}$, where $\mathbf{g} = [200 \quad 40]^T$ is constant, and $\mathbf{J} = [I_1 \quad I_2]^T$ is the vector of the currents. The method of undetermined coefficients applies. Since $\mathbf{g}$ is constant, you can choose a constant $\mathbf{I}^{(p)} = \mathbf{u} = [u_1 \quad u_2]^T = \mathbf{const}$ and substitute it into the the system, obtaining, since $\mathbf{u}' = \mathbf{0}$,

$$\mathbf{I}^{(p)'} = \mathbf{0} = \mathbf{A}\mathbf{u} + \mathbf{g} = \begin{bmatrix} -2 & 2 \\ -0.4 & 0.2 \end{bmatrix} \begin{bmatrix} u_1 \\ u_2 \end{bmatrix} + \begin{bmatrix} 200 \\ 40 \end{bmatrix} = \begin{bmatrix} -2u_1 + 2u_2 + 200 \\ -0.4u_1 + 0.2u_2 + 40 \end{bmatrix}.$$

Hence you can determine u_1 and u_2 from the system

$$-2u_1 + 2u_2 = -200$$
$$-0.4u_1 + 0.2u_2 = -40.$$

The solution is $u_1 = 100$, $u_2 = 0$. The answer is

$$\mathbf{J} = \mathbf{I}^{(h)} + \mathbf{I}^{(p)}.$$

CHAPTER 4. Series Solutions of Differential Equations. Special Functions

Sec. 4.2   Theory of the Power Series Method

Problem Set 4.2. Page 204

1. Power series solution. The equation $y' = -2xy$ can readily be solved by separating variables,

$$dy/y = -2x\,dx, \qquad \ln|y| = -x^2 + C, \qquad y = ce^{-x^2}.$$

If for some reason a Maclaurin series of this solution is wanted, you can obtain it by substituting $-x^2$ for x in the familiar series for e^x. Hence this problem (as well as the others) just serves to explain the techniques in a simple case (in which you would not need them), as a preparation for equations, such as Legendre's, Bessel's, and the hypergeometric equations, to name the most important ones, where you do need these techniques. Start from the series

$$y = a_0 + a_1 x + a_2 x^2 + a_3 x^3 + \dots \tag{a}$$

and differentiate it termwise, obtaining

$$y' = a_1 + 2a_2 x + 3a_3 x^2 + 4a_4 x^3 + \dots . \tag{b}$$

Leave space behind each series on your sheet, so that you can add terms if necessary. From (a) obtain the right side of the given differential equation (write corresponding powers below each other; this will facilitate your further work)

$$-2xy = \qquad -2a_0 x - 2a_1 x^2 - 2a_2 x^3 - 2a_3 x^4 - \dots$$

For each power $x^0, x, x^2, x^3, \dots$ equate the two corresponding terms. Denote these equations for determining the coefficients by [0] (constant terms), [1] (first power of x), etc. This looks as follows.

$$a_1 = 0 \qquad\qquad\qquad\qquad\qquad\qquad\qquad\qquad\qquad\qquad [0]$$

$$2a_2 = -2a_0, \quad \text{hence} \quad a_2 = -a_0, \; a_0 \text{ arbitrary} \qquad\qquad [1]$$

$$3a_3 = -2a_1 = 0 \qquad\qquad\qquad\qquad\qquad\qquad\qquad\qquad\qquad [2]$$

$$4a_4 = -2a_2, \quad \text{hence} \quad a_4 = (-2/4)a_2 = (1/2)a_0 \qquad\qquad [3]$$

$$5a_5 = -2a_3 = 0 \qquad\qquad\qquad\qquad\qquad\qquad\qquad\qquad\qquad [4]$$

$$6a_6 = -2a_4, \quad \text{hence} \quad a_6 = (-2/6)a_4 = -(1/3!)a_0 \qquad\qquad [5]$$

and so on. With a little more skill, you may use power series notation and write

$$y' + 2xy = \sum_{n=1}^{\infty} n a_n x^{n-1} + 2x \sum_{m=0}^{\infty} a_m x^m = 0.$$

In the second series multiply each term by $2x$. Then you have the general power x^{m+1}. To get the same power in the first series, set $n = m + 2$; this gives $x^{n-1} = x^{m+1}$ (this was the reason for choosing different summation letters in the two series). Also pull out the first term a_1 of the first series; then both summations begin with $m = 0$ and you can take the two series together, obtaining

$$a_1 + \sum_{m=0}^{\infty} (m+2) a_{m+2} x^{m+1} + \sum_{m=0}^{\infty} 2 a_m x^{m+1}$$

$$= a_1 + \sum_{m=0}^{\infty} [(m+2) a_{m+2} + 2 a_m] x^{m+1} = 0.$$

You see now that $a_1 = 0$ and get the recursion

$$(m+2)a_{m+2} + 2a_m = 0 \quad \text{or} \quad a_{m+2} = -2a_m/(m+2), \quad m = 0, 1, \dots$$

Choosing $m = 0, 1, \dots$, you obtain successively

$$a_2 = -\frac{2}{2}a_0, \quad a_3 = 0, \quad a_4 = -\frac{4}{2}a_2 = \frac{1}{2}a_0, \quad a_5 = 0, \dots$$

as before.

17. Radius of convergence. A power series in powers of x may converge for all x (this is the best possible case) or within an interval with the center x_0 as midpoint (in the complex plane: within a disk with center x_0) or only at the center (the practically useless case). In the second case the interval of convergence has length $2R$, where R is called the *radius of convergence* (it is a radius in the complex case, as has just been said) and is given by (7a) or (7b) of Sec. 4.2. Here it is assumed that the limits in these formulas exist. This will be the case in most applications. (For help when this is not the case, see Sec. 14.2.) The convergence radius is important whenever you want to use series for computing values, exploring properties of functions represented by series, or proving relations between functions, tasks of which you will gain a first impression in Secs. 4.3-4.7 and corresponding problems. In Prob. 17 you may set $x^2 = t$. Then you have a power series in t with coefficients of absolute value $|a_m| = 1/|k|^m$. Hence the root in (7a) is $1/|k|$, so that the radius of convergence of the power series in t is $|k|$. That is, the series converges for $|t| < |k|$. This implies $|x| = \sqrt{|t|} < \sqrt{|k|}$. Hence the given series has the convergence radius $\sqrt{|k|}$. Confirm this by using (7b). The quotient in (7b) is

$$|a_{m+1}/a_m| = 1/|k|.$$

This leads to the same result as before. Note that the problem is special; in general, the sequences of those roots and quotients in (7) will not be constant, that is, the terms of such a sequence of quotients (or roots) will not be all the same.

Sec. 4.3   Legendre's Equation. Legendre Polynomials $P_n(x)$

Problem Set 4.3. Page 209

1. Legendre functions for $n = 0$. The power series and Frobenius methods were instrumental in establishing large portions of the very extensive theory of special functions (see, for instance, Refs. [1], [11], [12] in Appendix 1), as needed in engineering, physics (astronomy!), and other areas, simply because many special functions appeared first in the form of power series solutions of differential equations. In general, this concerns properties and relationships between higher transcendental functions. The point of Prob. 1 is an illustration that sometimes such functions may reduce to elementary functions known from calculus. If you set $n = 0$ in (7), it was observed in the problem that $y_1(x)$ becomes $\frac{1}{2}\ln((1+x)/(1-x))$. In this case, the answer suggests using

$$\ln(1+x) = x - \frac{1}{2}x^2 + \frac{1}{3}x^3 - + \dots .$$

Replacing x by $-x$ and multiplying by -1 on both sides gives

$$-\ln(1-x) = \ln\frac{1}{1-x} = x + \frac{1}{2}x^2 + \frac{1}{3}x^3 + \dots .$$

Addition of these two series and division by 2 verifies the last equality sign in the formula of Prob. 1. You are requested to obtain this result directly by solving the Legendre equation (1) with $n = 0$, that is,

$$(1-x^2)y'' - 2xy' = 0 \quad \text{or} \quad (1-x^2)z' = 2xz, \quad \text{where} \quad z = y'.$$

Separation of variables and integration gives

$$\frac{dz}{z} = \frac{2x}{1-x^2}dx, \quad \ln|z| = -\ln|1-x^2| + c, \quad z = C_1/(1-x^2).$$

y is now obtained by another integration, using partial fractions

$$\frac{1}{1-x^2} = \frac{1}{2}\left(\frac{1}{x+1} - \frac{1}{x-1}\right).$$

This gives

$$y = \int z\, dx = \frac{1}{2}C_1(\ln(x+1) - \ln(x-1)) + c = \frac{1}{2}C_1 \ln \frac{x+1}{x-1} + c.$$

Since $y_1(x)$ in (6), Sec. 4.3, reduces to 1 if $n = 0$, you can now readily express your solution obtained in terms of the standard functions y_1 and y_2 in (6) and (7), namely,

$$y = cy_1(x) + C_1 y_2(x).$$

7. Differential equation. Set $x = az$ and apply the chain rule, according to which

$$\frac{d}{dx} = \frac{d}{dz}\frac{dz}{dx} = \frac{1}{a}\frac{d}{dz} \quad \text{and} \quad \frac{d^2}{dx^2} = \frac{1}{a^2}\frac{d^2}{dz^2}.$$

Substitution now gives

$$(a^2 - a^2 z^2)\frac{d^2 y}{dz^2}\frac{1}{a^2} - 2az\frac{dy}{dz}\frac{1}{a} + 3\cdot 4y = 0.$$

The factors a cancel and you are left with

$$(1 - z^2)y'' - 2zy' + 3\cdot 4y = 0.$$

Hence the solution is $P_3(z) = P_3(x/a)$, as claimed in Appendix 2.

Sec. 4.4 Frobenius Method

Problem Set 4.4. Page 216

5. Basis of solutions. Substitute y, y', and y'', given by (2) and (2*) in Sec. 4.4, into the differential equation $xy'' + 2y' + xy = 0$. This gives

$$\sum_{m=0}^{\infty}(m+r)(m+r-1)a_m x^{m+r-1} + \sum_{m=0}^{\infty}2(m+r)a_m x^{m+r-1} + \sum_{n=0}^{\infty}a_n x^{n+r+1} = 0.$$

The first two series have the same general power, and you can take them together. In the third series set $n = m - 2$ to get the same general power. $n = 0$ then gives $m = 2$. You obtain

$$\sum_{m=0}^{\infty}(m+r)(m+r+1)a_m x^{m+r-1} + \sum_{m=2}^{\infty}a_{m-2}x^{m+r-1} = 0. \tag{A}$$

For $m = 0$ this gives the indicial equation

$$r(r+1) = 0.$$

The roots are $r = 0$ and -1. They differ by an integer. This is Case 3 of Theorem 2 in Sec. 4.4. Consider the larger root $r = 0$. Then (A) takes the form

$$\sum_{m=0}^{\infty}m(m+1)a_m x^{m-1} + \sum_{m=2}^{\infty}a_{m-2}x^{m-1} = 0.$$

$m = 1$ gives $2a_1 = 0$. This implies $a_3 = a_5 = \dots = 0$, as is seen by taking $m = 3, 5, \dots$. Furthermore,

$$m = 2 \text{ gives } 2\cdot3 a_2 + a_0 = 0, \text{ hence } a_0 \text{ arbitrary,} \quad a_2 = -a_0/3!$$

$$m = 4 \text{ gives } 4\cdot5 a_4 + a_2 = 0, \text{ hence } a_4 = -a_2/(4\cdot5) = +a_0/5!$$

and so on. Since you want a basis and a_0 is arbitrary, you can take $a_0 = 1$. Recognize that you then have the Maclaurin series of

$$y_1 = (\sin x)/x.$$

Now determine an independent solution y_2. Since in Case 3 one would have to assume a term involving the logarithm (which may turn out to be zero), reduction of order (Sec. 2.1) seems to be simpler. This begins by writing the equation in standard form (divide by x):

$$y'' + (2/x)y' + y = 0.$$

In (2) of Sec. 2.1 you then have $p = 2/x$, $-\int p\,dx = -2\ln|x| = \ln(1/x^2)$, hence $\exp(-\int p\,dx) = 1/x^2$. Insertion of this and y_1^2 into (9) and cancellation of a factor x^2 gives

$$U = 1/\sin^2 x, \qquad u = \int U\,dx = -\cot x, \qquad y_2 = uy_1 = -\frac{\cos x}{x}.$$

9. Basis of solutions. Try the substitution $x + 2 = t$. Can you see why?

13. Euler-Cauchy equation. The point of this problem is that you should recognize how Euler-Cauchy equations fit into the Frobenius theory.

Sec. 4.5   Bessel's Equation. Bessel Functions $J_\nu(x)$

Problem Set 4.5. Page 226

3. Reduction to Bessel's equation. Bessel's equation gains additional importance by the fact that numerous other differential equations can be reduced to this equation, so that the extensive theory of Bessel functions becomes applicable to the solutions of those other equations and their engineering uses. The corresponding transformations involve applications of the chain rule in transforming derivatives. From $x^2 = z$ you obtain $y' = 2x\,(dy/dx)$ and y'' by differentiating this; in detail

$$y' = \frac{dy}{dx} = \frac{dy}{dz}\frac{dz}{dx} = 2x\frac{dy}{dz} = 2\sqrt{z}\,\frac{dy}{dz},$$

$$y'' = \frac{d^2 y}{dx^2} = 2\frac{dy}{dz} + 4x^2\frac{d^2 y}{dz^2} = 2\frac{dy}{dz} + 4z\frac{d^2 y}{dz^2}.$$

By substitution,

$$x^2 y'' + xy' + (4x^4 - \tfrac{1}{4})y = z(4z\frac{d^2 y}{dz^2} + 2\frac{dy}{dz}) + 2z\frac{dy}{dz} + (4z^2 - \tfrac{1}{4})y = 0.$$

Division by 4 gives the Bessel equation

$$z^2 \frac{d^2 y}{dz^2} + z\frac{dy}{dz} + (z^2 - \tfrac{1}{16})y = 0.$$

Hence a general solution is

$$y = AJ_{1/4}(z) + BJ_{-1/4}(z) = AJ_{1/4}(x^2) + BJ_{-1/4}(x^2). \tag{A}$$

The figures show the two Bessel functions that form the basis in (A), plotted as functions of z (their usual appearance) and as functions of x^2 over the x-axis, in which case the oscillations become more and more rapid with increasing x.

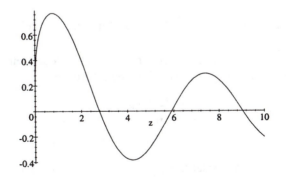

Section 4.5. Problem 3. Bessel function $J_{1/4}(z)$

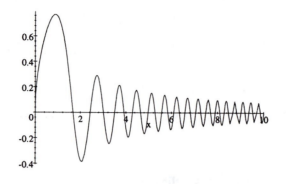

Section 4.5. Problem 3. Bessel function $J_{1/4}(x^2)$

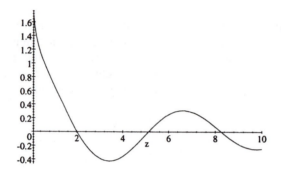

Section 4.5. Problem 3. Bessel function $J_{-1/4}(z)$

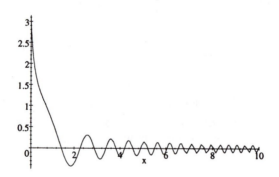

Section 4.5. Problem 3. Bessel function $J_{-1/4}(x^2)$

9. **Reduction to Bessel's equation.** Here you may first transform the dependent variable (the unknown function y) by the given transformation $y = x^{1/3}u$. You need the derivatives

$$y' = \frac{1}{3}x^{-2/3}u + x^{1/3}u',$$

$$y'' = -\frac{2}{9}x^{-5/3}u + \frac{2}{3}x^{-2/3}u' + x^{1/3}u''.$$

Substitute this into the given equation, order terms, and drop a common factor $x^{1/3}$. This gives

$$81x^2u'' + 81xu' + (9x^{2/3} - 1)u = 0. \qquad\qquad \textbf{(B)}$$

Now comes the second step: introduce $z = x^{1/3}$ as the new independent variable. You have to transform the derivatives

$$u' = \frac{du}{dz}\frac{dz}{dx} = \frac{du}{dz}\left(\frac{1}{3}\right)x^{-2/3} = \frac{1}{3}z^{-2}\frac{du}{dz}$$

$$u'' = \frac{d^2u}{dz^2}\left(\frac{1}{9}\right)x^{-4/3} + \frac{du}{dz}\left(-\frac{2}{9}\right)x^{-5/3} = \frac{1}{9}z^{-4}\frac{d^2u}{dz^2} - \frac{2}{9}z^{-5}\frac{du}{dz}.$$

Substitution into (B) and collection of terms gives

$$9\frac{z^6}{z^4}\frac{d^2u}{dz^2} - 18\frac{z^6}{z^5}\frac{du}{dz} + 81\frac{z^3}{z^2}\left(\frac{1}{3}\right)\frac{du}{dz} + (9z^2 - 1)u = 0.$$

Dividing this by 9 gives the Bessel equation with parameter 1/3 and unknown function $u(z)$, whose solution is

$$u(z) = AJ_{1/3}(z) + BJ_{-1/3}(z).$$

Replacing z by $x^{1/3}$ and multiplying by $x^{1/3}$ gives $y = x^{1/3}u(x^{1/3})$, as shown in Appendix 2.

Sec. 4.6   Bessel Functions of the Second Kind $Y_\nu(x)$

Problem Set 4.6. Page 232

7. **Reduction to Bessel's equation.** You have to transform the independent variable x by setting $z = kx^2/2$ as well as the unknown function y by setting $y = \sqrt{x}\,u$. Using the chain rule, you can perform the two transformations one after another – this would be similar to Prob. 9 in Problem Set 4.5 – or simultaneously, as we shall now explain. You will need $dz/dx = kx$. Differentiation with respect to x gives

$$\frac{dy}{dx} = \frac{1}{2}x^{-1/2}u + x^{1/2}\frac{du}{dz}kx$$

$$= \frac{1}{2}x^{-1/2}u + kx^{3/2}\frac{du}{dz}.$$

Differentiating this again, you obtain the second derivative

$$\frac{d^2y}{dx^2} = -\frac{1}{4}x^{-3/2}u + \frac{1}{2}x^{-1/2}\frac{du}{dz}kx + \frac{3}{2}kx^{1/2}\frac{du}{dz} + kx^{3/2}\frac{d^2u}{dz^2}kx$$

$$= -\frac{1}{4}x^{-3/2}u + 2kx^{1/2}\frac{du}{dz} + k^2x^{5/2}\frac{d^2u}{dz^2}.$$

Substituting this expression for y'' as well as y into the given equation and dividing the whole equation by $k^2x^{5/2}$ gives

$$\frac{d^2u}{dz^2} + \frac{2}{kx^2}\frac{du}{dz} + (1 - \frac{1}{4k^2x^4})u = 0.$$

Now recall that $kx^2/2 = z$. Hence $kx^2 = 2z$. Substitute this into the last equation to get

$$\frac{d^2u}{dz^2} + \frac{1}{z}\frac{du}{dz} + (1 - \frac{1}{16z^2})u = 0.$$

This is Bessel's equation with parameter $\nu = 1/4$. Hence a general solution of the given equation is

$$y = x^{1/2}u(z) = x^{1/2}(AJ_{1/4}(z) + BY_{1/4}(z)) = x^{1/2}(AJ_{1/4}(kx^2/2) + BY_{1/4}(kx^2/2)).$$

11. **Y_0 for small x.** Use (6). Neglect the series in (6), which is 0 for $x = 0$; hence solve $\ln(x/2) = -\gamma$. This gives $x = 1.1$, approximately. The actual 2S-value of the zero is 0.89; see Ref. [1], p. 410.

Sec. 4.7   Sturm-Liouville Problems. Orthogonal Functions

Example 5 and Theorem 2. In (13), n is fixed. The smallest n is $n = 0$. Hence then (13) concerns J_0. It then takes the form

$$\int_0^R x J_0(k_{m0} x) J_0(k_{j0} x)\, dx = 0 \qquad (j \neq m, \text{ both integer}).$$

If $n = 0$ were the only possible value of n, you could simply write k_m and k_j instead of k_{m0} and k_{j0}; write it down for youself to see what (13) then looks like. Recall that k_{m0} is related to the zero α_{m0} of J_0 by $k_{m0} = \alpha_{m0}/R$. In applications (vibrating drumhead in Sec. 11.10, for instance) the number R can have any value depending on the problem (in Sec. 11.10 it is the radius of the drumhead); this is the reason for introducing the arbitrary k near the beginning of the example; it gives us the flexibility needed in practice.

Problem Set 4.7. Page 238

1. **Case 3 of Theorem 1.** In this case the proof runs as follows. By assumption, $r(a) = 0$ and $r(b) \neq 0$. The starting point of the proof is (8), as before. Since $r(a) = 0$, you see that (8) reduces to

$$r(b)[y_n'(b)y_m(b) - y_m'(b)y_n(b)]$$

and you have to show that this is zero. Now from (2b) you have (we write L instead of l, to avoid confusion with the number 1)

$$L_1 y_n(b) + L_2 y_n'(b) = 0$$
$$L_1 y_m(b) + L_2 y_m'(b) = 0.$$

At least one of the two coefficients is different from zero, by assumption, say, $L_2 \neq 0$. Now multiply the first equation by $y_m(b)$ and the second by $-y_n(b)$ and add, obtaining

$$L_2[y_n'(b)y_m(b) - y_m'(b)y_n(b)] = 0.$$

Since L_2 is not zero, the expression in the brackets must be zero. But this expression is identical with that in the brackets in the first line of (8), which we have written above. The second line of (8) is zero because of the assumption $r(a) = 0$. Hence (8) is zero, and from (7) you obtain the relationship (9) to be proved. (For $L_1 \neq 0$ the proof is similar. Supply the details; this will show you whether you really understand the present proof.)

3. **Sturm-Liouville problem.** The given equation and boundary conditions do constitute a Sturm-Liouville problem. The equation is of the form (1) with $r = 1, q = 0, p = 1$. The interval in which solutions are sought is given by $a = 0$ and $b = 1$ as its endpoints. In the boundary conditions, $k_1 = 1, k_2 = 0, l_1 = 0, l_2 = 1$. You first have to find a general solution. In Prob. 3 it is

$$y = A \cos kx + B \sin kx \qquad \text{where} \quad k = \sqrt{\lambda}. \tag{A}$$

You obtain eigenvalues and functions by using the boundary conditions. The first condition gives $y(0) = A = 0$. Differentiation of the remaining part of equation (A) gives

$$y'(x) = kB \cos kx, \qquad \text{hence} \quad y'(1) = kB \cos k = 0,$$

thus $\cos k = 0$. This yields $k = k_n = (2n+1)\pi/2$, where $n = 0, 1, \ldots$; these are the positive values for which the cosine is zero. You need not consider negative values of n because the cosine is even, so that you would get the same eigenfunctions. The eigenvalues are $\lambda = \lambda_n = k_n^2$. The corresponding eigenfunctions are

$$y(x) = y_n(x) = \sin(k_n x) = \sin((2n+1)\pi x/2).$$

The figure shows the first few eigenfunctions. All of them start at 0 and have a horizontal tangent at the other end of the interval from 0 to 1. This is the geometric meaning of the boundary conditions. y_1 has no zero in the interior of this interval. Its graph shown corresponds to 1/4 of the period of the cosine. y_2 has one such zero (at 2/3), and its graph shown corresponds to 3/4 of that period. y_3 has two such zeros (at 0.4 and 0.8). y_4 has three, and so on.

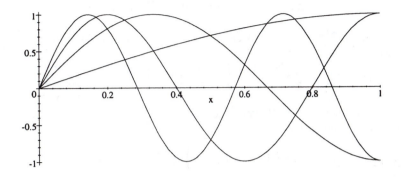

Section 4.7. Problem 3. First four eigenfunctions of the Sturm-Liouville problem

13. Change of x. Equate $ct + k$ to the endpoints of the given interval and solve for t to get the new interval on which you can prove orthogonality.

Sec. 4.8   Orthogonal Eigenfunction Expansions

Example 2. Answers to the questions near the end. $a_{13} P_{13}$ is the next term. $a_2 = a_4 = \ldots = 0$ because $\sin \pi x$ is odd. $P_3(x)$ resembles $-\sin \pi x$ more closely than any other term does; see Fig.101 in Sec. 4.3.

Problem Set 4.8. Page 246

1. Fourier-Legendre series. In Example 2 of the text we had to determine the coefficients by integration. In the present case this would be possible, but the method of undetermined coefficients is much simpler. The given function
$$f(x) = 70x^4 - 84x^2 + 30$$
is of degree 4, hence you need only $P_0, P_1, \ldots, P_4$. Since f is an even function, you actually need only P_0, P_2, P_4. Write
$$f(x) = a_4 P_4(x) + a_2 P_2(x) + a_0 P_0(x) = 70x^4 - 84x^2 + 30.$$
Begin by determining a_4 so that the x^4-terms on both sides agree. Since $P_4(x) = \frac{1}{8}(35x^4 - 30x^2 + 3)$ (see Sec. 4.3), you have the condition
$$a_4(35/8) = 70, \quad \text{hence} \quad a_4 = 70 \cdot 8/35 = 16.$$
Calculate the remaining function
$$f_1(x) = f(x) - 16P_4(x) = 70x^4 - 84x^2 + 30 - (16/8)(35x^4 - 30x^2 + 3)$$
$$= -84x^2 + 30 + 60x^2 - 6 = -24x^2 + 24.$$
Now determine a_2 by the same process so that the x^2-terms on both sides of the last equation agree. Using $P_2(x) = (1/2)(3x^2 - 1)$, you obtain
$$a_2(3/2) = -24, \quad \text{hence} \quad a_2 = -24 \cdot 2/3 = -16.$$
Calculate the remaining function (a constant!)
$$f_2(x) = f_1(x) - (-16)P_2(x) = -24x^2 + 24 - (-16/2)(3x^2 - 1) = 16.$$
Hence $a_0 = 16$ because $P_0(x) = 1$. The answer is
$$f = 16(P_0 - P_2 + P_4).$$

5. Fourier-Legendre series. The coefficients are given by (7) in the form

$$a_m = (m + 1/2) \int_{-1}^{1} \cos\left(\frac{\pi x}{2}\right) P_m(x)\, dx.$$

For $m = 0$ this gives $a_0 = 2/\pi = 0.6366$ by calculus. For even m you obtain by two successive integrations by parts

$$\int_{-1}^{1} x^m \cos\left(\frac{\pi x}{2}\right) dx = \frac{4}{\pi} - \frac{4m(m-1)}{\pi^2} \int_{-1}^{1} x^{m-2} \cos\left(\frac{\pi x}{2}\right) dx.$$

Thus for $m = 2$ this gives $4/\pi - (8/\pi^2)(4/\pi)$, where $4/\pi \; (= 2a_0)$ is the value of the integral of $\cos(\pi x/2)$ just calculated. From this you further obtain

$$a_2 = \frac{5}{2} \int_{-1}^{1} \cos\left(\frac{\pi x}{2}\right) \frac{1}{2}(3x^2 - 1)\, dx$$

$$= \frac{5}{4}\left(3\left(\frac{4}{\pi} - \frac{32}{\pi^3}\right) - \frac{4}{\pi}\right) = -0.687085,$$

and similarly for the further coefficients as given in the answer in Appendix 2 (which were calculated by a CAS).

CHAPTER 5.   Laplace Transforms

Sec. 5.1   Laplace Transform. Inverse Transform. Linearity. Shifting

Problem Set 5.1. Page 257

7. Laplace transform. Use the addition formula for the sine function.

13. Use of the defining integral. Problems 9-16 can be done by direct integration of the defining integral, with the proper lower and upper limits of integration. In some cases you will need integration by parts. In Prob. 13,

$$\int te^{-st} dt = \frac{te^{-st}}{-s} - \frac{1}{-s} \int e^{-st} dt$$

$$= \frac{te^{-st}}{-s} - \frac{1}{s^2} e^{-st}.$$

The lower limit of integration is 0, and the first term in the second line is 0 at $t = 0$. The upper limit of integration is k (see the figure in Problem Set 5.1). Hence by evaluating your result at the limits you obtain $ke^{-ks}/(-s)$ from the first term and $-(e^{-ks} - 1)/s^2$ from the second.

17. Inverse transform. The basic formulas are contained in Table 5.1 of Sec. 5.1. From the denominator of the given expression you see that formulas 7 and 8 of that table are needed. Since $s^2 + 3.24 = s^2 + 1.8^2$, you will obtain cos 1.8t and sin 1.8t. From formula 7 you may conclude that the cosine term is 0.1 cos 1.8t. From formula 8 you see that the sine term is 0.5 sin 1.8t because $0.9 = 1.8 \cdot 0.5$. The answer is the sum of these two expressions obtained.

23. Inverse transform. Factor out $1/L^2$.

27. Partial fraction reduction and formula 6 in Table 5.1 are needed. The two partial fractions will give you a sum of two exponential functions.

33. First shifting theorem. sinh t cos $t = (1/2)(e^t - e^{-t})$ cos $t = (1/2) e^t \cos t - (1/2) e^{-t} \cos t$ and formula 7 in Table 5.1 together with the first shifting theorem give $(1/2)(s - 1)/[(s - 1)^2 + 1]$ as the transform of the first term and $-(1/2)(s + 1)/[(s + 1)^2 + 1]$ as the transform of the second term. The sum of the two transforms has the common denominator

$$(s^2 - 2s + 2)(s^2 + 2s + 2) = (s^2 + 2)^2 - 4s^2 = s^4 + 4$$

and the numerator

$$(1/2)(s - 1)(s^2 + 2s + 2) - (1/2)(s + 1)(s^2 - 2s + 2) = s^2 - 2.$$

This gives the answer shown in Appendix 2.

39. First shifting theorem. You can write $s/D = [(s + 1/2) - 1/2]/D$, where D is the given denominator. Now by the shifting theorem, $(s + 1/2)/D$ has the inverse exp $(-t/2)$ cos t, and $(-1/2)/D$ has the inverse $(-1/2)$ exp $(-t/2)$ sin t.

Sec. 5.2   Transforms of Derivatives and Integrals. Differential Equations

Example 2. The question at the end. $f(t) = \sin \omega t, f''(t) = -\omega^2 f(t)$. Now take the transform and use (2) and $f'(0) = \omega \cos 0 = \omega$, obtaining

$$\mathcal{L}(f) = -\omega^2 \mathcal{L}(f) = s^2 \mathcal{L}(f) - \omega.$$

Now solve algebraically for $\mathcal{L}(f)$.

Problem Set 5.2. Page 264

1. **Initial value problem**. Denote the Laplace transform of the unknown function y (the solution sought) by Y. By (1) in Sec. 5.2 the transform of the derivative y' is $sY - y(0) = sY$; here the given initial condition is taken into account. From Table 5.1 in Sec. 5.1 you see that the transform of the right side of the given differential equation is $10/(s^2 + 1)$, Together this gives the subsidiary equation

$$sY + 3Y = 10/(s^2 + 1).$$

Algebraically solving for Y and writing the result in terms of partial fractions gives

$$Y = \frac{10}{(s+3)(s^2+1)} = \frac{A}{s+3} + \frac{Bs+C}{s^2+1}.$$

For determining the constants A, B, C use your favorite method. For instance, taking the common denominator on the right, by equating the numerators on both sides, you have

$$10 = A(s^2 + 1) + (s+3)(Bs+C). \tag{a}$$

Setting $s = -3$ gives

$$10 = A[(-3)^2 + 1], \quad \text{hence } A = 1.$$

Equating the terms in s^2 on both sides of (a) and using $A = 1$, you obtain

$$0 = 1 + B, \quad \text{hence } B = -1.$$

Equating the constant terms on both sides of (a) and again using $A = 1$ gives

$$10 = 1 + 3C, \quad \text{hence } C = 3.$$

$1/(s+3)$ has the inverse transform e^{-3t} (see Table 5.1). The inverse transform of the next term $Bs/(s^2+1) = -s/(s^2+1)$ is $-\cos t$. The last term $C/(s^2+1) = 3/(s^2+1)$ has the inverse transform $3 \sin t$. Together this gives the answer on p. A14 of Appendix 2. The figure shows that the solution is practically a harmonic oscillation; the influence of the transition term e^{-3t} is hardly visible.

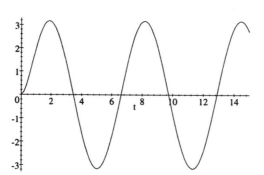

Section 5.2. Problem 1. Solution of the initial value problem

11. **Different derivations**. (a) Use $\cos^2 t = 1 - \sin^2 t$. (b) Set $f = \cos^2 t$. Then $f(0) = 1$ and $f' = -2\cos t \sin t = -\sin 2t$. From (1) in Sec. 5.2 you now obtain

$$\mathcal{L}(-\sin 2t) = -2/(s^2 + 4) = s\mathcal{L}(f) - 1.$$

Solving for $\mathcal{L}(f)$ gives the answer

$$\mathcal{L}(f) = \frac{1}{s}\left(1 - \frac{2}{s^2+4}\right) = \frac{1}{s}\left(\frac{s^2+2}{s^2+4}\right).$$

19. Application of Theorem 3. The expression in the parentheses has the inverse transform $\cos 3t + \frac{1}{3} \sin 3t$. The factor s^2 in the denominator suggests two successive integrations. The first gives

$$\frac{1}{3} \sin 3t - \frac{1}{9}(\cos 3t - 1).$$

Integrate this again from 0 to t (see Theorem 3; don't forget the contribution from the lower limit of integration). This gives

$$-\frac{1}{9}(\cos 3t - 1) - \frac{1}{9}(\frac{1}{3} \sin 3t - t).$$

Multiply the result by the factor 9 shown in the enunciation of the problem. This gives the answer in Appendix 2.

Sec. 5.3   Unit Step Function. Second Shifting Theorem. Dirac's Delta Function

Example 1 involves the term $u(t - 2\pi) \sin t$. You must convert it to $u(t - 2\pi) \sin (t - 2\pi)$ before you can apply (3) because in (3) both f and u have the same argument $t - a$. Here these two expressions are equal because of periodicity. In most cases two such expressions will not be equal. For instance, if you have $tu(t - 4)$ (sketch it to see what it looks like!), before you apply (3), you have to write it as

$$tu(t - 4) = (t - 4 + 4)u(t - 4) = (t - 4)u(t - 4) + 4u(t - 4).$$

Think this over before you go on.

Problem Set 5.3. Page 273

3. Unit step function. The given function $(t - 1)u(t - 1)$ is of the form $(2) f(t - a)u(t - a)$ with $f(t) = t$ and $a = 1$. The transform of $f(t)$ is $F(s) = 1/s^2$ (see Table 5.1 in Sec. 5.1). Hence (3) in the second shifting theorem shows that the given function has the transform $e^{-s} F(s) = e^{-s}/s^2$.

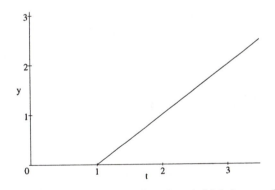

Section 5.3. Problem 3. Given function (which is zero for $t < 1$)

11. Laplace transform. Since $1 - u(t - 1)$ is 1 for t from 0 to 1 and 0 otherwise, you can represent the given function by $[1 - u(t - 1)]e^{-t} = e^{-t} - u(t - 1)e^{-t}$. The first term has the transform $1/(s - 1)$. Consider the second term and write it in the form (2) of this section,

$$-u(t - 1)e^t = -u(t - 1)e^{t-1}e.$$

Keep this step in mind, which is always needed in using the second shifting theorem (Theorem 1). Formula (3) with $F(s) = 1/(s - 1)$ and $a = 1$ now gives the transform

$$-e \cdot e^{-s}/(s - 1) = -e^{1-s}/(s - 1).$$

Hence the answer is $(1 - e^{1-s})/(s - 1)$.

19. Inverse transform. $F(s) = s/(s^2 + \pi^2)$ has the inverse $f(t) = \cos \pi t$ and is multiplied by e^{-2s}, which corresponds to the multiplication of $f(t)$ by $u(t-2)$ and a shift of t in the cosine to $t-2$. But $\cos(\pi(t-2)) = \cos t$ by periodicity. Hence the answer is $u(t-2)\cos \pi t$, a cosine curve that begins at $t = 2$, the first part of it between 0 and 2 being cut off (see the figure).

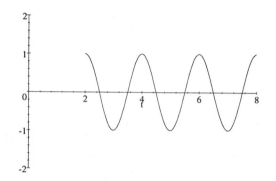

Section 5.3. Problem 19. Inverse $u(t-2)\cos \pi t$

23. Initial value problem. Express the right side by a unit step function, obtaining

$$y'' + 9y = 8[1 - u(t - \pi)]\sin t$$

$$= 8\sin t + 8u(t - \pi)\sin(t - \pi).$$

Here $\sin t = -\sin(t - \pi)$ has been used and the form of the right side has been chosen in order to prepare for the use of (2) in this section. Now use the initial conditions $y(0) = 0$, $y'(0) = 4$ and (2) in Sec. 5.2 as well as (3) in this section. This gives the subsidiary equation

$$s^2 Y - s \cdot 0 - 4 + 9Y = 8/(s^2 + 1) + 8e^{-\pi s}/(s^2 + 1).$$

Solving algebraically for Y and using partial fractions gives

$$Y = 4/(s^2 + 9) + 8/[(s^2 + 1)(s^2 + 9)] + 8e^{-\pi s}/[(s^2 + 1)(s^2 + 9)]$$

$$= \frac{4}{s^2 + 9} + \frac{1}{s^2 + 1} - \frac{1}{s^2 + 9} + \left(\frac{1}{s^2 + 1} - \frac{1}{s^2 + 9} \right) e^{-\pi s}.$$

The inverse is

$$\frac{4}{3}\sin 3t + \sin t - \frac{1}{3}\sin 3t \qquad\qquad\qquad\qquad (a)$$

$$+ u(t - \pi)\left[\sin(t - \pi) - \frac{1}{3}\sin(3t - 3\pi) \right].$$

This gives $\sin 3t + \sin t$ if $0 < t < \pi$. In the second line,

$$\sin(t - \pi) - \frac{1}{3}\sin(3t - 3\pi) = -\sin t + \frac{1}{3}\sin 3t.$$

These two terms cancel two terms in the first line of (a), so that for $t > \pi$ the solution is $(4/3)\sin 3t$. The figure shows the solution thus obtained.

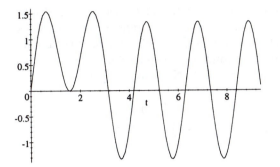

Section 5.3. Problem 23. Particular solution whose formula changes at π

37. RC-circuit. $100\,i + 10\int_0^t i(\tau)\,d\tau = 100[u(t-1) - u(t-2)]$. Now $q(0) = 0$ because of the limits of integration, and $i(0) = 0$ follows from $v(0) = 0$ and the equation. Using Theorem 3 in Sec. 5.2, obtain the subsidiary equation

$$100I + (10/s)I = 100(e^{-s}/s - e^{-2s}/s).$$

Division by 100, multiplication by s, and algebraic solution for I gives

$$I = e^{-s}/(s + 0.1) - e^{-2s}/(s + 0.1).$$

Now use the second shifting theorem to obtain the solution

$$i = e^{-0.1(t-1)}u(t-1) - e^{-0.1(t-2)}u(t-2).$$

Hence

$$i = \begin{cases} 0 & \text{if } t < 1, \\ e^{-0.1(t-1)} & \text{if } 1 < t < 2, \\ e^{-0.1(t-1)} - e^{-0.1(t-2)} & \text{if } t > 2. \end{cases}$$

Sec. 5.4   Differentiation and Integration of Transforms

Example 3. Proceeding as in Example 2 gives

$$-\frac{d}{ds}\left[\ln\left(1 - \frac{a^2}{s^2}\right)\right] = -\left(1 - \frac{a^2}{s^2}\right)^{-1}(-2)\left(-\frac{a^2}{s^3}\right)$$

$$= -2a^2/[s(s^2 - a^2)]$$

$$= 2/s - 2s/(s^2 - a^2).$$

The inverse is $f(t) = 2 - 2\cosh at$. Now divide by t.

Problem Set 5.4. Page 278

7. Inverse transform. $\sin t$ has the transform $F(s) = 1/(s^2 + 1)$; see Sec. 5.1 if necessary. The multiplication by e^{-t} corresponds to replacing s by $s + 1$, according to the first shifting theorem (Sec. 5.1), that is, $e^{-t}\sin t$ has the transform $F(s + 1) = 1/[(s + 1)^2 + 1] = 1/(s^2 + 2s + 2)$. Finally, the multiplication by t corresponds to the differentiation of the transform and multiplication by -1, so that you obtain the answer

$$\frac{1}{(s^2 + 2s + 2)^2}(2s + 2) = \frac{2s + 2}{(s^2 + 2s + 2)^2}.$$

9. **Inverse transform.** $1/s^3$ has the inverse transform $t^2/2$ (see Table 5.1 in Sec. 5.1). Hence by the first shifting theorem, the given function has the inverse transform $t^2e^{3t}/2$. Check the result by two successive applications of (1) in this section, as follows. $1/(s-3)$ has the inverse transform e^{3t}. Its derivative is $-1/(s-3)^2$ and has the inverse transform $-te^{3t}$, by (1). Similarly, its second derivative is $+2/(s-3)^3$ and has the inverse transform $-t(-te^{3t}) = +t^2e^{3t}$. Now divide by 2.

13. **Inverse transform.** Let $f(t)$ be the inverse transform of the given

$$\ln [(s^2 + 1)/(s-1)^2)] = \ln (s^2 + 1) - 2 \ln (s-1). \tag{A}$$

The derivative of (A) is

$$\frac{2s}{s^2 + 1} - \frac{2}{s-1}. \tag{B}$$

The inverse transform of (B) is $2\cos t - 2e^t$. From (1*) you see that this is the inverse of $-tf(t)$. Hence the answer is

$$f(t) = -(2 \cos t - 2e^t)/t = 2(e^t - \cos t)/t.$$

Sec. 5.5   Convolution. Integral Equations

Problem Set 5.5. Page 283

Write all the calculations in these problems, in particular, the occurring integrals and their evaluation, very orderly step by step on your worksheet, as explained in these solutions.

3. **Convolution by evaluating the integral (1).** In (1) you have $f(t) = e^t$, $g(t) = e^{-t}$, hence $f(\tau) = e^\tau$, $g(t-\tau) = e^{-(t-\tau)} = e^{-t+\tau}$. The function e^{-t} can be taken out from under the integral sign because you integrate with respect to τ. The remaining integrand is $e^\tau e^\tau = e^{2\tau}$. Integration gives $e^{2\tau}/2$, hence $e^{2t}/2$ at the upper limit of integration and $1/2$ at the lower limit of integration. This must now be subtracted and the result must be multiplied by the function e^{-t}, which was pulled out; thus,

$$e^{-t}(e^{2t} - 1)/2 = (1/2)(e^t - e^{-t}) = \sinh t,$$

where the last equality is a consequence of the definition of the hyperbolic sine.

11. **Inverse transform.** $1/s^2$ has the inverse transform t, and $1/(s-1)$ has the inverse transform e^t. Hence the integrand of the corresponding convolution integral (1) is $\tau e^{t-\tau}$. (Of course, write this integral and all the following calculations on your worksheet!) Integration by parts gives $-\tau e^{t-\tau}$ plus the integral of $e^{t-\tau}$, the first minus sign being a consequence of the fact that you integrate with respect to τ, not t! The integrated part gives $-t$ from the upper limit and 0 from the lower. The remaining integral equals $-e^{t-\tau}$, so that its upper limit gives -1 and its lower contributes e^t. Hence the answer is $-t - 1 + e^t$.

19. **Initial value problem.** The subsidiary equation is

$$s^2Y + Y = 3s/(s^2 + 4);$$

here the initial conditions $y(0) = 0$, $y'(0) = 0$ have been used. Solving algebraically for Y gives

$$Y = \left(\frac{1}{s^2 + 1}\right)\left(\frac{3s}{s^2 + 4}\right).$$

The two functions on the right have the inverse transforms $\sin t$ and $3\cos 2t$, respectively. The integrand of the convolution of these two functions of t is

$$3 \sin \tau \cos (2t - 2\tau).$$

Convert this to a sum of sines by using (12) in Appendix A3.1. Choose $u = 2t - \tau$ and $v = 3\tau - 2t$. The result is

$$\frac{3}{2}\left[\sin(2t-\tau)+\sin(3\tau-2t)\right].$$

Integration with respect to τ gives

$$\frac{3}{2}\left[\cos(2t-\tau)-\frac{1}{3}\cos(3\tau-2t)\right].$$

Evaluating at the upper limit $\tau=t$ and subtracting the value at the lower limit $\tau=0$ gives

$$\frac{3}{2}\left[\cos t-\frac{1}{3}\cos t-\left(\cos 2t-\frac{1}{3}\cos(-2t)\right)\right].$$

Here $\cos(-2t)=\cos 2t$. Simplification gives

$$\frac{3}{2}\left[\frac{2}{3}\cos t-\frac{2}{3}\cos 2t\right]=\cos t-\cos 2t.$$

Sec. 5.6   Partial Fractions. Differential Equations

Problem Set 5.6. Page 289

3. Unrepeated factors. The partial fraction representation of the given function is

$$1/s+(3/2)/(s-3)-(3/2)/(s+3).$$

You can find the inverse transform of each of these three terms in Table 5.1 (Sec. 5.1). Recalling the definition of the hyperbolic sine function, you obtain

$$1+(3/2)e^{3t}-(3/2)e^{-3t}=1+3\sinh 3t.$$

13. Verification by working backward. $\cosh at \cos at=(1/2)e^{at}\cos at+(1/2)e^{-at}\cos at$. Now $\cos at$ has the transform $s/(s^2+a^2)$. Hence, by the first shifting theorem, $e^{at}\cos at$ has the transform

$$(s-a)/[(s-a)^2+a^2]=(s-a)/(s^2-2as+2a^2)\tag{A}$$

and $e^{-at}\cos at$ has the transform

$$(s+a)/[(s+a)^2+a^2]=(s+a)/(s^2+2as+2a^2).\tag{B}$$

Now take the common denominator of the expressions in (A) and (B),

$$[(s^2+2a^2)-2as][(s^2+2a^2)+2as]=(s^2+2a^2)^2-4a^2s^2=s^4+4a^4.$$

The corresponding numerator of the sum of (A) and (B) is $2s^3$ because all the other terms cancel in pairs. Multiply this by 1/2 (the factors at the very beginning of this calculation) to get the formula in the problem.

Sec. 5.7   Systems of Differential Equations

Example 1. The subsidiary equations are obtained by transforming the derivatives by means of (1) in **Sec. 5.2,** just as in the case of a single differential equation. For the first equation this gives

$$sY_1-y_1(0)=-0.08Y_1+0.02Y_2+6/s.$$

For the second equation you obtain

$$sY_2-y_2(0)=0.08Y_1+0.08Y_2.$$

Inserting $y_1(0)=0$ and $y_2(0)=150$ and collecting terms in Y_1 and in Y_2 gives the subsidiary equations shown on p. 291.

Problem Set 5.7. Page 294

1. System of differential equations. The subsidiary equations are

$$sY_1 - 1 = -Y_1 + Y_2$$
$$sY_2 - 0 = -Y_1 - Y_2.$$

Note that the given initial conditions have been used, as in the example just explained, the notations being those used in the text. The next step is ordering these equations, taking all the Y-terms to the left. This gives

$$(s+1)Y_1 - \qquad Y_2 = -(-1)$$
$$Y_1 + (s+1)Y_2 = 0.$$

In the case of a single equation the subsidiary equation is solved algebraically. The same step must now be done with the system of subsidiary equations. Elimination or Cramer's rule (Sec. 6.6) gives

$$Y_1 = \frac{s+1}{(s+1)^2 + 1}$$

$$Y_2 = -\frac{1}{(s+1)^2 + 1}.$$

$s/(s^2+1)$ has the inverse transform $\cos t$, and $1/(s^2+1)$ has the inverse transform $\sin t$. This and the first shifting theorem give the answer

$$y_1 = e^{-t}\cos t, \qquad y_2 = -e^{-t}\sin t.$$

19. Electrical network. From Fig. 132 and Kirchhoff's laws you obtain the model

$$2i_1 + 4(i_1 - i_2) + i_1' = 195\sin t$$
$$4i_2 + 4(i_2 - i_1) + 2i_2' = 0$$

and $i_1(0) = 0$, $i_2(0) = 0$. The subsidiary equations are

$$2I_1 + 4(I_1 - I_2) + sI_1 = 195/(s^2+1)$$
$$4I_2 + 4(I_2 - I_1) + 2sI_2 = 0.$$

Collecting terms in I_1 and I_2 and dividing the second equation by 2 gives

$$(s+6)I_1 - \qquad 4I_2 = 195/(s^2+1)$$
$$-2I_1 + (s+4)I_2 = 0.$$

The solution by elimination or Cramer's rule is

$$I_1 = 195\frac{s+4}{(s^2+10s+16)(s^2+1)}$$

$$I_2 = 2\cdot195\frac{1}{(s^2+10s+16)(s^2+1)}.$$

In terms of partial fractions, this becomes

$$I_1 = \frac{2}{s+8} + \frac{13}{s+2} - 3\frac{5s-14}{s^2+1}$$

$$I_2 = -\frac{1}{s+8} + \frac{13}{s+2} - 6\frac{2s-3}{s^2+1}.$$

All the terms obtained have inverse transforms that should by now be well known to you. The last term in each line is a mixture of transforms of cosines and sines that you can easily take apart and determine. This gives the answer

$$i_1 = 2e^{-8t} + 13e^{-2t} - 15\cos t + 42\sin t$$

$$i_2 = -e^{-8t} + 13e^{-2t} - 12\cos t + 18\sin t.$$

The figure shows the exponential terms in i_1 and i_2 and (separately) the steady-state solutions, which are harmonic oscillations. It shows that the transition period is rather short and of the same length of time for both currents, and that the initial conditions seem to be satisfied (as you can readily check from the formulas of the answer). Compare this figure with Fig. 132 in the text and comment. Why do the two curves cross the t-axis at different points?

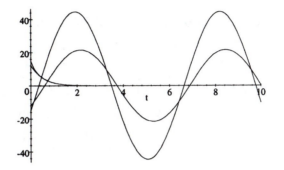

Section 5.7. Problem 19. Exponential terms and steady-state solutions

PART B.   LINEAR ALGEBRA, VECTOR CALCULUS

CHAPTER 6.   Linear Algebra: Matrices, Vectors, Determinants. Linear Systems of Equations

Sec. 6.1   Basic Concepts. Matrix Addition, Scalar Multiplication

Problem Set 6.1. Page 309

5. Matrix addition, scalar multiplication. First multiply $\mathbf{D}$ by 5, then $\mathbf{C}$ by 3. This gives

$$5\mathbf{D} = 5 \begin{bmatrix} 4 & 0 & -4 \\ -3 & 4 & 9 \end{bmatrix} = \begin{bmatrix} 20 & 0 & -20 \\ -15 & 20 & 45 \end{bmatrix}$$

and

$$3\mathbf{C} = 3 \begin{bmatrix} 6 & 0 & 3 \\ 1 & 0 & -5 \end{bmatrix} = \begin{bmatrix} 18 & 0 & 9 \\ 3 & 0 & -15 \end{bmatrix}.$$

The resulting matrices have the same size as the given ones because scalar multiplication does not alter the size of a matrix. Hence these matrices both have the size 2×3, so that the operations of addition and subtraction are defined for these matrices. Subtraction gives

$$5\mathbf{D} - 3\mathbf{C} = \begin{bmatrix} 20-18 & 0-0 & -20-9 \\ -15-3 & 20-0 & 45-(-15) \end{bmatrix} = \begin{bmatrix} 2 & 0 & -29 \\ -18 & 20 & 60 \end{bmatrix}.$$

The second task is very simple. Use (6) and conclude that

$$5\mathbf{D}^T - 3\mathbf{C}^T = (5\mathbf{D} - 3\mathbf{C})^T.$$

That is, you obtain the answer simply by taking the transpose of the matrix just obtained,

$$\begin{bmatrix} 2 & -18 \\ 0 & 20 \\ -29 & 60 \end{bmatrix}.$$

The first row (the second row) of that matrix becomes the first colum (the second column) of the transpose, which has size 3×2.

11. Vectors are special matrices (having a single column or a single row), and operations with them are the same as those for general matrices (and somewhat simpler). $\mathbf{c}$ and $\mathbf{d}$ are column vectors and they have the same number of components. These two properties are preserved under scalar multiplication. Hence $3(\mathbf{c} - 4\mathbf{d})$ is defined. You obtain

$$\mathbf{c} = \begin{bmatrix} 9 \\ 5 \\ 7 \end{bmatrix}, \quad 4\mathbf{d} = 4 \begin{bmatrix} 2 \\ -2 \\ 6 \end{bmatrix} = \begin{bmatrix} 8 \\ -8 \\ 24 \end{bmatrix}, \quad \text{hence} \quad \mathbf{c} - 4\mathbf{d} = \begin{bmatrix} 1 \\ 13 \\ -17 \end{bmatrix}, \quad 3(\mathbf{c} - 4\mathbf{d}) = \begin{bmatrix} 3 \\ 39 \\ -51 \end{bmatrix}.$$

17. Proof of (4a). **A** and **B** are assumed to be general 2×3 matrices, that is,

$$\mathbf{A} = \begin{bmatrix} a_{11} & a_{12} & a_{13} \\ a_{21} & a_{22} & a_{23} \end{bmatrix}, \quad \mathbf{B} = \begin{bmatrix} b_{11} & b_{12} & b_{13} \\ b_{21} & b_{22} & b_{23} \end{bmatrix}.$$

By the definition of matrix addition,

$$\mathbf{A} + \mathbf{B} = \begin{bmatrix} a_{11} + b_{11} & a_{12} + b_{12} & a_{13} + b_{13} \\ a_{21} + b_{21} & a_{22} + b_{22} & a_{23} + b_{23} \end{bmatrix}$$

and

$$\mathbf{B} + \mathbf{A} = \begin{bmatrix} b_{11} + a_{11} & b_{12} + a_{12} & b_{13} + a_{13} \\ b_{21} + a_{21} & b_{22} + a_{22} & b_{23} + a_{23} \end{bmatrix}.$$

Now comes the idea of the proof. Use the *commutativity of the addition of numbers* to show that the two sums of **A** and **B** are equal. Indeed, $a_{11} + b_{11} = b_{11} + a_{11}$ and similarly for the other five entries of $\mathbf{A} + \mathbf{B}$ and $\mathbf{B} + \mathbf{A}$. Hence corresponding entries of the matrix $\mathbf{A} + \mathbf{B}$ and of the matrix $\mathbf{B} + \mathbf{A}$ are equal. By the definition of the equality of matrices this proves (4a) for any 2×3 matrices **A** and **B**, that is,

$$\mathbf{A} + \mathbf{B} = \mathbf{B} + \mathbf{A}.$$

Sec. 6.2   Matrix Multiplication

Problem Set 6.2. Page 319

3. Matrix multiplication. $\mathbf{C}^2$ is obtained from **C** by a straightforward application of the definition of matrix multiplication,

$$\mathbf{C}^2 = \begin{bmatrix} 4 & 6 & 2 \\ 6 & 0 & 3 \\ 2 & 3 & -1 \end{bmatrix} \begin{bmatrix} 4 & 6 & 2 \\ 6 & 0 & 3 \\ 2 & 3 & -1 \end{bmatrix} = \begin{bmatrix} 56 & 30 & 24 \\ 30 & 45 & 9 \\ 24 & 9 & 14 \end{bmatrix}.$$

Sample calculation: Denoting the entries of $\mathbf{C}^2$ by d_{jk}, you obtain

$$d_{11} = c_{11}^2 + c_{12} c_{21} + c_{13} c_{31} = 4^2 + 6^2 + 2^2 = 56$$

$$d_{12} = c_{11} c_{12} + c_{12} c_{22} + c_{13} c_{32} = 4 \cdot 6 + 6 \cdot 0 + 2 \cdot 3 = 30.$$

Furthermore, $\mathbf{C}^T \mathbf{C} = \mathbf{C} \mathbf{C}^T = \mathbf{C}^2$ because **C** is symmetric (definition in Sec. 6.1).

5. Vectors. The product $\mathbf{a}^T \mathbf{d}$ is not defined because **a** is a column vector, so that $\mathbf{a}^T$ is a row vector, and so is **d**, but the product of two row vectors (with more than one component) is not defined. (Neither is the product of two column vectors.) $\mathbf{a}^T \mathbf{d}^T$ is defined because **a** is a column vector, so that $\mathbf{a}^T$ is a row vector, and $\mathbf{d}^T$ is a column vector, and in this product you multiply row times column, as usual in matrix multiplication,

$$\mathbf{a}^T \mathbf{d}^T = \begin{bmatrix} 1 & 4 & 3 \end{bmatrix} \begin{bmatrix} 4 \\ 3 \\ 0 \end{bmatrix} = 1 \cdot 4 + 4 \cdot 3 + 3 \cdot 0 = 16.$$

d a is defined and equals 16. This follows by calculation or by taking the transpose of the product just computed and noting that transposition of a 1×1 matrix (a number) has no effect. Products of vectors as just discussed are of great practical importance. Products, such as **a d**, which will give a 3×3 matrix, occur once in a while; they are of minor practical importance, but may add to your understanding of matrix multiplication. In this you multiply row times column, but here rows and columns have just a single entry.

This looks as follows

$$\mathbf{ad} = \begin{bmatrix} 1 \\ 4 \\ 3 \end{bmatrix} \begin{bmatrix} 4 & 3 & 0 \end{bmatrix} = \begin{bmatrix} 4 & 3 & 0 \\ 16 & 12 & 0 \\ 12 & 9 & 0 \end{bmatrix}.$$

The first entry of the product is $1 \cdot 4 = 4$, the next $1 \cdot 3 = 3$, the third $1 \cdot 0 = 0$, and so on. Note that the first row of the product matrix is $\mathbf{d}$, the second is $4\mathbf{d}$, and the third is $3\mathbf{d}$.

13. **Markov process.** The matrix of the Markov process is given as

$$\begin{array}{cc} \text{To I} & \text{To II} \end{array}$$

$$\mathbf{A} = \begin{bmatrix} 0.5 & 0.5 \\ 0.2 & 0.8 \end{bmatrix} \begin{array}{c} \text{From I} \\ \text{From II} \end{array}$$

For good understanding, give the process an interpretation. For instance, I and II are two kinds of soap that initially sell equally well, as expressed by the starting vector [0.7 0.7], measured in millions of bars sold per month. Then the first entry 0.5 shows that somebody who is using brand I will continue to use it with probability 0.5, hence he or she will switch to brand II with probability 0.5. A person who is using II will again use II with the high probability 0.8 and will try out I with the much smaller probability of 0.2. From this interpretation you can already see what will happen in the long run, namely, the sale of II will constantly increase with time. The calculation is as follows. (The notation is that of Example 14 in the book.)

$$\mathbf{y}^T = \mathbf{x}^T\mathbf{A} = [0.7 \quad 0.7] \begin{bmatrix} 0.5 & 0.5 \\ 0.2 & 0.8 \end{bmatrix} = [0.49 \quad 0.91].$$

In the next step calculate $\mathbf{y}^T\mathbf{A} = [0.427 \quad 0.973]$. Note that by substituting $\mathbf{y}^T = \mathbf{x}^T\mathbf{A}$ into $\mathbf{y}^T\mathbf{A}$ you get

$$\mathbf{y}^T\mathbf{A} = \mathbf{x}^T\mathbf{A}^2,$$

but this would not be a practical way of calculating the vectors in this iteration. And so on for the further steps.

Sec. 6.3   Linear Systems of Equations. Gauss Elimination

Problem Set 6.3. Page 329

13. **Gauss elimination.** It is, of course, perfectly acceptable to do the Gauss elimination in terms of equations rather than in terms of the augmented matrix. However, as soon as you feel sufficiently acquainted with matrices, you may wish to save work of writing by operating on matrices. The unknowns are eliminated in the order in which they occur in each equation. Hence start with w. Since w does not occur in the first equation, you must **pivot**. Take the second equation as your pivot equation in this first step. This gives the new augmented matrix

$$\begin{bmatrix} 2 & -3 & -3 & 6 & | & 2 \\ 0 & 5 & 5 & -10 & | & 0 \\ 4 & 1 & 1 & -2 & | & 4 \end{bmatrix}.$$

It is practical to indicate the operations after the corresponding row, as shown in the book. You obtain the next matrix row by row. Copy Row 1. This is the pivot row in the first step. w does not occur in Eq. 2, so you need not operate on Row 2 and simply copy it. To eliminate w from Eq. 3, subtract twice Row 1 from Row 3. Mark this after Row 3 of the next matrix, which, accordingly, takes the form

$$\begin{bmatrix} 2 & -3 & -3 & 6 & \bigm| & 2 \\ 0 & 5 & 5 & -10 & \bigm| & 0 \\ 0 & 7 & 7 & -14 & \bigm| & 0 \end{bmatrix} \quad \text{Row } 3 - 2\,\text{Row } 1.$$

Note well that rows may be left unlabeled if you do not operate on them. And the row numbers occurring in labels always refer to the *previous* matrix just as in the book. w has now been eliminated. Turn to the next unknown, x. Copy the first two rows of the present matrix and operate on Row 3 by subtracting from it 1.4 times Row 2 because this will eliminate x from Eq. 3. The result is surprising – not really if you have already noticed that you have produced two proportional rows. The calculation gives

$$\begin{bmatrix} 2 & -3 & -3 & 6 & \bigm| & 2 \\ 0 & 5 & 5 & -10 & \bigm| & 0 \\ 0 & 0 & 0 & 0 & \bigm| & 0 \end{bmatrix} \quad \text{Row } 3 - 1.4\,\text{Row } 2.$$

Note that if the last entry were not 0, the system would have no solution. Now solve the second equation,

$$5x + 5y - 10z = 0.$$

Solving for x, you obtain

$$x = \tfrac{1}{5}(10z - 5y) = 2z - y$$

where y and z remain arbitrary. Had you solved for y, you would have obtained the answer given in Appendix 2, namely, $y = 2z - x$ with arbitrary x and z. As a third possibility, you could solve for z and leave x and y arbitrary, $z = (x + y)/2$. Each of these three possibilities is acceptable. Finally, find w from the first equation of the system after pivoting, which is the second equation of the system in the form given, namely, $2w - 3x - 3y + 6z = 2$. This gives

$$w = \tfrac{1}{2}(2 + 3(2z - y) + 3y - 6z) = 1.$$

17. **Electrical network.** The elements of the circuits (batteries and Ohm's resistors in the present case) are given. The first step is the introduction of letters and directions for the unknown currents, which you want to determine. This has already been done in the figure of the network as shown. You do not know the directions of the currents. This does not matter. You make a choice, and if an unknown current comes out negative, this means that you have chosen the wrong direction and the current actually flows in the opposite direction. There are three currents I_1, I_2, I_3; hence you need three equations. An obvious choice is the right node, at which I_3 flows in and I_1 and I_2 flow out; thus, by KCL (Kirchhoff's current law, Sec. 1.7),

$$I_3 = I_1 + I_2.$$

The left node would do equally well. Can you see that you would get the same equation (except for a minus sign by which all three currents are now multiplied)? Two further equations are obtained from KVL (Kirchhoff's voltage law, Sec. 1.7), one for the upper circuit and one for the lower. In the upper circuit you have a voltage drop of $2I_1$ across the left resistor (which has resistance $R = 2$), a voltage drop of $1 \cdot I_3$ across the lower resistor (whose resistance is 1 and through which the current I_3 is flowing), and a voltage drop $2I_1$ across the right resistor. Hence the sum of the voltage drops is $2I_1 + I_3 + 2I_1 = 4I_1 + I_3$. By KVL this sum must equal the electromotive force 16 on the upper circuit; here resistance is measured in ohms and voltage in volts. Thus, the second equation for determining the currents is

$$4I_1 + I_3 = 16.$$

A third equation is obtained by KVL from the lower circuit. The voltage drop across the left resistor is $4I_2$ because this resistor has resistance 4 ohms and the current I_2 is flowing through it, causing the drop. A second voltage drop occurs across the upper (horizontal) resistor in the circuit, namely $1 \cdot I_3$, as before. The sum of these two voltage drops must equal the electromotive force of 32 volts in this circuit, again by KVL. This gives

$$4I_2 + I_3 = 32.$$

Hence the system of the three equations for the three unknowns, properly ordered, is

$$I_1 + I_2 - I_3 = 0$$
$$4I_1 \quad\quad + I_3 = 16$$
$$4I_2 + I_3 = 32.$$

In your further work use the corresponding augmented matrix (write a vertical bar before the entries on the right if you still need it)

$$\begin{bmatrix} 1 & 1 & -1 & 0 \\ 4 & 0 & 1 & 16 \\ 0 & 4 & 1 & 32 \end{bmatrix}.$$

Subtract 4 times Row 1 (your pivot row) from Row 2, obtaining

$$\begin{bmatrix} 1 & 1 & -1 & 0 \\ 0 & -4 & 5 & 16 \\ 0 & 4 & 1 & 32 \end{bmatrix} \quad \text{Row } 2 - 4\,\text{Row } 1.$$

Note well that the pivot row (or pivot equation) remains untouched as it is. The new pivot row is Row 2. Use it to eliminate I_2 (its coefficient 4) from Row 3. For this you must add it to Row 3, obtaining

$$\begin{bmatrix} 1 & 1 & -1 & 0 \\ 0 & -4 & 5 & 16 \\ 0 & 0 & 6 & 48 \end{bmatrix} \quad \text{Row } 3 + \text{Row } 2.$$

The system has now reached triangular form. Back substitution begins. You may perhaps first write the transformed system in terms of equations,

$$I_1 + I_2 - I_3 = 0$$
$$-4I_2 + 5I_3 = 16$$
$$6I_3 = 48.$$

From Eq. 3 you obtain $I_3 = 48/6 = 8$. With this and $5I_3 = 40$, Eq. 2 gives

$$I_2 = \frac{1}{-4}(16 - 40) = 6.$$

Finally, Eq. 1 gives

$$I_1 = -I_2 + I_3 = 2.$$

Sec. 6.4   Rank of a Matrix. Linear Independence. Vector Space

Problem Set 6.4. Page 336

1. Linear independence and dependence are concepts of general importance, for instance, in dropping unnecessary vectors from a set of vectors, keeping only the essential ones that make up a possibly much smaller and therefore simpler set to work with. To show linear independence or dependence in the problem, use (1), with a simpler notation for the scalars, say, a, b, c. Then (1) takes the form

$$a\begin{bmatrix} 1 & 0 & 0 \end{bmatrix} + b\begin{bmatrix} 1 & 1 & 0 \end{bmatrix} + c\begin{bmatrix} 1 & 1 & 1 \end{bmatrix} = \mathbf{0}.$$

This vector equation is equivalent to three equations for the three components. You can first do the three scalar multiplications and then add the three resulting vectors. This gives

$$\begin{bmatrix} a & 0 & 0 \end{bmatrix} + \begin{bmatrix} b & b & 0 \end{bmatrix} + \begin{bmatrix} c & c & c \end{bmatrix}$$
$$= \begin{bmatrix} a+b+c & b+c & c \end{bmatrix} = \mathbf{0}.$$

The equation for the third component gives $c = 0$. This reduces the equation for the second component to $b = 0$. Finally, the equation for the first component is now reduced to $a = 0$. Hence the only solution of (1) is $a = 0$, $b = 0$, $c = 0$, which means linear independence.

15. **Rank of a matrix.** To determine the rank of a matrix, you can use the definition and work with rows, or Theorem 1 and work with columns, and reduce that matrix as shown in Example 6 in the text. The determination of a rank by inspection will hardly be possible in practice. The given matrix is 4×3. Hence its rank can be 3 at most, but may be less. Whether using rows or using columns will be better, this is hard to say; so start with rows, that is, consider the matrix as given and reduce it as in the Gauss elimination, first taking Row 1 as pivot row and then Row 2. This looks as follows. The given matrix is

$$\begin{bmatrix} 3 & 1 & 4 \\ 0 & 5 & 8 \\ -3 & 4 & 4 \\ 1 & 2 & 4 \end{bmatrix}$$

Generate zeros in the first column:

$$\begin{bmatrix} 3 & 1 & 4 \\ 0 & 5 & 8 \\ 0 & 5 & 8 \\ 0 & 5/3 & 8/3 \end{bmatrix} \begin{array}{l} \\ \\ \text{Row 3 + Row 1} \\ \text{Row 4 − 1/3 Row 1} \end{array}$$

(Can you see already here that rank $\mathbf{A} = 2$? But go on, to see how the algorithm works.) Now take the second row as pivot row and generate zeros in the second column:

$$\begin{bmatrix} 3 & 1 & 4 \\ 0 & 5 & 8 \\ 0 & 0 & 0 \\ 0 & 0 & 0 \end{bmatrix} \begin{array}{l} \\ \\ \text{Row 3 − Row 2} \\ \text{Row 4 − 1/3 Row 2} \end{array}$$

Your task is finished; you see that the number of linearly independent rows of this matrix, hence of the given matrix, is 2. Hence it has rank 2.

    You may be curious to see whether you could have saved work by taking columns instead of rows. This question seems natural since the matrix has only 3 columns (and its rank cannot exceed 3), whereas it has 4 rows. Write the matrix in transposed form, for convenient work. Use the first row of the transpose (the first column of the given matrix) as pivot row and then the second row in the second step.

$$\begin{bmatrix} 3 & 0 & -3 & 1 \\ 1 & 5 & 4 & 2 \\ 4 & 8 & 4 & 4 \end{bmatrix}$$

Generate zeros in the first column of this transposed matrix.

$$\begin{bmatrix} 3 & 0 & -3 & 1 \\ 0 & 5 & 5 & 5/3 \\ 0 & 8 & 8 & 8/3 \end{bmatrix} \begin{array}{l} \\ \text{Row 2 − 1/3 Row 1} \\ \text{Row 3 − 4/3 Row 1} \end{array}$$

Now generate a zero in the second column of this matrix, and this will automatically reduce the last row to zero.

$$\begin{bmatrix} 3 & 0 & -3 & 1 \\ 0 & 5 & 5 & 5/3 \\ 0 & 0 & 0 & 0 \end{bmatrix} \quad \text{Row } 3 - 8/5 \text{ Row } 2$$

The matrix has a maximum number of 2 linearly independent rows. Theorem 1 now implies that the given matrix has rank 2. This confirms your previous result.

27. Row space. By definition, a matrix **B** is row-equivalent to a matrix **A** if **B** is obtained from **A** by elementary row operations (p. 326), that is, by

    (a) Interchanging two rows of **A**,

    (b) Adding a constant multiple of one row of **A** to another row of **A**,

    (c) Multiplying a row by a nonzero constant .

By definition, the row space of **A** is the span of the row vectors of **A**; similarly for **B**, and you have to show that these two spans agree.

    Clearly, operation (a) does not alter the span because the order in which one writes the vectors does not affect their span.

    Consider the effect of (b). The row space of **A**, being the span of the row vectors of **A**, is also obtained as the span of the maximum number of *linearly independent* row vectors of **A**. But this maximum number (which equals rank **A**) is not altered by (b).

    Finally, (c) has no influence on that span since multiplication by a constant that is not zero multiplies the terms of a linear combination by that constant, converting it to another linear combination already present because that span consists of *all* linear combinations of the row vectors of **A**. This completes the proof.

Sec. 6.6   Determinants. Cramer's Rule

Problem Set 6.6. Page 349

9. Determinant. The simplest way seems a development by any of the rows or columns, for instance, by the first row.

13. Determinant. In older books, determinants were evaluated "unsystematically" by looking for rows or columns that already had zeros and then increasing their number by suitable operations. In the light of programmability, this method has become obsolete and has been replaced by reduction to triangular form, as shown in Example 7 of the text. For special determinants this can be modified. If a square matrix is of the form

$$\begin{bmatrix} \mathbf{A} & \mathbf{0} \\ \mathbf{0} & \mathbf{B} \end{bmatrix}$$

then you may perform the reduction only to the extent that you do not "destroy" the zeros in the two zero matrices. Accordingly, use Row 1 as pivot row, but reduce only Row 2, obtaining

$$\begin{vmatrix} 3 & 2 & 0 & 0 \\ 6 & 8 & 0 & 0 \\ 0 & 0 & 4 & 7 \\ 0 & 0 & 2 & 5 \end{vmatrix} = \begin{vmatrix} 3 & 2 & 0 & 0 \\ 0 & 4 & 0 & 0 \\ 0 & 0 & 4 & 7 \\ 0 & 0 & 2 & 5 \end{vmatrix} \quad \text{Row } 2 - 2 \text{ Row } 1.$$

Then use Row 3 as pivot row for reducing Row 4. This gives

$$\begin{vmatrix} 3 & 2 & 0 & 0 \\ 0 & 4 & 0 & 0 \\ 0 & 0 & 4 & 7 \\ 0 & 0 & 0 & 1.5 \end{vmatrix} \quad \text{Row 4} - (1/2)\,\text{Row 3}.$$

This determinant can now be developed by columns, first by the first column, the resulting third order determinant by *its* first column, and so on. The result is $3 \cdot 4 \cdot 4 \cdot 1.5 = 72$.

15. **Rank by determinants.** Theorem 3 shows immediately that the rank must be at least 2 because the 2×2 submatrix in the left upper corner has determinant -4. To find out by determinants whether the rank is 3, you must compute the determinant of the matrix. The development by the first row is

$$\begin{vmatrix} 0 & 2 & -3 \\ 2 & 0 & 5 \\ -3 & 5 & 0 \end{vmatrix} = -2 \begin{vmatrix} 2 & 5 \\ -3 & 0 \end{vmatrix} - 3 \begin{vmatrix} 2 & 0 \\ -3 & 5 \end{vmatrix} = -2 \cdot 15 - 3 \cdot 10 = -60.$$

Since this determinant is not zero, Theorem 3 implies that the matrix has rank 3. To check this result, apply row reduction to the matrix obtained by pivoting (by intechanging Rows 1 and 2).

19. **Cramer's rule.** The determinant of the system is

$$D = \begin{vmatrix} 3 & 7 & 8 \\ 2 & 0 & 9 \\ -4 & 1 & -26 \end{vmatrix} = 101.$$

The determinants D_1, D_2, D_3 needed in (13) are

$$D_1 = \begin{vmatrix} -13 & 7 & 8 \\ -5 & 0 & 9 \\ 2 & 1 & -26 \end{vmatrix} = -707$$

$$D_2 = \begin{vmatrix} 3 & -13 & 8 \\ 2 & -5 & 9 \\ -4 & 2 & -26 \end{vmatrix} = 0$$

$$D_3 = \begin{vmatrix} 3 & 7 & -13 \\ 2 & 0 & -5 \\ -4 & 1 & 2 \end{vmatrix} = 101.$$

Hence the solution is

$$x = -707/101 = -7, \quad y = 0, \quad z = 101/101 = 1.$$

Sec. 6.7   Inverse of a Matrix. Gauss-Jordan Elimination

Problem Set 6.7. Page 357

9. **Inverse.** The inverse of a square matrix **A** is obtained by the Gauss-Jordan elimination as shown in Example 1 in the text, and this needs hardly any further comments. In particular, Example 1 shows that the entries of the inverse will in general be fractions, even if the entries of **A** are integers. If **A** is special

(symmetric, triangular, etc.), its inverse may be special. The given matrix is upper triangular, and you start from

$$\left[\begin{array}{ccc|ccc} 1 & 8 & -7 & 1 & 0 & 0 \\ 0 & 1 & 3 & 0 & 1 & 0 \\ 0 & 0 & 1 & 0 & 0 & 1 \end{array}\right]$$

Since the given matrix is upper triangular, the Gauss part of the Gauss-Jordan method is not needed and you can begin immediately with the Jordan elimination of 8, −7, and 3 above the main diagonal, which will reduce the given matrix to the unit matrix. Using Row 2 as the pivot row and working "upward", calculate

$$\left[\begin{array}{ccc|ccc} 1 & 0 & -31 & 1 & -8 & 0 \\ 0 & 1 & 3 & 0 & 1 & 0 \\ 0 & 0 & 1 & 0 & 0 & 1 \end{array}\right] \quad \text{Row } 1 - 8\,\text{Row } 2$$

Using Row 3 as the pivot row, eliminate 3 and −31, obtaining

$$\left[\begin{array}{ccc|ccc} 1 & 0 & 0 & 1 & -8 & 31 \\ 0 & 1 & 0 & 0 & 1 & -3 \\ 0 & 0 & 1 & 0 & 0 & 1 \end{array}\right] \quad \begin{array}{l} \text{Row } 1 + 31\,\text{Row } 3 \\ \text{Row } 2 - 3\,\text{Row } 3 \end{array}$$

The right half of this 3×6 matrix is the inverse of the given matrix. Since the latter has 1 1 1 as the main diagonal, you needed no multiplications, as they would usually be necessary (see of the first matrix on p. 353, for example).

We mention that for a (nonsingular) triangular matrix the entries of the inverse can be determined one-by-one, without solving any system of equations. Can you show this for the present problem (which thus merely has the purpose of illustrating the Gauss-Jordan method for determining the inverse)?

13. **Inverse of a symmetric matrix.** Let $\mathbf{A}$ be symmetric, $\mathbf{A} = \mathbf{A}^T$, and nonsingular. Denote its inverse by $\mathbf{B}$. Then by the definition of the inverse,

$$\mathbf{AB} = \mathbf{I}.$$

Transposition and the use of the symmetry of $\mathbf{A}$ give

$$(\mathbf{AB})^T = \mathbf{B}^T\mathbf{A}^T = \mathbf{B}^T\mathbf{A} = \mathbf{I}^T = \mathbf{I}.$$

Multiply this equation by $\mathbf{B}$ from the right, obtaining

$$\mathbf{B}^T\mathbf{AB} = \mathbf{IB} = \mathbf{B}.$$

But $\mathbf{AB} = \mathbf{I}$, so that this equation reduces to

$$\mathbf{B}^T = \mathbf{B}.$$

This shows that $\mathbf{B}$ is symmetric and completes the proof.

By Prob. 12 you would simply have $(\mathbf{A}^{-1})^T = (\mathbf{A}^T)^{-1} = \mathbf{A}^{-1}$.

15. **Formula for the inverse.** Problems 15-20 should aid in understanding the use of minors and cofactors. The given matrix is

$$\mathbf{A} = \left[\begin{array}{ccc} 2 & 0 & -1 \\ 5 & 1 & 0 \\ 0 & 1 & 3 \end{array}\right].$$

Calculate its determinant, $D = \det \mathbf{A} = 1$. Denote the inverse of $\mathbf{A}$ simply by $\mathbf{B} = [b_{jk}]$. From (4) you thus obtain

$$b_{11} = A_{11} = \begin{bmatrix} 1 & 0 \\ 1 & 3 \end{bmatrix} = 3$$

$$b_{12} = A_{21} = -\begin{bmatrix} 0 & -1 \\ 1 & 3 \end{bmatrix} = -1$$

$$b_{13} = A_{31} = \begin{bmatrix} 0 & -1 \\ 1 & 0 \end{bmatrix} = 1$$

$$b_{21} = A_{12} = -\begin{bmatrix} 5 & 0 \\ 0 & 3 \end{bmatrix} = -15$$

$$b_{22} = A_{22} = \begin{bmatrix} 2 & -1 \\ 0 & 3 \end{bmatrix} = 6$$

$$b_{23} = A_{32} = -\begin{bmatrix} 2 & -1 \\ 5 & 0 \end{bmatrix} = -5$$

$$b_{31} = A_{13} = \begin{bmatrix} 5 & 1 \\ 0 & 1 \end{bmatrix} = 5$$

$$b_{32} = A_{23} = -\begin{bmatrix} 2 & 0 \\ 0 & 1 \end{bmatrix} = -2$$

$$b_{33} = A_{33} = \begin{bmatrix} 2 & 0 \\ 5 & 1 \end{bmatrix} = 2.$$

Hence, since $\det \mathbf{A} = 1$, the inverse is

$$\mathbf{A}^{-1} = \begin{bmatrix} 3 & -1 & 1 \\ -15 & 6 & -5 \\ 5 & -2 & 2 \end{bmatrix}.$$

Check your calculations by verifying that $\mathbf{A}\mathbf{A}^{-1} = \mathbf{I}$, the unit matrix.

Sec. 6.8   Vector Spaces, Inner Product Spaces, Linear Transformations.   *Optional*

Problem Set 6.8. Page 364

1.**Vector space.** To see whether the vectors satisfying

$$v_1 - 3v_2 + 2v_3 = 0 \tag{A}$$

form a vector space V, you have to find out whether for any two vectors $\mathbf{v}$ and $\mathbf{w}$ satisfying (A) and

$$w_1 - 3w_2 + 2w_3 = 0 \tag{B}$$

a linear combination

$$\mathbf{u} = a\mathbf{v} + b\mathbf{w} = [u_1 \quad u_2 \quad u_3] = [av_1 + bw_1 \quad av_2 + bw_2 \quad av_3 + bw_3]$$

also satisfies the corresponding relation

$$u_1 - 3u_2 + 2u_3 = 0. \tag{C}$$

This is true, as you can see by inserting the components of $\mathbf{u}$ into (C), obtaining

$$av_1 + bw_1 - 3(av_2 + bw_2) + 2(av_3 + bw_3) = a(v_1 - 3v_2 + 2v_3) + b(w_1 - 3w_2 + 2w_3)$$

because from (A) and (B) it follows that each of the expressions in the two pairs of parentheses on the right is zero.

V has dimension 2 because R^3 has dimension 3 and the vectors of V satisfy one relation, namely, (A), which may be written

$$v_1 = 3v_2 - 2v_3. \tag{A*}$$

You may use (A*) in determining a basis $\mathbf{p}, \mathbf{q}$ as follows. $\mathbf{p}$ in V with $p_2 = 1$ and $p_3 = 0$ must have $p_1 = 3$, because of (A*). Hence $\mathbf{p} = \begin{bmatrix} 3 & 1 & 0 \end{bmatrix}^T$. A vector $\mathbf{q}$ in V with $q_2 = 0$ and $q_3 = 1$ must have $q_1 = -2$, again because of (A*). Hence $\mathbf{q} = \begin{bmatrix} -2 & 0 & 1 \end{bmatrix}^T$. The answer in Appendix 2 gives $\mathbf{p}$ and $-\mathbf{q}$. Recall that you have infinitely many choices for a basis for V. Indeed, another basis is $\mathbf{u} = a\mathbf{p} + b\mathbf{q}$ and $\mathbf{w} = c\mathbf{p} + k\mathbf{q}$ with any scalars a, b, c, k such that $\mathbf{u}$ and $\mathbf{w}$ are linearly independent.

5. **Vector space.** A 2×2 skew-symmetric matrix has main diagonal 0   0 and $a_{21} = -a_{12}$. This gives dimension 1 and explains the form of the basis given in the answer, consisting of a single element (a single matrix).

21. **Orthogonality** is a concept of basic importance; for instance, the choice of orthogonal vectors for a basis simplifies many calculations. The inner product of the given $\mathbf{a} = \begin{bmatrix} 1 & 2 & 0 \end{bmatrix}^T$ and any $\mathbf{v} = \begin{bmatrix} v_1 & v_2 & v_3 \end{bmatrix}^T$ is

$$\mathbf{a} \bullet \mathbf{v} = \mathbf{a}^T \mathbf{v} = v_1 + 2v_2$$

and is zero if and only if $v_2 = -v_1/2$ and v_3 is arbitrary. Geometrically, these are all the vectors in R^3 whose orthogonal projections in the xy-plane are orthogonal to $\begin{bmatrix} 1 & 2 \end{bmatrix}^T$ (perpendicular to this vector or zero). Make a sketch. These vectors $\mathbf{v}$ form a vector space. This can be proved as in Prob. 1 above.

25. **Linear transformation.** In vector form you have $\mathbf{y} = \mathbf{A}\mathbf{x}$, where

$$\mathbf{A} = \begin{bmatrix} 3 & 2 \\ 4 & 1 \end{bmatrix}.$$

The inverse is $\mathbf{x} = \mathbf{A}^{-1}\mathbf{y}$. Hence Probs. 25-30 are solved by determining the inverse of the coefficient matrix $\mathbf{A}$ of the given transformation (if it exists, that is, if $\mathbf{A}$ is nonsingular).

CHAPTER 7.   Linear Algebra: Matrix Eigenvalue Problems

Sec. 7.1   Eigenvalues, Eigenvectors

Problem Set 7.1. Page 375

1. **Eigenvalues and eigenvectors.** For a diagonal matrix the eigenvalues are the main diagonal entries because the characteristic equation is

$$\det (\mathbf{A} - \lambda \mathbf{I}) = \begin{vmatrix} a_{11} - \lambda & 0 \\ 0 & a_{22} - \lambda \end{vmatrix} = (a_{11} - \lambda)(a_{22} - \lambda) = 0.$$

For the given matrix you obtain from this $\lambda_1 = 4, \lambda_2 = -6$. Now determine an eigenvector of $\mathbf{A}$ corresponding to $\lambda_1 = 4$. In components, $(\mathbf{A} - \lambda_1 \mathbf{I})\mathbf{x} = \mathbf{0}$ is

$$(a_{11} - \lambda_1)x_1 + \quad a_{12}x_2 \quad = (4-4)x_1 + 0x_2 \ = 0$$
$$a_{21}x_1 \quad + (a_{22} - \lambda_1)x_2 = 0x_1 + (-6-4)x_2 = 0.$$

The first equation gives no condition. The second gives $x_2 = 0$. Hence an eigenvector of $\mathbf{A}$ corresponding to $\lambda_1 = 4$ is $[x_1 \quad 0]^T$. Since an eigenvector is determined only up to a nonzero constant, you can simply take $[1 \quad 0]^T$ as an eigenvector. For $\lambda_2 = -6$ the procedure is similar and leads to $[0 \quad 1]^T$.

13. **Eigenvalues and eigenvectors.** Ordinarily one would expect that a 3×3 matrix has 3 linearly independent eigenvectors. For symmetric, skew-symmetric and many other matrices this is true. A simple example is the 3×3 unit matrix, which has but one eigenvalue, 1, but every (nonzero) vector is an eigenvector, so that you can choose, for instance, $[1 \quad 0 \quad 0]^T, [0 \quad 1 \quad 0]^T, [0 \quad 0 \quad 1]^T$. The given matrix has the characteristic equation

$$\begin{vmatrix} 2 - \lambda & 0 & -1 \\ 0 & 1/2 - \lambda & 0 \\ 1 & 0 & 4 - \lambda \end{vmatrix} = (2 - \lambda)(\tfrac{1}{2} - \lambda)(4 - \lambda) - (-(\tfrac{1}{2} - \lambda))$$

$$= -\lambda^3 + 6.5\lambda^2 - 12\lambda + 4.5 = 0.$$

The solutions are 1/2 and 3. In product form, the characteristic equation is

$$- (\lambda - 0.5)(\lambda - 3)^2 = 0;$$

hence $\lambda = 3$ has algebraic multiplicity 2. Now determine an eigenvector for $\lambda = 0.5$ from

$$(2 - 0.5)x_1 - x_3 = 0, \quad 0 = 0, \quad x_1 + (4 - 0.5)x_3 = 0.$$

This gives $x_1 = 0, x_3 = 0, x_2$ arbitrary. Hence you can take $[0 \quad 1 \quad 0]^T$. Similarly for $\lambda = 3$

$$(2 - 3)x_1 - x_3 = 0, \quad (0.5 - 3)x_2 = 0, \quad x_1 + (4 - 3)x_3 = 0.$$

Hence $x_1 = -x_3, x_2 = 0$, and you can take as an eigenvector $[-1 \quad 0 \quad 1]^T$, but you cannot obtain another eigenvector such that the three eigenvectors are linearly independent.

19. **Orthogonal projection.** This matrix has no inverse. This is geometrically obvious because all the points (x, y_0) on the horizontal line $y = y_0$ are projected onto the same point $(0, y_0)$ on the y-axis.

Sec. 7.2   Some Applications of Eigenvalue Problems

Problem Set 7.2. Page 379

3. **Elastic membrane.** Problems 1-6 amount to the determination of the eigenvalues (giving the extension or contraction factors) and eigenvectors (giving the principal directions) and a sketch of the latter. The

characteristic equation of the given matrix is

$$\begin{vmatrix} 3.0 - \lambda & 1.5 \\ 1.5 & 3.0 - \lambda \end{vmatrix} = (3.0 - \lambda)^2 - 1.5^2 = (\lambda - 1.5)(\lambda - 4.5) = 0.$$

You see that $\lambda = 1.5$ is an eigenvalue. A corresponding eigenvector is obtained from one of the two equations

$$(3.0 - 1.5)x_1 + 1.5x_2 = 0, \quad 1.5x_1 + (3.0 - 1.5)x_2 = 0$$

which both give $x_1 = -x_2$, so that you can take $[1 \quad -1]^T$, the vector from the origin $(0, 0)$ to the point $(1, -1)$ in the fourth quadrant, making a 45 degree angle with the x-axis. In this direction the membrane is stretched by a factor 1.5. Similarly, the other eigenvalue is $\lambda = 4.5$, and an eigenvector is obtained from

$$(3.5 - 4.5)x_1 + 1.5x_2 = 0, \quad \text{thus} \quad x_1 = x_2,$$

or from the other equation, which gives the same result. Hence you can take $[1 \quad 1]^T$ as an eigenvector, which you can graph as an arrow from the origin to the point $(1, 1)$ in the first quadrant. In this direction the membrane is stretched by a factor 4.5. The figure shows a circle of radius 1 and its image under stretching, which is an ellipse. A formula of the latter can be obtained by first stretching, leading from $x_1^2 + x_2^2 = 1$ (circle) to $x_1^2/4.5^2 + x_2^2/1.5^2 = 1$ (an ellipse whose axes coincide with the x_1 and x_2 axes) and then applying a 45 degree rotation (rotation through an angle $\alpha = \pi/4$) given by

$$u = x_1 \cos \alpha - x_2 \sin \alpha = (x_1 - x_2)/\sqrt{2},$$

$$v = x_1 \cos \alpha + x_2 \sin \alpha = (x_1 + x_2)/\sqrt{2}.$$

This problem is very similar to Example 1 on p. 376.

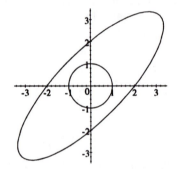

Section 7.2. Problem 3. Circular elastic membrane stretched to an ellipse

15. **Open Leontief input-output model.** For reasons explained in the enunciation of the problem you have to solve $\mathbf{x} - A\mathbf{x} = \mathbf{y}$ for $\mathbf{x}$, where A and $\mathbf{y}$ are given. With the given data you thus have to solve

$$\mathbf{x} - A\mathbf{x} = (I - A)\mathbf{x} = \begin{bmatrix} 1 - 0.1 & -0.4 & -0.2 \\ -0.5 & 1 & -0.1 \\ -0.1 & -0.4 & 1 - 0.4 \end{bmatrix} \begin{bmatrix} x_1 \\ x_2 \\ x_3 \end{bmatrix} = \mathbf{y} = \begin{bmatrix} 0.1 \\ 0.3 \\ 0.1 \end{bmatrix}.$$

For this you can apply the Gauss elimination to the augmented matrix of the system

$$\begin{bmatrix} 0.9 & -0.4 & -0.2 & 0.1 \\ -0.5 & 1.0 & -0.1 & 0.3 \\ -0.1 & -0.4 & 0.6 & 0.1 \end{bmatrix}.$$

If you use 6 decimals in your calculation, you will obtain the solution

$$x_1 = 0.55, \quad x_2 = 0.64375, \quad x_3 = 0.6875.$$

Sec. 7.3   Symmetric, Skew-Symmetric, and Orthogonal Matrices

Example 1. For a skew-symmetric matrix, $a_{kj} = -a_{jk}$. Hence for the main diagonal entries a_{jj} this gives $a_{jj} = -a_{jj} = 0$.

Problem Set 7.3. Page 384

3. **A common mistake.** This matrix is *NOT* skew-symmetric because a skew-symmetric matrix (which by definition is *real*) must have all diagonal entries zero. Hence you cannot expect its spectrum to lie on the y-axis. You obtain the eigenvalues from the characteristic equation

$$\begin{vmatrix} 1-\lambda & 4 \\ -4 & 1-\lambda \end{vmatrix} = (1-\lambda)^2 + 16 = \lambda^2 - 2\lambda + 17 = 0.$$

By the usual formula for the solutions of a quadratic equation you obtain

$$\lambda_1 = 1 + \sqrt{1-17} = 1 + 4i, \quad \lambda_2 = 1 - 4i \quad (i = \sqrt{-1}).$$

You see that a real matrix may very well have complex eigenvalues. However, note that if λ is a complex eigenvalue of such a matrix, so is the complex conjugate number; $1 + 4i$ and $1 - 4i$ are complex conjugates. It is interesting that $\mathbf{A} = \mathbf{B} + \mathbf{I}$, where

$$\mathbf{B} = \begin{bmatrix} 0 & 4 \\ -4 & 0 \end{bmatrix}$$

is skew-symmetric. The characteristic equation of $\mathbf{B}$ is $\lambda^2 + 16 = 0$. The roots (eigenvalues of $\mathbf{B}$) are $-4i$ and $4i$; they are pure imaginary. Hence $\mathbf{A} = \mathbf{B} + \mathbf{I}$ must have the eigenvalues $1 + 4i$ and $1 - 4i$, according to the spectral shift explained in Project 16d of Sec. 7.2.

11. **Inverse of a skew-symmetric matrix.** Let $\mathbf{A}$ be skew-symmetric, that is,

$$\mathbf{A}^T = -\mathbf{A}. \tag{1}$$

Let $\mathbf{A}$ be nonsingular. Let $\mathbf{B}$ be its inverse. Then

$$\mathbf{AB} = \mathbf{I}. \tag{2}$$

Transposition of (2) and the use of the skew symmetry (1) of $\mathbf{A}$ give

$$\mathbf{I} = \mathbf{I}^T = (\mathbf{AB})^T = \mathbf{B}^T\mathbf{A}^T = \mathbf{B}^T(-\mathbf{A}) = -\mathbf{B}^T\mathbf{A}. \tag{3}$$

Now multiply (3) by $\mathbf{B}$ from the right and use (2), obtaining

$$\mathbf{B} = -\mathbf{B}^T\mathbf{AB} = -\mathbf{B}^T.$$

This proves that $\mathbf{B} = \mathbf{A}^{-1}$ is skew-symmetric.

Sec. 7.4   Complex Matrices: Hermitian, Skew-Hermitian, Unitary

Example 1. In $\mathbf{A}$ the diagonal entries are real, hence equal to their conjugates. $a_{21} = 1 + 3i$ is the complex conjugate of $a_{12} = 1 - 3i$, as it should be for a Hermitian matrix. In $\mathbf{B}$ you have $\bar{b}_{11} = -3i = -b_{11}$, $\bar{b}_{22} = i = -b_{22}$, $\bar{b}_{21} = -2 - i = -b_{12}$. Hence $\mathbf{B}$ is skew-Hermitian. The complex conjugate transpose of $\mathbf{C}$ is

$$\begin{bmatrix} -i/2 & \sqrt{3}/2 \\ \sqrt{3}/2 & -i/2 \end{bmatrix}.$$

Multiplying this by **C**, you obtain the unit matrix. This verifies the defining relation of a unitary matrix.

Problem Set 7.4. Page 390

3. **Complex matrix**. The determination of eigenvalues and eigenvectors is the same in principle as for a real matrix. The matrix

$$\mathbf{B} = \begin{bmatrix} 3i & 2+i \\ -2+i & -i \end{bmatrix}$$

is skew-Hermitian, as has just been shown. The characteristic equation is

$$\begin{vmatrix} 3i - \lambda & 2+i \\ -2+i & -i-\lambda \end{vmatrix} = (3i - \lambda)(-i - \lambda) - (2+i)(-2+i)$$

$$= \lambda^2 + (-3i + i)\lambda + 3 - (-4 + 2i - 2i - 1)$$

$$= \lambda^2 - 2i\lambda + 8 = 0.$$

The roots (eigenvalues of **B**) are obtained by the usual formula for solving a quadratic equation

$$\lambda_1 = i + \sqrt{-1-8} = i + 3i = 4i, \quad \lambda_2 = i - 3i = -2i,$$

as given on p. 387 of the book. Observe that the eigenvalues need no longer be complex conjugates because the matrix is no longer real. To determine an eigenvector corresponding to $\lambda_1 = 4i$, substitute $\lambda = \lambda_1$ into the two equations, obtaining

$$(3i - 4i)x_1 + (2+i)x_2 = 0, \qquad (-2+i)x_1 + (-i - 4i)x_2 = 0.$$

Simplification gives

$$-ix_1 + (2+i)x_2 = 0,$$

$$(-2+i)x_1 - 5ix_2 = 0.$$

The first equation suggests choosing $x_1 = 2 + i, x_2 = i$. Check your result as follows. For that choice the second equation gives

$$(-2+i)(2+i) - 5ii = -4 - 2i + 2i - 1 + 5 = 0,$$

as expected. An eigenvector corresponding to $\lambda_2 = -2i$ is obtained in the same way from

$$(3i + 2i)x_1 + (2+i)x_2 = 0, \quad \text{that is,} \quad 5ix_1 + (2+i)x_2 = 0.$$

You see that you can choose $x_1 = 2 + i, x_2 = -5i$, as indicated in the solution on p. A19 in Appendix 2.

11. **Decomposition**. Let $\mathbf{A} = [a_{jk}]$ be arbitrary complex. Then $\bar{\mathbf{A}}^T = [\bar{a}_{kj}]$. The sum multiplied by 1/2 is

$$\mathbf{H} = \frac{1}{2}(\mathbf{A} + \bar{\mathbf{A}}^T) = \frac{1}{2}[a_{jk} + \bar{a}_{kj}].$$

Its conjugate transpose is

$$\bar{\mathbf{H}}^T = \frac{1}{2}[\bar{a}_{kj} + a_{jk}] = \mathbf{H}.$$

Hence **H** is Hermitian. Similarly, the difference multiplied by 1/2 is

$$\mathbf{S} = \frac{1}{2}(\mathbf{A} - \bar{\mathbf{A}}^T) = \frac{1}{2}[a_{jk} - \bar{a}_{kj}].$$

Its conjugate transpose is

$$\bar{\mathbf{S}}^T = \frac{1}{2}[\bar{a}_{kj} - a_{jk}] = -\mathbf{S}.$$

Hence **S** is skew-Hermitian. The sum $\mathbf{H} + \mathbf{S}$ equals the given matrix **A**. This completes the derivation of this representation.

Sec. 7.5   Similarity of Matrices. Basis of Eigenvectors. Diagonalization

Problem Set 7.5. Page 397

1. **Similar matrices.** The solutions of Probs. 1-6 are obtained by calculating the inverse and performing straightforward matrix multiplication. The importance of similarity transformations, for instance, in designing numerical methods for eigenvalue problems, justifies these problems. Given

$$\mathbf{A} = \begin{bmatrix} 1 & 2 \\ 2 & 4 \end{bmatrix} \quad \text{and} \quad \mathbf{P} = \begin{bmatrix} 1 & 3 \\ 3 & 6 \end{bmatrix}.$$

First calculate the inverse of **P**, which is best done by (4*) in Sec. 6.7. Then calculate

$$\hat{\mathbf{A}} = \mathbf{P}^{-1}\mathbf{A}\mathbf{P} = \begin{bmatrix} -2 & 1 \\ 1 & -1/3 \end{bmatrix} \begin{bmatrix} 1 & 2 \\ 2 & 4 \end{bmatrix} \begin{bmatrix} 1 & 3 \\ 3 & 6 \end{bmatrix}$$

$$= \begin{bmatrix} -2 & 1 \\ 1 & -1/3 \end{bmatrix} \begin{bmatrix} 7 & 15 \\ 14 & 30 \end{bmatrix} = \begin{bmatrix} 0 & 0 \\ 7/3 & 5 \end{bmatrix}.$$

The eigenvalues of both **A** and $\hat{\mathbf{A}}$ are 0 and 5. Eigenvectors **y** of $\hat{\mathbf{A}}$ are $[15 \ -7]^T$ and $[0 \quad 1]^T$, respectively. Multiplying these by **P** yields the vectors **x** given in the answer on p. A19 of Appendix 2.

15. **Diagonalization** is done by (5). For this you need the matrix **X** whose columns are eigenvectors of **A**. The characteristic equation of **A** gives the eigenvalues 15, −15, and 0. Eigenvectors are $[1 \quad 1 \quad 0]^T$, $[0 \quad 1 \quad 1]^T$, and $[2 \quad 0 \quad 1]^T$, respectively. Using these, form **X** and calculate its inverse. This gives

$$\mathbf{X} = \begin{bmatrix} 1 & 0 & 2 \\ 1 & 1 & 0 \\ 0 & 1 & 1 \end{bmatrix}, \quad \mathbf{X}^{-1} = \frac{1}{3}\begin{bmatrix} 1 & 2 & -2 \\ -1 & 1 & 2 \\ 1 & -1 & 1 \end{bmatrix}.$$

Now obtain the diagonal matrix **D** from (5) with main diagonal 15, −15, 0. This differs from the answer on p. A20 in Appendix 2 because the columns of **X** were chosen in a different order, which corresponds to the order of the eigenvalues on the main diagonal.

17. **Principal axes transformation.** The symmetric coefficient matrix of the given form is

$$\mathbf{A} = \begin{bmatrix} 7 & 3 \\ 3 & 7 \end{bmatrix}.$$

The eigenvalues 4 and 10 of **A** are obtained from the characteristic equation
$$(7 - \lambda)(7 - \lambda) - 9 = \lambda^2 - 14\lambda + 40 = (\lambda - 4)(\lambda - 10) = 0.$$
Eigenvectors are $[1 \quad -1]^T$ and $[1 \quad 1]^T$, respectively. From (10) you thus obtain the principal axes form
$$Q = 4y_1^2 + 10y_2^2 = 200.$$
This is an ellipse. The orthonormal matrix **X** in (9) is

$$\mathbf{X} = \frac{1}{\sqrt{2}}\begin{bmatrix} 1 & 1 \\ -1 & 1 \end{bmatrix}.$$

It gives the transformation $\mathbf{x} = \mathbf{X}\mathbf{y}$ shown in the answer on p. A20 in Appendix 2. Note that the eigenvectors (the columns of **X**) are determined only up to a minus sign; hence another **X** is obtained if you take $[-1 \quad 1]^T$ instead of $[1 \quad -1]^T$, giving another correct answer.

CHAPTER 8.   Vector Differential Calculus. Grad, Div, Curl

Sec. 8.1   Vector Algebra in 2-Space and 3-Space

Problem Set 8.1. Page 407

1. **Components.** According to the definition of components you have to calculate the differences of the coordinates of the terminal point Q minus the corresponding coordinates of the initial point P of the vector. Thus, $v_1 = 4 - 1 = 3$, etc. Since the z-coordinates of P and Q are zero, the vector $\mathbf{v}$ is a vector in the xy-plane; it has no component in the z-direction. Sketch the vector, so that you see what it looks like as an arrow in the xyz-coordinate system in space.

15. **Vector addition** has properties quite similar to those of the addition of numbers because it is defined in terms of components and the latter are numbers. Prob. 15 illustrates the commutativity and Prob. 17 the associativity of vector addition.

19. **Addition and scalar multiplication.** It makes no difference whether you first multiply and then add, or whether you first add the given vectors and then multiply their sum by the scalar 4. This problem and Example 2 in the text illustrate formula (6b).

29. **Forces** were foremost among the applications that have suggested the concept of a vector, and forming the resultant of forces has motivated vector addition to a large extent. Thus, each of Problems 24–28 amounts to the addition of three vectors. "Equilibrium" means that the resultant of the given forces is the zero vector. Hence in Prob. 29 you must determine $\mathbf{p}$ such that

$$\mathbf{p} + \mathbf{q} + \mathbf{u} = \mathbf{0}.$$

    Hence

$$\mathbf{p} = -\mathbf{q} - \mathbf{u} = -[3, \quad 2, \quad 0] - [-2, \quad 4, \quad 0] = [-3, \quad -2, \quad 0] + [2, \quad -4, \quad 0] = [-1, \quad -6, \quad 0].$$

33. **Force in a rope.** This is typical of many problems in mechanics. Choose an xy-coordinate system with the x-axis pointing horizontally to the right and the y-axis pointing vertically downward. Then the given weight is a force $\mathbf{w} = [w_1, w_2] = [0, w]$ pointing vertically downward. You have to determine the unknown force in the left rope, call it $\mathbf{u} = [u_1, u_2]$, and the unknown force $\mathbf{v} = [v_1, v_2]$ in the right rope. The three forces are in equilibrium (they have the resultant $\mathbf{0}$) because the system does not move. Thus,

$$\mathbf{u} + \mathbf{v} + \mathbf{w} = \mathbf{0}. \tag{A}$$

    You have two choices of giving $\mathbf{u}$ a direction, and similarly for $\mathbf{v}$. The choice is up to you. Suppose you choose the vectors to point from the point where the weight is attached upward to the points where the ropes are fixed. Then (see the figure on p. 406)

$$\begin{aligned} \mathbf{u} &= [u_1, \quad u_2] = [-|\mathbf{u}|\cos\alpha, \quad -|\mathbf{u}|\sin\alpha] \\ \mathbf{v} &= [v_1, \quad v_2] = [|\mathbf{v}|\cos\alpha, \quad -|\mathbf{v}|\sin\alpha] \\ \mathbf{w} &= [w_1, \quad w_2] = [0, \quad w], \end{aligned} \tag{B}$$

    where the components preceded by a minus sign are those that point in the negative direction of the corresponding coordinate axis (to the left or upward). From (A) and (B) you obtain for the horizontal components

$$-|\mathbf{u}|\cos\alpha + |\mathbf{v}|\cos\alpha = 0, \quad \text{hence} \quad |\mathbf{u}| = |\mathbf{v}|. \tag{C}$$

    Obviously, this could have been concluded from the symmetry of the figure. From (A) and (B) you obtain for the vertical components

$$-|\mathbf{u}|\sin\alpha - |\mathbf{v}|\sin\alpha + w = 0.$$

    From this and (C) conclude

$$|\mathbf{u}|\sin\alpha = |\mathbf{v}|\sin\alpha = \frac{1}{2}w.$$

Hence $\mathbf{u}$ and $\mathbf{v}$ have equal vertical components. Also this is a consequence of the symmetry.

Sec. 8.2   Inner Product (Dot Product)

Problem Set 8.2. Page 413

5. Inner product. This problem illustrates the rule

$$(p\mathbf{b})\bullet(q\mathbf{c}) = (pq)(\mathbf{b} \bullet \mathbf{c}) \tag{A}$$

which can be used to simplify calculations.

For the given vectors you obtain from (2), which gives the inner product in terms of components, for the first expression (corresponding to the left side of (A))

$$(2\mathbf{b})\bullet(5\mathbf{c}) = (2[2, 0, -5])\bullet(5[4, -2, 1]) = [4, 0, -10]\bullet[20, -10, 5]$$
$$= 4\cdot20 + 0 + (-10)\cdot5 = 30$$

and for the second expression, again from (2), (corresponding to the right side of (A))

$$10(\mathbf{b} \bullet \mathbf{c}) = 10([2, 0, -5]\bullet[4, -2, 1]) = 10(8 + 0 + (-5)) = 30.$$

The general formula (A) can be proved as follows. Let $\mathbf{b} = [b_1, b_2, b_3]$ and $\mathbf{c} = [c_1, c_2, c_3]$. Then $p\mathbf{b} = [pb_1, pb_2, pb_3]$ by (5) in Sec. 8.1, and similarly for $q\mathbf{c}$. From this and (2) in this section,

$$(p\mathbf{b}) \bullet (q\mathbf{c}) = [pb_1, pb_2, pb_3]\bullet[qc_1, qc_2, qc_3] = pb_1 qc_1 + pb_2 qc_2 + pb_3 qc_3.$$

Hence the last expression contains a common factor pq, which you can pull out, obtaining

$$pq(b_1 c_1 + b_2 c_2 + b_3 c_3).$$

From (2) (with $\mathbf{b}$ and $\mathbf{c}$ instead of $\mathbf{a}$ and $\mathbf{b}$) you see that this equals $(pq)(\mathbf{b} \bullet \mathbf{c})$. But this is the right side of (A), and the proof is complete.

Looking back and comparing, you may find that this general proof was not much more difficult than the calculation in the case of the given special vectors. This is an interesting experience worth remembering because it may happen in other cases, too. And, as a general rule, if a general proof seems too difficult, start with a simple special case.

13. Work. This is a major motivation and application of inner products. In this problem, the force $\mathbf{p} = [2, 6, 6]$ acts in the displacement from the point A with coordinates $(3, 4, 0)$ to the point B with coordinates $(5, 8, 0)$. Both points lie in the xy-plane. Hence the same is true for the segment AB, which represents a "displacement vector" $\mathbf{d}$, whose components are obtained as differences of corresponding coordinates of B minus A (endpoint minus initial point). Thus,

$$\mathbf{d} = [d_1, d_2, d_3] = [5 - 3, \quad 8 - 4, \quad 0 - 0] = [2, 4, 0].$$

This gives the work as an inner product (a dot product) as in Example 2 in the text,

$$W = \mathbf{p} \bullet \mathbf{d} = [2, 6, 6]\bullet[2, 4, 0] = 2\cdot2 + 6\cdot4 + 6\cdot0 = 28.$$

19. Angle. Use (4).

25. Normal to a plane. Angle between planes. A **normal** to a plane is a straight line perpendicular to the plane. See Example 6 in the text. The angle γ between the given planes P_1 and P_2 is the angle between their normals. (Equivalently, if P is a plane perpendicular to the line of intersection between P_1 and P_2, then γ is the angle between the lines along which P_1 and P_2 intersect P.) Normal vectors of the given $P_1 : x + y + z = 1$ and $P_2 : x + 2y + 3z = 6$ are

$$\mathbf{n}_1 = [1, 1, 1] \quad \text{and} \quad \mathbf{n}_2 = [1, 2, 3],$$

respectively. Now use (4). You need $\mathbf{n}_1 \bullet \mathbf{n}_2 = 1 + 2 + 3 = 6$ and in the denominator $|\mathbf{n}_1| = \sqrt{3}$ and

$|n_2| = \sqrt{1+4+9} = \sqrt{14}$. This gives

$$\cos\gamma = 6/\sqrt{3\cdot 14} = 6/6.48074 = 0.92582, \quad \gamma = 22.2 \text{ degrees.}$$

Sec. 8.3   Vector Product (Cross Product)

Problem Set 8.3. Page 421

1. **Vector product.** Both the dot product and the cross product are concepts created for reasons of applications. The dot product is a scalar, as it occurs, for instance, in connection with work, as you have seen in Problem Set 8.2. The cross product is a vector perpendicular to the two vectors (or is the zero vector in the exceptional cases described in the definition). Examples 4-6 in the text illustrate important applications that helped to motivate this kind of product. Given $\mathbf{a} = [1, 2, 0]$ and $\mathbf{b} = [-3, 2, 0]$, you obtain the cross product $\mathbf{v} = \mathbf{a} \times \mathbf{b}$ from (2). The vectors $\mathbf{a}$ and $\mathbf{b}$ lie in the xy-plane (more precisely, they are parallel to this plane) since their components in the z-direction are zero. Hence their cross product must be perpendicular to the xy-plane, as follows directly from the definition. See Fig. 167. Much more easily to remember than (2) is (2**), which implies (2*), giving the components by second-order determinants. If you need help with determinants, look up the beginning of Sec. 6.6. From (2*) you obtain

$$v_1 = \begin{vmatrix} 2 & 0 \\ 2 & 0 \end{vmatrix} = 0, \quad v_2 = \begin{vmatrix} 0 & 1 \\ 0 & -3 \end{vmatrix} = 0, \quad v_3 = \begin{vmatrix} 1 & 2 \\ -3 & 2 \end{vmatrix} = 2+6 = 8.$$

Hence $\mathbf{v} = [0, 0, 8]$. From this you have $\mathbf{b} \times \mathbf{a} = -\mathbf{v} = [0, 0, -8]$. This illustrates that cross multiplication is not commutative but *anticommutative*; see (6) and Fig. 171. Hence always observe the order of the factors carefully.

17. **Scalar triple product.** This is the most useful of products with three or more factors. The reason is its geometric interpretation (see Figs. 175 and 176). Using (10) and developing the determinant by the third column gives

$$(\mathbf{b}\,\mathbf{a}\,\mathbf{d}) = \mathbf{b}\cdot(\mathbf{a}\times\mathbf{d}) = \begin{vmatrix} -3 & 2 & 0 \\ 1 & 2 & 0 \\ 6 & -7 & 2 \end{vmatrix} = 2\begin{vmatrix} -3 & 2 \\ 1 & 2 \end{vmatrix} = 2(-6-2) = -16.$$

The three vectors make up the three rows of the determinant, as you see. Interchanging two rows multiplies the determinant by -1; this gives the second result.

21. **Rotations** can be conveniently handled by vector products, as Example 5 of the text shows. For a rotation about the y-axis with $\omega = 9\,\text{sec}^{-1}$ the rotation vector $\mathbf{w}$, which always lies in the axis of rotation (if you choose a point on the axis as the initial point of $\mathbf{w}$), is

$$\mathbf{w} = \begin{bmatrix} 0, 9, 0 \end{bmatrix}$$

or $\mathbf{w} = \begin{bmatrix} 0, -9, 0 \end{bmatrix}$, depending on the sense of the rotation, which is not given in the problem. Also, $\mathbf{r} = [3, 4, 8]$, the position vector of the point P at which you want to find the velocity vector $\mathbf{v}$ (see Fig. 174). From these data the basic formula (9) gives

$$\mathbf{v} = \mathbf{w}\times\mathbf{r} = \begin{vmatrix} \mathbf{i} & \mathbf{j} & \mathbf{k} \\ 0 & 9 & 0 \\ 3 & 4 & 8 \end{vmatrix} = (9\cdot 8)\mathbf{i} - 0\mathbf{j} + (-(9\cdot 3))\mathbf{k} = [72, 0, -27]$$

or $-[72, 0, -27] = [-72, 0, 27]$ if you take the other possible $\mathbf{w}$. The speed is the length of the velocity vector $\mathbf{v}$, that is,

$$|\mathbf{v}| = \sqrt{(-72)^2 + 0 + 27^2} = \sqrt{5913} = 76.896.$$

29. Area of a triangle. Given the vertices $A : (1,3,2)$, $B : (3,-4,2)$, $C : (5,0,-5)$, you can find the area from a vector product. (Try it without vectors.) All you have to do is to derive two vectors that form two sides of the triangle. Obviously, there are essentially three possibilities for doing this. For instance, derive **b** and **c** with common inital point A and terminal points B and C. This gives

$$\mathbf{b} = [3-1, \ -4-3, \ 2-2] = [2,-7, \ 0], \quad \mathbf{c} = [5-1, \ 0-3, \ -5-2] = [4, \ -3, \ -7].$$

Their vector product is

$$\mathbf{v} = \mathbf{b} \times \mathbf{c} = \begin{vmatrix} \mathbf{i} & \mathbf{j} & \mathbf{k} \\ 2 & -7 & 0 \\ 4 & -3 & -7 \end{vmatrix} = (-7)\cdot(-7)\mathbf{i} - (2\cdot(-7))\mathbf{j} + (2\cdot(-3) - (-7)\cdot 4)\mathbf{k}$$

$$= 49\mathbf{i} + 14\mathbf{j} + 22\mathbf{k}.$$

By the definition of vector product, $|\mathbf{v}|$ gives the area of the parallelogram determined by these vectors,

$$|\mathbf{v}| = \sqrt{49^2 + 14^2 + 22^2} = \sqrt{3081} = 55.507.$$

Hence the answer is $(1/2)\sqrt{3081} = 27.753$.

Sec. 8.4   Vector and Scalar Functions and Fields. Derivatives

Problem Set 8.4. Page 427

1. Value at a point. $f(2,4) = 9\cdot 2^2 + 4\cdot 4^2 = 100$. Etc.

5. Level curves. arctan $(y/x) = k = const$ if and only if $y/x = c = const$, $y = cx$. Straight lines passing through the origin.

13. Level surfaces. $4x^2 + y^2 + 9z^2 = const$. Division by 36 gives

$$\frac{x^2}{3^2} + \frac{y^2}{6^2} + \frac{z^2}{2^2} = c = const.$$

These are ellipsoids. For $c = 1$ the ellipsoid has the semi-axes $3, 6, 2$ in the x-, y-, and z-directions, respectively. For other (positive) values you obtain an ellipsoid with semi-axes proportional to 3, 6, and 2, respectively.

19. Vector field. The vector function $\mathbf{v} = [y^2, \ 1]$ is constant on each horizontal straight line $y = const$. Hence along each such line the arrows representing $\mathbf{v}$ have the same length and the same direction (they are parallel and their arrow heads point in the same direction).

23(e). Vector field. The vectors $\mathbf{v} = [\cos x, \ \sin x]$ are all unit vectors because $v_1^2 + v_2^2 = \cos^2 x + \sin^2 x = 1$. On each vertical straight line $x = const$ these vectors are constant (represented by parallel arrows with heads pointing in the same direction). For $x = 0$ they are horizontal (why?). As x increases from 0, their slope increases until x reaches the value $\pi/2$, at which $\cos x = 0$ and the arrows are vertical. Make a sketch corresponding to this discussion and continue it for values of $x > \pi/2$ until you reach 2π. From there on the vector field repeats itself, for reasons of periodicity.

27. Partial derivatives. Since vectors are differentiated componentwise, this is as in calculus, and you obtain the answer given on p. A21 of Appendix 2 of the book.

Sec. 8.5   Curves. Tangents. Arc Length

Problem Set 8.5. Page 433

5. Straight line. Eq. (4) gives the general parametric representation of a straight line

$$\mathbf{r}(t) = \mathbf{a} + t\mathbf{b} = [a_1 + tb_1,\ a_2 + tb_2,\ a_3 + tb_3].$$

Here, $\mathbf{a}$ is the position vector of a point P through which the line passes; hence this point corresponds to the parameter value $t = 0$. In this problem you can take $\mathbf{a} = [2,\ 3,\ 0]$, the position vector of A. The vector $\mathbf{b}$ gives the direction of the line. If $\mathbf{b}$ is a unit vector, then $|t|$ is the distance of the point with position vector $\mathbf{r}(t)$ from P. In the problem, $\mathbf{b}$ is not given, but must be determined, and you can take the vector with initial point A and terminal point B. Its components are the differences of corresponding coordinates, that is,

$$\mathbf{b} = [5 - 2,\ -1 - 3,\ 0 - 0] = [3,\ -4,\ 0].$$

Make sure that you understand this step. This gives the answer

$$\mathbf{r}(t) = [2,\ 3,\ 0] + t[3,\ -4,\ 0] = [2 + 3t,\ 3 - 4t,\ 0].$$

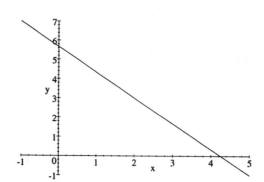

Section 8.5. Problem 5. Parametric representation of a straight line

11. Parametric representation of a circle. Circles occur often, and it is important that you fully understand this representation

$$\mathbf{r}(t) = [0,\ 5\cos t,\ 5\sin t].$$

In components,

$$x(t) = 0,\quad y(t) = 5\cos t,\quad z(t) = 5\sin t.$$

This gives

$$y^2 + z^2 = 25\cos^2 t + 25\sin^2 t = 25.$$

The circle lies in the yz-plane. Its center is the origin. Its radius is 5. A circle of radius c and center (a, b) in the xy-plane has the representation

$$\mathbf{r}(t) = [a + c\cos t,\ b + c\sin t,\ 0]. \tag{A}$$

From this you obtain

$$(x - a)^2 + (y - b)^2 = c^2\cos^2 t + c^2\sin^2 t = c^2, \tag{B}$$

the familiar representation often used in calculus. Before going on, make sure that you understand the relationship between (A) and (B).

23. Tangent to an ellipse. The ellipse is given by

$$\mathbf{r}(t) = [\cos t,\ 2\sin t]$$

where t varies from 0 to 2π. You have to find the tangent at the point $P : (1/2, \sqrt{3}\,)$. Now $\cos t = 1/2$

when $t = \pi/3 = 60$ degrees. Then $2 \sin t = 2 \sin (\pi/3) = 2(1/2) \sqrt{3} = \sqrt{3}$. Hence $\mathbf{r}(\pi/3) = [1/2, \ \sqrt{3}]$, so $t = \pi/3$ is in fact the parametric value of P. Now the slope of the tangent is $\mathbf{r}'(t) = [-\sin t, \ 2 \cos t]$. At P this becomes

$$\mathbf{r}'(\pi/3) = [-\sin (\pi/3), \ 2 \cos (\pi/3)] = [-(1/2) \sqrt{3}, \ 1].$$

From these data and (9) you obtain a parametric representation of the tangent at P in the form

$$\mathbf{q}(w) = [1/2, \ \sqrt{3}] + w[-(1/2) \sqrt{3}, \ 1] \ = [0.500 - 0.866w, \ 1.732 + w].$$

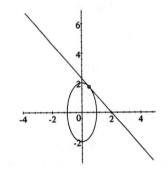

Section 8.5. Problem 23. Ellipse and tangent

33. Famous curves.

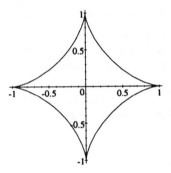

Section 8.5. CAS Project 33. Astroid

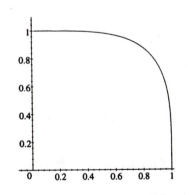

Section 8.5. CAS Project 33. Arc of Lamé's curve

Sec. 8.6   Curves in Mechanics. Velocity and Acceleration

Problem Set 8.6. Page 439

5. Acceleration. The path

$$\mathbf{r}(t) = [b \cos t, \quad b \sin t, \quad c] \quad (b > 0)$$

is a circle of radius b in the horizontal plane $z = c$. The velocity is

$$\mathbf{v}(t) = \mathbf{r}'(t) = [-b \sin t, \quad b \cos t, \quad 0].$$

This vector is horizontal; there is no velocity in z-direction, as had to be expected for physical reasons. The speed is

$$|\mathbf{v}(t)| = \sqrt{(-b \sin t)^2 + (b \cos t)^2} = b.$$

Hence the speed is constant. This does **not** imply that the acceleration is zero because the velocity is changing direction. Indeed, the acceleration is

$$\mathbf{a}(t) = \mathbf{v}'(t) = \mathbf{r}''(t) = [-b \cos t, \quad -b \sin t, \quad 0].$$

You see that $\mathbf{a}(t) = -\mathbf{r}(t)$. This is the centripetal acceleration. It points from the moving body to the center of the circle (the origin). All these results are in agreement with Example 1 of the text, with $R = b$ and $\omega = 1$. Note that the special $\omega = 1$ makes the centripetal acceleration *equal* to $-\mathbf{r}(t)$, whereas for general ω it will just be *proportional* to it.

9. Cycloid. This curve is shown in the figure. From the given formula

$$\mathbf{r}(t) = [R \sin \omega t + \omega R t, \quad R \cos \omega t + R]$$

you see that $t = 0$ gives $\mathbf{r}(0) = [0, 2R]$; this corresponds to the first maximum of the curve in the figure. For $t = \pi/\omega$ you obtain $\mathbf{r}(\pi/\omega) = [\pi R, 0]$; this corresponds to the first cusp of the curve in the figure. Differentiating $\mathbf{r}(t)$ twice and substituting those values of t, you obtain the answer on p. A21 in Appendix 2 of the book. If the wheel rolls without slipping on a circle (instead of the x-axis), one obtains an epicycloid or a hypocycloid. These curves are discussed in most engineering handbooks, for instance, on p. 290 of O. V. Eshbach, *Handbook of Engineering Fundamentals*, 3rd ed., New York: Wiley, 1975.

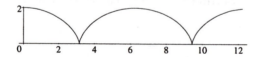

Section 8.6. Problem 9. Cycloid with $R = 1$.

Sec. 8.7   Curvature and Torsion of a Curve.   *Optional*

Problem Set 8.7. Page 443

7. Curvature. The given curve

$$\mathbf{r}(t) = [a \cos t, \quad a \sin t, \quad ct]$$

is a helix, right-handed when $c > 0$ (Fig. 184 in Sec. 8.5), left-handed when $c < 0$ (Fig. 185), and degenerated to a circle when $c = 0$. In each case the curve lies on a circular cylinder of radius a and the z-axis as the axis of rotation. In Example 1 of the text it is shown how the curvature can be obtained from (1). This, however, presupposes that the curve is represented with the arc length s as parameter or the given representation can be easily converted to such a representation. This will hardly be the case in practice. For this reason, from a practical point of view, formula (1') is more important than formula (1) (which gives a better geometrical characterization of the curvature in principle). In the present case, an application of (1') proceeds as follows.

$$\mathbf{r} = [a \cos t, \quad a \sin t, \quad ct]$$
$$\mathbf{r}' = [-a \sin t, \quad a \cos t, \quad c]$$
$$\mathbf{r}'' = [-a \cos t, -a \sin t, \quad 0]$$
$$\mathbf{r}' \bullet \mathbf{r}' = a^2 \sin^2 t + a^2 \cos^2 t + c^2 = a^2 + c^2$$
$$\mathbf{r}' \bullet \mathbf{r}'' = a^2 \sin t \cos t - a^2 \sin t \cos t + 0 = 0$$
$$\mathbf{r}'' \bullet \mathbf{r}'' = a^2 \cos^2 t + a^2 \sin^2 t = a^2.$$

Hence the numerator in (1') is

$$\sqrt{(a^2 + c^2) a^2} = a \sqrt{a^2 + c^2}$$

and the denominator is $(a^2 + c^2)^{3/2}$. The quotient is the curvature

$$\kappa = a/(a^2 + c^2).$$

For a circle of radius a you have $c = 0$ and get the curvature $\kappa = a/a^2 = 1/a$, the reciprocal of the radius, as expected. The amount of work was not much more than that in Example 1. In general the expression for s will be cumbersome or will even be given by a nonelementary integral, so that (1') must be used instead of (1).

13. Torsion. Use that for a plane curve, $\mathbf{u}$ and thus $\mathbf{u}'$ in (2) and (3) lie in the plane of the curve (when you choose a point of the curve as the initial point of these vectors). Use what this implies with respect to the cross product in (3), and then take a look at (4), from which you will see the result without any calculation.

15. Torsion. Use (4''') and straightforward calculation. By repeated differentiation you get for the determinant

$$\begin{vmatrix} 1 & 2t & 3t^2 \\ 0 & 2 & 6t \\ 0 & 0 & 6 \end{vmatrix} = 12.$$

The denominator requires more effort, but the expression

$$(1 + 4t^2 + 9t^4)(4 + 36t^2) - (4t + 18t^3)^2$$

first obtained simplifies nicely by cancellation of terms.

Sec. 8.8   Review from Calculus in Several Variables.   *Optional*

Problem Set 8.8. Page 446

7. Chain rule. The result is obtained by applying the chain rule and simplifying. From $w = xy$ and $x = e^u \cos v, y = e^u \sin v$ you obtain

$$w_u = x_u y + x y_u = e^u \cos v \, e^u \sin v + e^u \cos v \, e^u \sin v$$
$$= 2e^{2u} \cos v \sin v = e^{2u} \sin 2v$$

and similarly for the partial derivative of w with respect to v. In the present case, substitution and differentiation is much simpler; you obtain

$$w = e^{2u} \cos v \sin v = \frac{1}{2} e^{2u} \sin 2v$$

and from this by differentiation and use of the chain rule

$$w_u = e^{2u} \sin 2v, \qquad w_v = e^{2u} \cos 2v.$$

Sec. 8.9   Gradient of a Scalar Field. Directional Derivative

Problem Set 8.9. Page 452

7. Gradient. Heat flows from higher to lower temperatures. If the temperature is given by $T(x,y,z)$, the isotherms are the surfaces $T = const$, and the direction of heat flow is the direction of $-\text{grad}\, T$. For $T = z/(x^2 + y^2)$ you obtain, using the chain rule,

$$- \text{grad}\, T = -[-2xz/(x^2 + y^2)^2, \quad -2yz/(x^2 + y^2)^2, \quad 1/(x^2 + y^2)]$$
$$= (1/(x^2 + y^2)^2)[2xz, \quad 2yz, \quad -(x^2 + y^2)].$$

The given point P has the coordinates $x = 0, y = 1, z = 2$. Hence at P,

$$- \text{grad}\, T(P) = [0, \quad 4, \quad -1].$$

(In the first printing of the textbook, the last minus sign is missing.) The figure shows the isotherms in the plane $z = 2$, that is, the circles of intersection of the paraboloids $T = c = const$ (thus $z = c(x^2 + y^2)$) with the horizontal plane $z = 2$. The point P is marked by a small circle on the vertical y-axis.

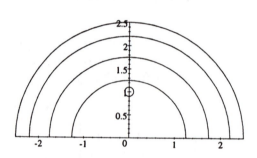

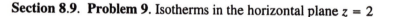

Section 8.9. Problem 9. Isotherms in the horizontal plane $z = 2$

17. Normal of a curve. $f(x, y) = x^2 + y^2 = 25$ represents a circle of radius 5 with the center at the origin. A normal vector to the circle is

$$\mathbf{N} = \text{grad}\, f = \begin{bmatrix} 2x, & 2y \end{bmatrix}.$$

At $P : (3,4)$ this becomes

$$\mathbf{N}(P) = \begin{bmatrix} 6, & 8 \end{bmatrix}.$$

This normal vector has the length 10. Hence a unit normal vector of the circle at P is

$$\mathbf{n} = [0.6, \quad 0.8]$$

(and the other is $-\mathbf{n} = -[0.6, \quad 0.8]$).

23. Potential. Not every vector function (vector field) has a potential, that is, not every vector function $\mathbf{v}(x, y, z)$ can be obtained as the gradient of a scalar function $f(x,y,z)$.

In the present case, $\mathbf{v}(x,y) = [xy, \quad 2xy]$. If $\mathbf{v}$ had a potential, you should have $\mathbf{v} = \text{grad}\, f = [f_x, \quad f_y]$. Give it a trial. From the first component you obtain the condition

$$f_x = xy.$$

By integration with respect to x,

$$f = \tfrac{1}{2}x^2 y + g(y).$$

To understand this step, take the partial derivative of f with respect to x. Similarly, from the second component you obtain the condition

$$f_y = 2xy.$$

By integration with respect to y,

$$f = xy^2 + h(x).$$

Again, check this by differentiation with respect to y.

Now, if v had a potential f, you should be able to find a function $g(y)$ depending only on y and a function $h(x)$ depending only on x such that the two expressions for f agree. You see that this is not possible. From this it follows that v has no potential. (In Sec. 8.11 we shall discover a systematic method that will make you independent of trial and error.)

29. **Directional derivative.** The directional derivative measures the rate of change of a function f in a given direction in the plane or in space. The direction is given by a vector which we call a, and the derivative is given by (6'), for reasons explained in the text. In the present problem, $f = x^2 + y^2$, $a = [2, \ -4]$, and $P : (1, \ 1)$. In (6') you need

$$|a| = \sqrt{20}, \quad \text{grad } f = [2x, \ 2y], \quad a \bullet \text{grad } f = 2 \cdot 2x - 4 \cdot 2y = 4x - 8y.$$

Hence at an arbitrary point the derivative of f in the direction of a is

$$(4x - 8y)/\sqrt{20} \quad \text{and at } P, \quad (4 - 8)/\sqrt{20} = -2/\sqrt{5} = -0.89.$$

The curves $f = const$ are concentric circles. The figure shows the circle through P, as well as P itself (the small circle) and the direction of a (the small segment inside the circle). At P the gradient is $[2, \ 2]$, pointing to the exterior of the circle. The angle between grad f and a is greater than 90 degrees; this gives the minus sign in -0.89. The value $|-0.89|$ is relatively small because a is not too much different from the tangent direction of the circle at P (for which the directional derivative is zero–explain!).

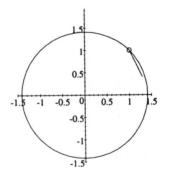

Section 8.9. Problem 29. Point P, contour line $f = const$ through P, and direction of the vector a

Sec. 8.10   Divergence of a Vector Field

Example 2. v is the velocity. ρ is the density. $u = \rho v$ has the components $u_1 = \rho v_1, u_2 = \rho v_2, u_3 = \rho v_3$. Hence ρv_2 on p. 455 is the mass flowing in the y-direction, and $(\rho v_2)_y$ is its value at some y, that is, the mass entering the box through the face $y = const$ (the left face in Fig. 199, which also shows the face $x + \Delta x = const$ as the right face and $z + \Delta z = const$ as the upper face.

Problem Set 8.10. Page 456

5. **Divergence.** The physical meaning and practical importance of the divergence of a vector function (a vector field) are explained in the text. div is applied to a vector function $\mathbf{v}$ and gives a scalar function div $\mathbf{v}$, whereas grad is applied to a scalar function f and gives a vector function grad f. Of course, grad and div are not inverses of each other; they are entirely different operations, created because of their applicability in physics, geometry, and elsewhere. The calculation of div $\mathbf{v}$ by (1) is partial differentiation as in calculus. In the present problem, $\mathbf{v} = [v_1, \ v_2]$ has as its first component

$$v_1 = -y/(x^2 + y^2), \quad \text{and you need} \quad \frac{\partial}{\partial x} v_1 = +2xy/(x^2 + y^2)^2$$

with the factor $2x$ resulting from the chain rule. The second component is

$$v_2 = x/(x^2 + y^2), \quad \text{and you need} \quad \frac{\partial}{\partial y} v_2 = -2xy/(x^2 + y^2)^2.$$

The sum of the two expressions on the right is div $\mathbf{v}$, and you see that div $\mathbf{v} = 0$.

11. **Incompressible flow.** The velocity vector is $\mathbf{v} = y\mathbf{i} = [y, \ 0, \ 0]$. Hence div $\mathbf{v} = 0$. This shows that the flow is incompressible; see (7) on p. 456. $\mathbf{v}$ is parallel to the x-axis. In the upper half-plane it points to the right and in the lower to the left. On the x-axis ($y = 0$) it is the zero vector. On each horizontal line $y = const$ it is constant. The speed is the larger the farther away from the x-axis you are. From $\mathbf{v} = y\mathbf{i}$ and the definition of a velocity vector you obtain

$$\mathbf{v} = [dx/dt, \ dy/dt, \ dz/dt] = [y, \ 0, \ 0].$$

This vector equation gives three equations for the corresponding components,

$$dx/dt = y, \quad dy/dt = 0, \quad dz/dt = 0.$$

Integration of $dz/dt = 0$ gives

$$z(t) = c_3 \quad \text{with} \quad c_3 = 0 \text{ for the face } z = 0, \quad c_3 = 1 \text{ for the face } z = 1$$

and $0 < c_3 < 1$ for particles inside the cube. Similarly, by integration of $dy/dt = 0$ you obtain $y(t) = c_2$ with $c_2 = 0$ for the face $y = 0$, $c_2 = 1$ for the face $y = 1$ and $0 < c_2 < 1$ for particles inside the cube.

Finally, $dx/dt = y$ with $y = c_2$ becomes $dx/dt = c_2$. By integration,

$$x(t) = c_2 t + c_1.$$

From this,

$$x(0) = c_1 \quad \text{with} \quad c_1 = 0 \text{ for the face } x = 0, \quad c_1 = 1 \text{ for the face } x = 1.$$

Also

$$x(1) = c_1 + c_2,$$

hence

$$x(1) = c_2 + 0 \text{ for the one face}, \quad x(1) = c_2 + 1 \text{ for the other face }.$$

This shows that the distance of these two parallel faces has remained the same, namely, 1. And since nothing happened in the y- and z-directions, this shows that the volume at time $t = 1$ is still 1, as it should be in the case of incompressibility.

13. **Formulas for the divergence.** These formulas help in simplifying calculations as well as in theoretical work. They follow by straightforward calculations directly from the definitions. For instance, by the definition of the divergence and by product differentiation you obtain

$$\text{div} (f\mathbf{v}) = (fv_1)_x + (fv_2)_y + (fv_3)_z$$
$$= f_x v_1 + f_y v_2 + f_z v_3 + f[(v_1)_x + (v_2)_y + (v_3)_z]$$
$$= (\text{grad} f) \bullet \mathbf{v} + f \, \text{div} \, \mathbf{v}.$$

15. Laplacian. Problems 14-20 are calculational exercises proposed because of the importance of the Laplacian and the Laplace equation in several branches of physics (see the Index of the book). In general, the Laplacian is itself a function of x, y, z. In Prob. 15 it is constant,

$$\nabla^2 f = 8 + 18 + 2 = 28,$$

since f is so simple. The case of two variables x, y is best handled by complex analysis, as is explained in Chaps. 12 and 16.

Sec. 8.11   Curl of a Vector Field

Example 1. curl $\mathbf{v} = \mathbf{i}(z_y - (3xz)_z) - \mathbf{j}(z_x - (yz)_z) + \mathbf{k}((3xz)_x - (yz)_y) = -3x\mathbf{i} + y\mathbf{j} + (3z - z)\mathbf{k}$.

Problem Set 8.11. Page 459

7. Calculation of the curl. curl is the third of the three operators grad, div, curl designed to meet practical needs in connection with vector and scalar functions and fields. These operators increase the usefulness of vector calculus very significantly, notably in connection with integrals, as will be shown in Chap. 9. For the calculation of the curl use (1), where the "symbolic determinant" helps to memorize the actual formulas for the components given in (1) below the determinant. Given

$$v_1 = x^2 yz, \quad v_2 = xy^2 z, \quad v_3 = xyz^2.$$

With this you obtain from (1) for the components of $\mathbf{u} = $ curl $\mathbf{v}$ the expressions

$$u_1 = (v_3)_y - (v_2)_z = (xyz^2)_y - (xy^2 z)_z = xz^2 - xy^2$$

$$u_2 = (v_1)_z - (v_3)_x = (x^2 yz)_z - (xyz^2)_x = x^2 y - yz^2$$

$$u_3 = (v_2)_x - (v_1)_y = (xy^2 z)_x - (x^2 yz)_y = y^2 z - x^2 z.$$

11. Fluid flow. Both div and curl characterize essential properties of flows, which are usually given in terms of the velocity vector field $\mathbf{v}(x, y, z)$. The present problem is two-dimensional, that is, in each plane $z = const$ the flow is the same. The given velocity is

$$\mathbf{v} = [y, \quad -x].$$

Hence div $\mathbf{v} = 0 + 0 = 0$. This shows that the flow is incompressible (see the previous section). Furthermore, from (1) in the present section you see that the first two components of the curl are zero because they consist of expressions involving v_3, which is zero, or involving the partial derivative with respect to z, which is zero because $\mathbf{v}$ does not depend on z. There remains

$$\text{curl } \mathbf{v} = ((v_2)_x - (v_1)_y)\mathbf{k} = (-1 - 1)\mathbf{k} = -2\mathbf{k}.$$

This shows that the flow is not irrotational. Now determine the paths of the particles of the fluid. From the definition of velocity you have

$$v_1 = dx/dt, \quad v_2 = dy/dt.$$

From this and the given velocity vector $\mathbf{v}$ you see that

$$dx/dt = y \tag{A}$$

$$dy/dt = -x. \tag{B}$$

This system of differential equations can be solved by a trick worth remembering. The right side of (B) times the left side of (A) is $-x\,dx/dt$. This must equal the right side of (A) times the left side of (B), which is $y\,dy/dt$. Hence

$$-x\,dx/dt = y\,dy/dt.$$

You can now integrate with respect to t on both sides and multiply by -2, obtaining

$$x^2 = -y^2 + c.$$

This shows that the paths of the particles (the streamlines of the flow) are concentric circles

$$x^2 + y^2 = c.$$

This agrees with the fact that (2) on p.458 relates the curl with the rotation vector. In the present case,

$$\mathbf{w} = \tfrac{1}{2}\operatorname{curl}\mathbf{v} = \mathbf{k},$$

which shows that the z-axis is the axis of rotation.

CHAPTER 9.   Vector Integral Calculus. Integral Theorems

Sec. 9.1   Line Integrals

Problem Set 9.1. Page 470

3. Line integral in the plane. This generalizes a definite integral of calculus. Instead of integrating along the x-axis you now integrate over a curve C, a quarter-circle, in the xy-plane.

Integrals of the form (3) have various applications, for instance, in connection with work done by a force in a displacement. The right side of (3) shows how such an integral is converted to a definite integral with t as the variable of integration. The conversion is done by using the representation of the path of integration C.

The problem parallels Example 1 in the text. The quarter-circle C has the radius 2 and can be represented by

$$\mathbf{r}(t) = [2\cos t, \quad 2\sin t], \quad \text{in components}, \quad x = 2\cos t, \quad y = 2\sin t, \tag{I}$$

where t varies from $t = 0$ (the initial point of C on the x-axis) to $t = \pi/2$ (the terminal point of C on the y-axis). The given function is a vector function

$$\mathbf{F} = [xy, \quad x^2 y^2]. \tag{II}$$

$\mathbf{F}$ defines a vector field in the xy-plane. At each point (x, y) it gives a certain vector, which you could draw as a little arrow. In particular, at each point of C the vector function $\mathbf{F}$ gives a vector. You can obtain these vectors simply by substituting x and y from (I) into (II). This gives

$$\mathbf{F}(\mathbf{r}(t)) \quad = [4\cos t \sin t, \quad 16\cos^2 t \sin^2 t]. \tag{III}$$

This is now a vector function of t defined on the quarter-circle C.

Now comes an important point to observe. You do not integrate $\mathbf{F}$ itself, but you integrate the dot product of $\mathbf{F}$ in (III) and the tangent vector $\mathbf{r}'(t)$ of C. This dot product $\mathbf{F} \bullet \mathbf{r}'$ can be "visualized" because it is the component of $\mathbf{F}$ in the direction of the tangent of C (times the factor $|\mathbf{r}'(t)|$), as you can see from (11) in Sec. 8.2 with $\mathbf{F}$ playing the role of $\mathbf{a}$ and $\mathbf{r}'$ playing the role of $\mathbf{b}$. (Note that if t is the arc length s, then $\mathbf{r}'$ is a unit vector, so that that factor equals 1 and you get exactly that tangential projection.) Think this over before you go on calculating.

Differentiation with respect to t gives the tangent vector

$$\mathbf{r}'(t) = [-2\sin t, \quad 2\cos t].$$

Hence the dot product is

$$\begin{aligned}\mathbf{F}(\mathbf{r}(t)) \bullet \mathbf{r}'(t) &= 4\cos t \sin t \,(-2\sin t) + 16\cos^2 t \sin^2 t \,(2\cos t) \\ &= -8\cos t \sin^2 t + 32\cos^3 t \sin^2 t \\ &= -8\cos t \sin^2 t + 32\cos t \,(1 - \sin^2 t) \sin^2 t.\end{aligned} \tag{IV}$$

Now by the chain rule,

$$(\sin^3 t)' = 3\sin^2 t \cos t, \quad (\sin^5 t)' = 5\sin^4 t \cos t.$$

This shows that the expression in the last line of (IV) can be readily integrated; you obtain

$$-\tfrac{8}{3}\sin^3 t + 32\left(\tfrac{1}{3}\sin^3 t - \tfrac{1}{5}\sin^5 t\right).$$

At 0 the sine is 0, and at the upper limit of integration $\pi/2$ it is 1. Hence the result is

$$-\tfrac{8}{3} + 32\left(\tfrac{1}{3} - \tfrac{1}{5}\right) = \tfrac{8}{5}.$$

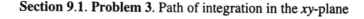

Section 9.1. Problem 3. Path of integration in the *xy*-plane

15. **Line integral (7) in space.** Line integrals in space are handled by the same method as line integrals in the plane. Quite generally, it is a great advantage of vector methods that methods in space and in the plane are very similar in most cases. Integrals (3), as just discussed, are suggested by work. Integrals (7) are conceptually a little simpler because the integrand is given directly (whereas in (3) it is obtained from a vector function by taking a dot product).

In the present problem, the path of integration C is a portion of a helix

$$C : \mathbf{r}(t) = [\cos t, \ \sin t, \ 2t]$$

with t varying from 0 to 4π. It lies on the circular cylinder of radius $a = 1$ with the z-axis as the axis, and has $c = 2$. Hence the arc length is (see Example 5 in Sec. 8.5)

$$s = t\sqrt{a^2 + c^2} = t\sqrt{5}.$$

It follows that with respect to s the limits of integration are 0 and $4\pi\sqrt{5}$. Furthermore, the integrand is

$$f = x^2 + y^2 + z^2 = \cos^2 t + \sin^2 t + 4t^2 = 1 + 4t^2 = 1 + \frac{4}{5}s^2.$$

You thus obtain the integral

$$\int_0^{4\pi\sqrt{5}} \left(1 + \frac{4s^2}{5}\right) ds = s + \frac{4s^3}{15} \Big|_0^{4\pi\sqrt{5}} = \sqrt{5}\left(4\pi + \frac{4}{3} \cdot 64\pi^3\right),$$

as given on p. A23 in Appendix 2.

Sec. 9.2   Line Integrals Independent of Path

Problem Set 9.2. Page 477

3. **Exactness and independence of path.** In general, a line integral will depend on the path along which you integrate from a given point A to a given point B. The reason for the importance of independence of path is explained at the beginning of the section.

In the present problem the form under the integral sign is

$$3z^2\, dx + 6zx\, dz.$$

The limits of integration in the xz-plane are $A : (-1, 5)$ and $B : (4, 3)$. Theorem 1 relates path independence to the existence of a function f such that

$$[3z^2, \ 6zx] = \operatorname{grad} f = [f_x, \ f_z],$$

thus

$$f_x = 3z^2, \quad \text{integrated with respect to } x : f = 3z^2 x + g(z)$$

$$f_z = 6zx, \quad \text{integrated with respect to } z : f = 3z^2 x + h(x).$$

Hence you can take $f(x, z) = 3z^2 x$. Then at the lower limit of integration, $f(-1, 5) = 3 \cdot 25 \cdot (-1) = -75$ and at the upper, $f(4, 3) = 108$, so that (3) gives the answer $108 - (-75) = 183$.

More systematically, you can check for the existence of a function f by (6) in Theorem 3. Now, since y does not occur in the integrand or in the limits of integration, F_2 and the partial derivatives of $\mathbf{F}$ with respect to y are zero, (6′) reduces to its second relationship, that is,

$$(F_1)_z = (3z^2)_z = 6z = (F_3)_x.$$

This shows that the integral is independent of path in the xz-plane. Its value is now obtained as just explained.

11. Integral in space. The method is the same as for an integral in the plane. It involves somewhat more work. In (6′) you may have to check all three relationships; of course, you can stop if you arrive at a relationship that is not satisfied; then you know that you have path dependence and cannot use (3).

In the present problem the form under the integral sign is

$$2xy^2 \, dx + 2x^2 y \, dy + dz.$$

Hence

$$F_1 = 2xy^2, \quad F_2 = 2x^2 y, \quad F_3 = 1.$$

In (6′) you thus obtain

$$(F_3)_y = 0 = (F_2)_z, \quad (F_1)_z = 0 = (F_3)_x, \quad (F_2)_x = 4xy = (F_1)_y.$$

Hence the differential form is exact in space, so that the integral is independent of path.

To evaluate the integral, you have to find f such that

$$\mathbf{F} = [F_1, \ F_2, \ F_3] = [2xy^2, \ 2x^2 y, \ 1] = \operatorname{grad} f = [f_x, \ f_y, \ f_z].$$

This gives the conditions

$$f_x = 2xy^2, \quad \text{integrated with respect to } x \quad f = x^2 y^2 + g(y, z)$$

$$f_y = 2x^2 y, \quad \text{integrated with respect to } y \quad f = x^2 y^2 + h(x, z)$$

$$f_z = 1, \quad \text{integrated with respect to } z \quad f = z + k(x, y).$$

You see that the three functions on the right agree if you choose

$$g(y, z) = h(x, z) = z, \quad k(x, y) = x^2 y^2.$$

Hence your result is $f = x^2 y^2 + z$. From (3) you now obtain the answer

$$f(a, b, c) - f(0, 0, 0) = a^2 b^2 + c.$$

13. Path dependence. The integrand is $F_1 \, dx + F_3 \, dz$, where

$$F_1 = z \sinh xz, \quad F_3 = -x \sinh xz.$$

Hence the second formula in (6′) gives on the left

$$(F_1)_z = \sinh xz + xz \cosh xz$$

but on the right

$$(F_3)_x = -\sinh xz - xz \cosh xz.$$

This shows path dependence. (The other two conditions in (6′) are satisfied because $F_2 = 0$, and F_1 and F_3 are independent of y.)

Sec. 9.3   From Calculus: Double Integrals.   *Optional*

Problem Set 9.3. Page 484

3. **Double integral**. Further work in this chapter will be concerned with basic integral theorems for transforming different kinds of integrals (line, double, surface, and triple integrals) into one another, and double integrals will occur in connection with line integrals (Sec. 9.3) and surface integrals (Secs. 9.6, 9.9).

The figure shows the triangular region of integration. Indeed, you first integrate with respect to x horizontally from $-y$ to y. The resulting integral is a function of y,

$$\int_{-y}^{y} (x^2 + y^2)\, dx = \frac{x^3}{3} + xy^2 \Big|_{-y}^{y} = \frac{y^3}{3} + y^3 - \left(\frac{(-y)^3}{3} + (-y)y^2 \right) = \frac{8y^3}{3}.$$

You have to integrate this with respect to y from 0 to 3, obtaining

$$\int_{0}^{3} \frac{8y^3}{3}\, dy = \frac{2y^4}{3} \Big|_{0}^{3} = 54.$$

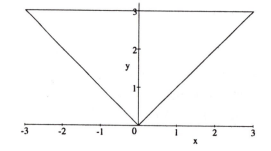

Section 9.3. Problem 3. Region of integration

15. **Center of gravity**. The circular disk of radius a has area πa^2. Hence $M = \pi a^2/4$ for the given region (a quarter of that disk). Since the mass density is 1, the formula in the text gives for the x-coordinate of the center of gravity

$$\bar{x} = \frac{1}{M} \iint x\, dx\, dy.$$

The form of the region of integration suggests the use of polar coordinates $x = r \cos \theta$, $y = r \sin \theta$. Then $dx\, dy = r\, dr\, d\theta$ and you have to integrate with respect to r from 0 to a and with respect to θ from 0 to $\pi/2$; this corresponds to the quarter of the circular disk. You obtain

$$\bar{x} = \frac{1}{M} \int_{0}^{\pi/2} \int_{0}^{a} r(\cos \theta)\, r\, dr\, d\theta.$$

The integral of r^2 is $r^3/3$, hence $a^3/3$ at the upper limit. The integral of $\cos \theta$ from 0 to $\pi/2$ gives 1. Hence, together, since $1/M = 4/(\pi a^2)$,

$$\bar{x} = (4/\pi a^2)(a^3/3) = 4a/(3\pi).$$

$\bar{y}$ has the same value, for reasons of symmetry.

Sec. 9.4   Green's Theorem in the Plane

Problem Set 9.4. Page 490

1. Transformation of a line integral into a double integral. Green's theorem in the plane transforms double integrals over a region R in the xy-plane into line integrals over the boundary curve C of R and conversely. These transformations are of practical and theoretical interest in both directions, depending on the purpose. In Probs. 1-10 the direct evaluation of the line integral would be much more involved than that of the double integral obtained by Green's theorem. Given

$$\mathbf{F} = [F_1, \ F_2] = [x^2 e^y, \ y^2 e^x].$$

In (1) you obtain on the left side

$$(F_2)_x - (F_1)_y = y^2 e^x - x^2 e^y. \tag{A}$$

Sketch the given rectangle (the region of integration R in the double integral). Then you see that you have to integrate over x from 0 to 2 and over y from 0 to 3. Integrating (A) over x gives

$$y^2 e^x - (x^3/3)e^y.$$

Substituting the upper limit $x = 2$ and then the lower limit $x = 0$ and taking the difference, you obtain

$$\left(y^2 e^2 - \tfrac{8}{3}e^y\right) - (y^2 \cdot 1 - 0) = y^2 e^2 - \tfrac{8}{3}e^y - y^2.$$

Integrating this over y gives

$$\tfrac{1}{3} y^3 e^2 - \tfrac{8}{3}e^y - \tfrac{1}{3}y^3.$$

Substituting the upper limit $y = 3$ and the lower limit $y = 0$, and taking the difference of the two expressions obtained you finally arrive at the answer

$$\left(\tfrac{27}{3}e^2 - \tfrac{8}{3}e^3 - \tfrac{27}{3}\right) - \left(0 - \tfrac{8}{3}\cdot 1 - 0\right) = 9e^2 - \tfrac{8}{3}e^3 - 9 + \tfrac{8}{3},$$

in agreement with the answer on p. A23 in Appendix 2 of the book.

13. Area. The given Pascal snail has the representation in polar coordinates

$$r = 1 + 2 \cos \theta.$$

In the formula (5) for the area you need

$$r^2 = (1 + 2 \cos \theta)^2 = 1 + 4 \cos \theta + 4 \cos^2\theta.$$

Integration of the three terms on the right from 0 to $\pi/2$ gives

$$\pi/2 + 4 + 4(\pi/4) = 4 + 3\pi/2.$$

Now multiply by 1/2 as shown in (5) to obtain the answer given on p. A23 of the book.

The figure shows the entire curve which you get if you let θ range over a full interval of periodicity of $\cos \theta$, say, from 0 to 2π. For $\theta = 0$ you get $r = 3$ on the x-axis. For $\theta = \pi/2$ you have $r = 1$ because $\cos (\pi/2) = 0$; this is the point where the curve intersects the y-axis. Hence our answer gives the area under the curve in the first quadrant. For what values of θ does the curve pass through the origin? Through the point 1 on the x-axis?

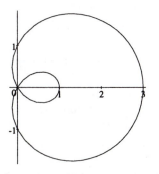

Section 9.4. Problem 13. Limaçon

15. Consequences of Green's theorem in the plane are given in formulas (9)-(12). Formula (9) shows that if you integrate the normal derivative of a function w over a closed curve and if w is harmonic, then you get zero.

For other functions you may use (9) to simplify the evaluation of integrals of the normal derivative. Such integrals occur, for instance, in connection with the flux of a fluid through a surface. For $w = \cosh x$ two partial differentiations with respect to x give $\nabla^2 w = w_{xx} = \cosh x$. The hypothenuse of the triangle has the representation

$$y = x/2. \qquad \text{Hence} \qquad x = 2y.$$

(Make a sketch.) The integral of $\cosh x$ is $\sinh x$, and you integrate from $x = 0$ horizontally to $x = 2y$. Since $\sinh 0 = 0$, this gives $\sinh 2y$. You now integrate over y, obtaining $(\cosh 2y)/2$. The limits are 0 and 2. This gives the answer $(\cosh 4 - 1)/2$.

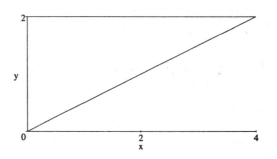

Section 9.4. Problem 15. Region of integration in the double integral

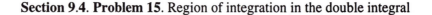

Sec. 9.5   Surfaces for Surface Integrals

Problem Set 9.5. Page 495

3. **Parametric surface representations** have the advantage that the components x, y, z of the position vector **r** play the same role in the sense that none of them is an independent variable (as it is the case when we use $z = f(x,y)$), but all three are functions of two variables ("**parameters**") u and v (we need two of them because a surface is two-dimensional). Thus, in the present problem,

$$\mathbf{r}(u,v) = [x(u,v), \quad y(u,v), \quad z(u,v)] = [u \cos v, \quad u \sin v, \quad cu].$$

In components,

$$x = u \cos v, \quad y = u \sin v, \quad z = cu \quad (c \text{ constant}). \tag{A}$$

If cos and sin occur, you can often use $\cos^2 v + \sin^2 v = 1$. At present,

$$x^2 + y^2 = u^2(\cos^2 v + \sin^2 v) = u^2.$$

From this and $z = cu$ you see that

$$z = c\sqrt{x^2 + y^2}.$$

This is a representation of the cone of the form $z = f(x,y)$.

If you set $u = const$, you see that $z = const$, so these curves are the intersections of the cone with horizontal planes $u = const$. They are circles.

If you set $v = const$, then $y/x = \tan v = const$ (since u drops out in (A)). Hence $y = kx$, where $k = \tan v = const$. These are straight lines through the origin in the xy-plane, hence they are planes through the z-axis in space, which intersect the cone along straight lines.

To find a surface normal, you first have to calculate the partial derivatives of **r**,

$$\mathbf{r}_u = [\cos v, \quad \sin v, \quad c],$$
$$\mathbf{r}_v = [-u \sin v, \quad u \cos v, \quad 0],$$

and then form their cross product $\mathbf{N}$ because this cross product is perpendicular to the two vectors, which span the tangent plane, so that $\mathbf{N}$ in fact is a normal vector. You obtain

$$\mathbf{N} = \mathbf{r}_u \times \mathbf{r}_v = \begin{vmatrix} \mathbf{i} & \mathbf{j} & \mathbf{k} \\ \cos v & \sin v & c \\ -u \sin v & u \cos v & 0 \end{vmatrix}$$

$$= \mathbf{i} \begin{vmatrix} \sin v & c \\ u \cos v & 0 \end{vmatrix} - \mathbf{j} \begin{vmatrix} \cos v & c \\ -u \sin v & 0 \end{vmatrix} + \mathbf{k} \begin{vmatrix} \cos v & \sin v \\ -u \sin v & u \cos v \end{vmatrix}$$

$$= [-cu \cos v, \quad -cu \sin v, \quad u].$$

In this calculation, the third component resulted by simplification,

$$(\cos v) u \cos v - (\sin v)(-u \sin v) = u (\cos^2 v + \sin^2 v) = u.$$

23. Representation $z = f(x, y)$. The simplest way to convert this to a parametric representation $\mathbf{r}(u, v)$ is to set $x = u, y = v$. Then $z = f(u, v)$, and by substituting this into $\mathbf{r}(u, v)$ you obtain the first of the two formulas in (6),

$$\mathbf{r}(u, v) = [u, \quad v, \quad f(u, v)].$$

A normal vector $\mathbf{N}$ is now obtained by first calculating the partial derivatives

$$\mathbf{r}_u = [1, \quad 0, \quad f_u],$$
$$\mathbf{r}_v = [0, \quad 1, \quad f_v]$$

and then their cross product

$$\mathbf{N} = \mathbf{r}_u \times \mathbf{r}_v = \begin{vmatrix} \mathbf{i} & \mathbf{j} & \mathbf{k} \\ 1 & 0 & f_u \\ 0 & 1 & f_v \end{vmatrix}$$

$$= \mathbf{i} \begin{vmatrix} 0 & f_u \\ 1 & f_v \end{vmatrix} - \mathbf{j} \begin{vmatrix} 1 & f_u \\ 0 & f_v \end{vmatrix} + \mathbf{k} \begin{vmatrix} 1 & 0 \\ 0 & 1 \end{vmatrix}$$

$$= [-f_u, \quad -f_v, \quad 1].$$

Sec. 9.6   Surface Integrals

Problem Set 9.6. Page 503

1. Surface integral over a plane in space. The surface S is given by

$$\mathbf{r}(u, v) = [u, \quad v, \quad 2u + 3v].$$

Hence $x = u, y = v, z = 2u + 3v = 2x + 3y$. This shows that this is a plane in space. The region of integration is a rectangle; u varies from 0 to 2 and v from -1 to 1. Since $x = u, y = v$, this is the same rectangle in the xy-plane.

     In (3) on the right you need the normal vector $\mathbf{N} = \mathbf{r}_u \times \mathbf{r}_v$. Now

$$\mathbf{r}_u = [1, \quad 0, \quad 2],$$
$$\mathbf{r}_v = [0, \quad 1, \quad 3],$$

so that

$$N = \begin{vmatrix} \mathbf{i} & \mathbf{j} & \mathbf{k} \\ 1 & 0 & 2 \\ 0 & 1 & 3 \end{vmatrix} = -2\mathbf{i} - 3\mathbf{j} + \mathbf{k} = [-2, \quad -3, \quad 1].$$

Next calculate $\mathbf{F}$ on the surface by substituting the components of $\mathbf{r}$ into $\mathbf{F}$. This gives

$$\mathbf{F} = [3x^2, \quad y^2, \quad 0] = [3u^2, \quad v^2, \quad 0].$$

Hence the dot product in (3) on the right is

$$\mathbf{F} \cdot \mathbf{N} = [3u^2, \quad v^2, \quad 0] \cdot [-2, \quad -3, \quad 1] = -6u^2 - 3v^2.$$

The following is quite interesting. Since $\mathbf{N}$ is a cross product, $\mathbf{F} \cdot \mathbf{N}$ is a scalar triple product (Sec. 8.3), given by the determinant

$$\begin{vmatrix} 3u^2 & v^2 & 0 \\ 1 & 0 & 2 \\ 0 & 1 & 3 \end{vmatrix} = 3u^2(-2) - v^2 \cdot 3 = -6u^2 - 3v^2.$$

In this way you have done two steps in one.

Now integrate $-6u^2 - 3v^2$. Integration over u gives

$$-6u^3/3 - 3uv^2 = -2u^3 - 3uv^2.$$

At the upper limit of integration $u = 2$ this equals $-16 - 6v^2$, and at the lower limit 0 it is zero. Integration of this result $-16 - 6v^2$ over v gives $-16v - 2v^3$. At the upper limit $v = 1$ this equals $-16 - 2 = -18$. At the lower limit $v = -1$ you obtain the value $16 + 2 = 18$. The difference of these two values gives the answer $-18 - 18 = -36$.

17. Surface integrals of the form (7). Integrate $G = z$ over the upper half of the sphere of radius 3 and center at the origin

$$x^2 + y^2 + z^2 = 9.$$

It seems best to use a parametric representation, say, (3) in Sec. 9.5 with $a = 3$, that is,

$$\mathbf{r} = [3 \cos v \cos u, \quad 3 \cos v \sin u, \quad 3 \sin v].$$

Here, u varies from 0 to 2π and v from 0 to $\pi/2$. Also,

$$G = z = 3 \sin v.$$

You need $\mathbf{N} = \mathbf{r}_u \times \mathbf{r}_v$. Differentiation gives

$$\mathbf{r}_u = [-3 \cos v \sin u, \quad 3 \cos v \cos u, \quad 0],$$
$$\mathbf{r}_v = [-3 \sin v \cos u, \quad -3 \sin v \sin u, \quad 3 \cos v].$$

The cross product is

$$\mathbf{N} = \mathbf{r}_u \times \mathbf{r}_v = [9 \cos^2 v \cos u, \quad 9 \cos^2 v \sin u, \quad 9 \cos v \sin v],$$

where the last component has been obtained by simplification, namely,

$$(-3 \cos v \sin u)(-3 \sin v \sin u) - (3 \cos v \cos u)(-3 \sin v \cos u)$$
$$= 9(\cos v \sin v \sin^2 u + \cos v \sin v \cos^2 u)$$
$$= 9 \cos v \sin v.$$

$\mathbf{N}$ has the length

$$|\mathbf{N}| = \sqrt{81 \cos^4 v \cos^2 u + 81 \cos^4 v \sin^2 u + 81 \cos^2 v \sin^2 v}$$
$$= 9\sqrt{\cos^4 v + \cos^2 v \sin^2 v}$$
$$= 9\sqrt{\cos^4 v + \cos^2 v (1 - \cos^2 v)}$$
$$= 9 \cos v.$$

Hence $G|\mathbf{N}| = 27 \cos v \sin v = 13.5 \sin 2v$. Integration of $\sin 2v$ gives $-(1/2) \cos 2v$. At the upper limit $v = \pi/2$ this is $-(1/2)(-1) = 1/2$. At the lower limit $v = 0$ it is $-1/2$. The difference of the two values is 1, so that your intermediate result is 13.5. Finally, integration over u from 0 to 2π produces a factor 2π, so that the answer is 27π.

21. **Moment of inertia.** The integrals for the moments of inertia in Prob. 19 involve the square of the distance of points (x, y, z) from the axis. In the present example, where the surface S is the cylinder $x^2 + y^2 = 1$ for z from 0 to h, that distance is simply $x^2 + y^2 = 1$. The density is $\sigma = 1$ by assumption. Integration over the angle from 0 to 2π gives 2π. Subsequent integration over z from 0 to h gives a factor h. You thus get the answer $2\pi h$.

Sec. 9.7   Triple Integrals. Divergence Theorem of Gauss

Problem Set 9.7. Page 509

5. **Triple integral.** The vertices of the region of integration are the unit points on the axes and the origin. The faces are triangles, three of them are portions of the coordinate planes and one of them is a portion of the plane $x + y + z = 1$, which intersects each of the coordinate axes at the point 1. Hence integration over z extends from 0 to $z = 1 - x - y$. The integrand is $12\,xy$ and is independent of z, so that the integration over z gives $12\,xyz$. At the upper limit of integration this equals

$$12xy(1 - x - y) = 12x(1 - x)y - 12xy^2. \tag{A}$$

At the lower limit $z = 0$ it is 0.

     That plane intersects the xy-plane along the straight line $y = 1 - x$. (Make a sketch to be sure.) Hence the integration over y extends from 0 to $1 - x$. It gives $6x(1 - x)y^2 - 4xy^3$. At the upper limit $y = 1 - x$ this equals

$$6x(1 - x)^3 - 4x(1 - x)^3 = 2x(1 - x)^3. \tag{B}$$

At the lower limit $y = 0$ it is 0. The function in (B) must finally be integrated over x from 0 to 1. (Use $u = 1 - x$ as a new variable of integration.) After simplification this gives the answer $1/10$.

19. **Divergence theorem.** In this problem you use the divergence theorem for evaluating the surface integral of the normal component $\mathbf{F} \cdot \mathbf{n}$ of

$$\mathbf{F} = [x^3, \quad y^3, \quad z^3]$$

over the sphere S of radius 3 and center at the origin. This integral is converted to a volume integral of the divergence

$$\operatorname{div}\mathbf{F} = 3x^3 + 3y^2 + 3z^2 = 3(x^2 + y^2 + z^2) = 3r^2.$$

Here $r^2 = x^2 + y^2 + z^2$ and you see that $\operatorname{div}\mathbf{F}$ is constant on any of these concentric spheres $r = const$. Hence the integral of $\operatorname{div}\mathbf{F}$ over such a sphere is simply $3r^2$ times the area $4\pi r^2$ of the sphere. This gives $12\pi r^4$. All you still have to do is integrating over r from 0 to 3 (the radius of S). This gives

$$12\pi\,3^5/5 = 2916\pi/5.$$

If you do not see this way, you can use a parametric representation, say, (3) in Sec. 9.5, that is,

$$\mathbf{r} = [r \cos v \cos u, \quad r \cos v \sin u, \quad r \sin v],$$

where you assume r to vary from 0 to 3. Then you have $\operatorname{div}\mathbf{F} = 3r^2$, as before. This must be multiplied by the volume element in spherical coordinates

$$dV = r^2 \cos v\, dr\,du\,dv, \quad \text{thus} \quad 3r^4 \cos v\, dr\,du\,dv,$$

and integrated. Integration over u from 0 to 2π gives a factor 2π. Integration of $\cos v$ over v from $-\pi/2$ to $\pi/2$ gives

$$\sin \pi/2 - (-\sin \pi/2) = 2.$$

Integration of $3r^4$ over r from 0 to 3 gives $3 \cdot 3^5/5$. Together, $12\pi \, 3^5/5$, as before.

Sec. 9.8   Divergence Theorem. Further Applications

Problem Set 9.8. Page 514

1. **Formula (9)** expresses a very remarkable property of harmonic functions stated in Theorem 1. Of course, the formula can also be used for other functions. The point of the problem is to gain confidence in the formula and to see how to organize more involved calculations so that errors are avoided as much as possible. The box has six faces $S_1, .., S_6$. The normals to them (which you need in connection with normal derivatives) have the direction of the axes; this makes the calculation of the surface integrals simple, in addition to the fact that the normal derivatives will turn out to be constant on each face. Thus to the faces $x = 0$ and $x = 1$ there correspond the negative and positive x-directions, respectively, as outer normal direction, and so on. For the given function $f = 2z^2 - x^2 - y^2$ you thus obtain

$$
\begin{array}{llll}
S_1 : x = 0 & f_x = -2x = 0 & \text{integrated} & 0 \\
S_2 : x = 1 & f_x = -2 & " & -2 \cdot 2 \cdot 4 = -16 \\
S_3 : y = 0 & f_y = -2y = 0 & " & 0 \\
S_4 : y = 2 & f_y = -4 & " & -4 \cdot 1 \cdot 4 = -16 \\
S_5 : z = 0 & f_z = 4z = 0 & " & 0 \\
S_6 : z = 4 & f_z = 16 & " & 16 \cdot 1 \cdot 2 = 32.
\end{array}
$$

In each line the value of the normal derivative is multiplied by the area of the corresponding face of the box. For a more general f you would have to evaluate double integrals over these six faces. You see that the six integrals add up to zero, confirming formula (9) for our special case. Indeed, the Laplacian of f is $4 - 2 - 2 = 0$, as differentiation shows; hence f is harmonic.

5. **Divergence theorem**. This concerns the surface integral of the outer normal component of
$\mathbf{F} = [9x, \quad y \cosh^2 x, \quad -z \sinh^2 x]$ over the ellipsoid

$$4x^2 + y^2 + 9z^2 = 36, \quad \text{thus} \quad x^2/9 + y^2/36 + z^2/4 = 1.$$

To evaluate the integral, use the divergence theorem for converting it to a volume integral of the divergence. The latter is found to be

$$\operatorname{div} \mathbf{F} = 9 + \cosh^2 x - \sinh^2 x.$$

The sum of the last two terms is 1. Hence the divergence is constant, $\operatorname{div} \mathbf{F} = 10$. It follows that the volume integral equals 10 times the volume of the region bounded by the ellipsoid, that is, $\frac{4}{3}\pi \, abc$, where $a = 3$, $b = 6$, and $c = 2$ are the lengths of the three semi-axes of the ellipsoid. This gives the answer $\frac{4}{3} \cdot 36 \cdot 10\pi = 480\pi$.

Sec. 9.9   Stokes's Theorem

Problem Set 9.9. Page 520

3. **Stokes's theorem** converts surface integrals into line integrals over the boundary of the (portion of the) surface and conversely. It will depend on the special problem which of the two integrals is simpler, the surface integral or the line integral. In the present problem, the application of Stokes's theorem proceeds as follows. Given are $\mathbf{F} = [\, e^z, \quad e^z \sin y, \quad e^z \cos y]$ and the surface $S : z = y^2$, with x varying from 0 to 4 and y from 0 to 2. (S is called a *cylinder*.) Using (1), calculate the curl of $\mathbf{F}$,

$$\text{curl } \mathbf{F} = \begin{vmatrix} \mathbf{i} & \mathbf{j} & \mathbf{k} \\ \partial/\partial x & \partial/\partial y & \partial/\partial z \\ e^z & e^z \sin y & e^z \cos y \end{vmatrix}$$

$$= [-e^z \sin y + e^z \sin y, \quad e^z - 0, \quad 0 - 0]$$

$$= [0, \quad e^z, \quad 0]$$

$$= [0, \quad \exp y^2, \quad 0] \qquad \text{on } S$$

In Stokes's theorem you further need a normal vector of S. To get it, write S in the form

$$S : \mathbf{r} = [x, \quad y, \quad y^2].$$

You could also write $\mathbf{r} = [u, \quad v, \quad v^2]$, so that $x = u, y = v$; this would not make any difference in what follows. The partial derivatives are

$$\mathbf{r}_x = [1, \quad 0, \quad 0],$$

$$\mathbf{r}_y = [0, \quad 1, \quad 2y].$$

Their cross product is the normal vector

$$\mathbf{N} = \mathbf{r}_x \times \mathbf{r}_y = [0, \quad -2y, \quad 1].$$

From it you obtain

$$(\text{curl } \mathbf{F}) \bullet \mathbf{n} \, dA = (\text{curl } \mathbf{F}) \bullet \mathbf{N} \, dx \, dy = (0 - 2y \exp y^2 + 0) \, dx \, dy = -2y \exp y^2 \, dx \, dy.$$

Integration over x from 0 to 4 gives a factor 4. Integration over y gives $-\exp y^2$. At the upper limit $y = 2$ this equals $-e^4$, and at the lower limit $y = 0$ it equals -1. Together with the factor 4 you thus obtain the answer

$$4(1 - e^4) \tag{A}$$

(or $-4(1 - e^4)$ if you reverse the direction of the normal vector).

   Confirm this result by Stokes's theorem, as follows. Sketch the surface S, so that you see what is going on. The boundary curve of S has four portions. The first, C_1, is the segment of the x-axis from the origin to $x = 4$. On it, $y = 0, z = 0, x$ varies from 0 to 4, $\mathbf{F} = [1, \quad 0, \quad 1], \mathbf{r} = [x, \quad 0, \quad 0]$ (because $s = x$ on C_1), $\mathbf{r}' = [1, \quad 0, \quad 0], \mathbf{F} \bullet \mathbf{r}' = 1$, integrated from 0 to 4 gives 4. The third, C_3, is the upper straight-line edge from $(0, 2, 4)$ to $(4, 2, 4)$. On it, $y = 2, z = 4, x$ varies from 4 to 0,

$$\mathbf{F} = [e^4, \quad e^4 \sin 2, \quad e^4 \cos 2], \qquad \mathbf{r} = [x, \quad 2, \quad 4], \qquad \mathbf{r}' = [1, 0, 0],$$

$\mathbf{F} \bullet \mathbf{r}' = e^4$, integrated from 4 to 0 gives $-4e^4$, the minus sign appearing because you integrate in the negative x-direction. The sum of the two integrals equals $4 - 4e^4$; this is your result in (A). Now show that the sum of the other two integrals over the portions of parabolas is zero. The second portion, C_2, is the parabola $z = y^2$ in the plane $x = 4$, which you can represent by $\mathbf{r} = [4, \quad y, \quad y^2]$. The derivative with respect to y is $\mathbf{r}' = [0, \quad 1, \quad 2y]$. Furthermore, $\mathbf{F}$ on C_2 is

$$\mathbf{F} = [\exp y^2, \quad (\exp y^2) \sin y, \quad (\exp y^2) \cos y].$$

This gives the dot product

$$\mathbf{F} \bullet \mathbf{r}' = 0 + (\exp y^2) \sin y + 2y (\exp y^2) \cos y.$$

This must be integrated over y from 0 to 2. But for the fourth portion, C_4, you obtain exactly the same expression because C_4 can be given by $\mathbf{r} = [0, \quad y, \quad y^2]$, so that $\mathbf{r}' = [0, \quad 1, \quad 2y]$ is exactly as for C_2, and so is $\mathbf{F} \bullet \mathbf{r}'$ because $\mathbf{F}$ does not involve x. Now on C_4 you have to integrate over y in the opposite sense, from 2 to 0, so that the two integrals do indeed cancel each other and their sum is zero. This remains true if the arc length s is introduced as a variable of integration. This completes the integration over the boundary of S, the result being the same as in (A).

9. Evaluation of a line integral. If you want to do this by Stokes's theorem, you have to find a surface S whose boundary is the path of integration C of the line integral. In Prob. 9 you can use the plane of the ellipse, given by

$$\mathbf{r} = [u, \quad v, \quad v+1].$$

For $\mathbf{F} = [4z, \quad -2x, \quad 2x]$ you get $\operatorname{curl}\mathbf{F} = [0, \quad -2+4, \quad -2] = [0, \quad 2, \quad -2]$. A normal vector to the plane is obtained from

$$\mathbf{r}_u = [1, \quad 0, \quad 0],$$
$$\mathbf{r}_v = [0, \quad 1, \quad 1]$$

in the form

$$\mathbf{N} = \mathbf{r}_u \times \mathbf{r}_v = [0, \quad -1, \quad 1].$$

(You get this more quickly by writing the plane as $0 \cdot x - y + z = -1$ and remembering Example 6 in Sec. 8.2.) From this you see that

$$(\operatorname{curl}\mathbf{F}) \cdot \mathbf{N} = [0, \quad 2, \quad -2] \cdot [0, \quad -1, \quad 1] = -4 = const.$$

Hence the answer is -4 times the area of the region of integration in the uv-plane, which is the interior of the circle $u^2 + v^2 = 1$ and has the area π. This gives the answer -4π.

Alternatively, instead of $\mathbf{N}\,du\,dv$ used you can use $\mathbf{n}\,dA$, where $\mathbf{n}$ is the unit normal vector

$$\mathbf{n} = [0, \quad -1/\sqrt{2}, \quad 1/\sqrt{2}].$$

Then $(\operatorname{curl}\mathbf{F}) \cdot \mathbf{n} = -4/\sqrt{2} = const.$ This must now be multiplied by the area of the ellipse in that plane, with semi-axes $\sqrt{2}$ and 1, which equals $\pi\sqrt{2}$, so that $\sqrt{2}$ cancels and the result is the same as before.

PART C.   FOURIER ANALYSIS AND PARTIAL DIFFERENTIAL EQUATIONS

CHAPTER 10.   Fourier Series, Integrals, and Transforms

Sec. 10.1   Periodic Functions. Trigonometric Series

Problem Set 10.1. Page 528

3. **Linear combinations of periodic functions.** Addition of periodic functions with the same period p gives a periodic function with that period p, and so does the multiplication of such a function by a constant. Thus all functions of period p form an important example of a vector space. This is what you are supposed to prove. Now, by assumption, given any periodic functions f and g that have period p, that is,

$$f(x+p) = f(x), \quad g(x+p) = g(x), \tag{A}$$

form a linear combination of them, say,

$$h = af + bg \quad (a, b \text{ constant})$$

and show that h is periodic with period p; that is, you must show that $h(x+p) = h(x)$. This follows by calculating and using the definition of h and then (A):

$$h(x+p) = af(x+p) + bg(x+p) = af(x) + bg(x) = h(x).$$

11. **Periodic functions** in applications (mechanics, electrical networks, etc.) are often given over portions of the interval of periodicity by several expressions, and you should work Probs. 7-18 with care, so that you get used to this kind of formulas. The function in Prob. 11 occurs as current in electrical rectifiers, as we shall see in more detail later (in Sec. 10.3).

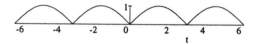

Section 10.1. Problem 9. Full-wave rectification of a sinusoidal current

Sec. 10.2   Fourier Series

Problem Set 10.2. Page 536

1. **Fourier series, Euler formulas.** This problem and the others amount to the calculation of Fourier coefficients by integration. In (6) you integrate over a period, from $-\pi$ to π. The figure in the problem

shows that the function is zero from $-\pi$ to $-\pi/2$ and from $\pi/2$ to π. Hence in this problem you integrate from $-\pi/2$ to $\pi/2$. In (6a) the integrand is 1 (the given function), so that the integral equals π. Division by 2π gives $a_0 = 1/2$. Note that this is the mean value of our $f(x)$ over the interval from $-\pi$ to π. The coefficient a_0 will always be the mean value of $f(x)$, so that in many cases, integration can be avoided. In (6b) the integrand is 1 times $\cos nx$. The integral gives $(1/n)\sin nx$ (if you don't see it right away, put $nx = u$; then $dx = du/n$ shows why $1/n$ arises). At $x = \pi/2$ this is $(1/n)\sin(n\pi/2)$. At the lower limit $x = -\pi/2$ it gives $(1/n)\sin(-n\pi/2) = -(1/n)\sin(n\pi/2)$. Hence the value at the upper limit minus the value at the lower limit is

$$(1/n)\sin(n\pi/2) - (-(1/n)\sin(n\pi/2)) = (2/n)\sin(n\pi/2).$$

Division by π, as indicated in (6b), gives

$$a_n = (2/n\pi)\sin(n\pi/2). \tag{A}$$

Now remember that $\sin x$ equals $0, 1, 0, -1, 0$ when $x = 0, \pi/2, \pi, 3\pi/2, 2\pi$, respectively. From (A) with $n = 1, 2, 3, \ldots$ thus conclude that

$$a_1 = 2/\pi$$
$$a_2 = 0/(2\pi) = 0$$
$$a_3 = -2/(3\pi)$$
$$a_4 = 0/(4\pi) = 0.$$

Hence the beginning of the Fourier series is

$$\frac{1}{2} + \frac{2}{\pi}\cos x + 0 - \frac{2}{3\pi}\cos 3x + 0.$$

The further terms should now be clear since n in $1/n$ keeps growing and $\sin x$ repeats its values $0, 1, 0, -1$. Also, all the b_n in (6c) are zero, as you should verify by integration. Hence the series has only cosine terms, it is a "Fourier cosine series". We shall later find the reason for this (f is an even function, $f(-x) = f(x)$). Your answer agrees with that on p. A25 in Appendix 2 of the book,

$$f(x) = \frac{1}{2} + \frac{2}{\pi}\left(\cos x - \frac{1}{3}\cos 3x + \frac{1}{5}\cos 5x - \frac{1}{7}\cos 7x + -\ldots\right).$$

11. **Breaking up the interval of integration.** The function is given by two different expressions for positive and negative x. Accordingly, you have to integrate $+1$ from $-\pi$ to 0 and then -1 from 0 to π. This is similar to Example 1. You obtain $a_0 = 0$ from (6a) (the mean value of the function is zero) and $a_n = 0$ from (6b) for all n. Do this integration. (We shall see the reason later: the function is odd, $f(-x) = -f(x)$.) For b_n you obtain from (6c)

$$b_n = \frac{1}{\pi}\int_{-\pi}^{\pi} f(x)\sin nx\,dx$$
$$= \frac{1}{\pi}\int_{-\pi}^{0} 1\sin nx\,dx + \frac{1}{\pi}\int_{0}^{\pi}(-1)\sin nx\,dx.$$

The integral of $\sin nx$ is $-(1/n)\cos nx$. In the first term this gives

$$\frac{1}{\pi}\left(-\frac{1}{n}\cos 0 + \frac{1}{n}\cos(-n\pi)\right). \tag{B}$$

In the second term it gives

$$\frac{1}{\pi}\left(\frac{1}{n}\cos n\pi - \frac{1}{n}\cos 0\right), \tag{C}$$

where a minus sign was contributed to each of these two terms from the function value -1 for x from 0 to π. Since $\cos 0 = 1$ and $\cos(-n\pi) = \cos n\pi$, the sum of (B) and (C) is

$$b_n = \frac{2}{n\pi}(\cos n\pi - 1).$$

Now $\cos \pi = -1$, $\cos 2\pi = +1$, $\cos 3\pi = -1$, and so on. Hence you get 0 when the cosine is positive, and

$-4/(n\pi)$ when it is negative, that is,

$$b_n = -4/(n\pi) \quad \text{for} \quad n = 1, 3, 5, ..., \quad b_n = 0 \quad \text{for} \quad n = 2, 4, 6, ...$$

This gives the Fourier series

$$f(x) = -\frac{4}{\pi}\left(\sin x + \frac{1}{3}\sin 3x + \frac{1}{5}\sin 5x + \frac{1}{7}\sin 7x + ...\right).$$

Finally, take a look at Example 1 in Sec.10.2 and compare. Do you see something? Can you explain it?

Sec. 10.3   Functions of Any Period $p = 2L$

Problem Set 10.3. Page 540

9. **Period $p = 2L$.** For the period 2π the formulas for Fourier series are simpler than for arbitrary periods. For this reason we began with period 2π. Of course, in applications, all kinds of different periods occur, so that we have to provide Fourier series for functions of arbitrary periods p. Our notation is

$$p = 2L$$

because in applications, L will be the length of a vibrating string or of a bar whose temperature we want to study, and so on. The corresponding Euler formulas (2) are obtained by a stretch of scale, chosen to let $x = -\pi$ and π correspond to $-L$ and L on the new scale, as explained in the text. In Prob. 9 you have $p = 2L = 2$, hence $L = 1$. Integration goes from 0 to $L = 1$ only because $f(x)$ is zero for x from -1 to 0. From (2a) you thus obtain

$$a_0 = \frac{1}{2}\int_0^1 x\,dx = \frac{1}{4}.$$

Note that this is the average (the mean value) of $f(x)$ over the interval of periodicity from -1 to 1. From (2b) with $L = 1$ you obtain for the cosine coefficients by integration by parts

$$a_n = \int_0^1 x\cos n\pi x\,dx$$
$$= \frac{x}{n\pi}\sin n\pi x\Big|_0^1 - \frac{1}{n\pi}\int_0^1 \sin n\pi x\,dx$$
$$= 0 + \frac{1}{n^2\pi^2}\cos n\pi x\Big|_0^1$$
$$= \frac{1}{n^2\pi^2}(\cos n\pi - 1)$$
$$= -\frac{2}{n^2\pi^2} \quad \text{if} \quad n = 1, 3, ...$$

and $a_n = 0$ if $n = 2, 4, ...$ This takes care of the cosine coefficients. Now turn to the sine coefficients. From (2c) you obtain, again by integration by parts,

$$b_n = \int_0^1 x\sin n\pi x\,dx$$
$$= -\frac{x}{n\pi}\cos n\pi x\Big|_0^1 + \frac{1}{n\pi}\int_0^1 \cos n\pi x\,dx.$$

The first expression equals 0 at the lower limit and

$$-(\cos n\pi)/n\pi = (-1)^{n+1}/(n\pi)$$

at the upper limit. The integral gives $(1/n^2\pi^2)\sin n\pi x$, which is zero at both limits $x = 0$ and $x = 1$. Hence $b_n = 1/(n\pi)$ for odd n and $-1/(n\pi)$ for even n. With these coefficients a_n and b_n the Fourier series becomes

$$f(x) = \frac{1}{4} - \frac{2}{\pi^2}\left(\cos \pi x + \frac{1}{9}\cos 3\pi x + \frac{1}{25}\cos 5\pi x + ...\right)$$

$$+ \frac{1}{\pi}\left(\sin \pi x - \frac{1}{2}\sin 2\pi x + \frac{1}{3}\sin 3\pi x - +...\right).$$

It has both cosine and sine terms because $f(x)$ is neither even nor odd.

17. **Changes of period**, as they may occur, for instance, if we switch from inches to centimeters or gallons to liters, can readily be taken care of in connection with Fourier series. Problem 17 describes the standard situation that you have a Fourier series for the period 2π and want derive from it a Fourier series for some other period p, in the present case, for $p = 2L = 4$. Now the function in Prob. 3 has a jump of height 2 at the origin. Hence you choose $k = 1$ in Example 1 in Sec. 10.2, so that the function in the example also has a jump of height 2. In that example, $\sin x$ has period 2π. Hence $\sin \pi x$ has period 2 and $\sin (\pi x/2)$ has period 4, as wanted in Prob. 3. Similarly for the other terms in Example 1, which upon this change take the form of the corresponding terms in the answer to Prob. 3 on p. A25 of Appendix 2. Finally, you have to lift up the function in Example 1 by 1 unit for obtaining the function in Prob. 3; that is, you have to add the constant term 1 to the Fourier series to get that in Prob. 3.

Sec. 10.4   Even and Odd Functions. Half-Range Expansions

Problem Set 10.4. Page 546

3. **Even or odd functions.** It is quite important that you observe carefully for what interval a periodic function is given. x^2 would be even if it were given for $-\pi$ to π, but it is not because in Prob. 3 the formula is given for the interval from 0 to 2π. Sketch the function in this interval and in a few neighboring intervals of periodicity, so that the point becomes completely clear.

13. **Odd function.** Even and odd functions were considered before, but in this section the formulas (4) and (6) for the Fourier coefficients have been adjusted, so that you integrate only over half of the interval of periodicity. In the problem the function is odd (see the figure) and of period 2π. For x from 0 to $\pi/2$ it is $f(x) = x$. For x from $\pi/2$ to π it is $f(x) = \pi - x$. Accordingly you have to split the integral in (6) from 0 to π into two integrals. Using integration by parts, you obtain

$$b_n = \frac{2}{\pi}\int_0^{\pi/2} x \sin nx\, dx + \frac{2}{\pi}\int_{\pi/2}^{\pi}(\pi - x)\sin nx\, dx$$

$$= -\frac{2}{n\pi}x\cos nx\Big|_0^{\pi/2} + \frac{2}{n\pi}\int_0^{\pi/2}\cos nx\, dx$$

$$-\frac{2}{n\pi}(\pi - x)\cos nx\Big|_{\pi/2}^{\pi} - \frac{2}{n\pi}\int_{\pi/2}^{\pi}\cos nx\, dx.$$

The minus sign in front of the last integral results from $(\pi - x)' = -1$. The first integral-free expression is zero at $x = 0$ and $-(1/n)\cos(n\pi/2)$ at the upper limit of integration $\pi/2$ (a factor 2π drops out) . The other integral-free expression is zero at $x = \pi$ (since $x - \pi = 0$ there) and equal to $-(1/n)\cos(n\pi/2)$ at the lower limit. You see that these two integral-free expressions are equal; hence the first expression minus the second (minus because it refers to the *lower* limit) equals zero.

Consider the two integrals. The integral of $\cos nx$ is $(1/n)\sin nx$; together with the factor $2/(n\pi)$ in front this gives $(2/(n^2\pi))\sin nx$. This is 0 at 0 (the lower limit of the first integral) and at π (the upper limit of the second integral). At the remaining two limits you obtain

$$\frac{2}{n^2\pi}\sin(n\pi/2) + \frac{2}{n^2\pi}\sin(n\pi/2) = \frac{4}{n^2\pi}\sin(n\pi/2).$$

For $n = 1, 2, 3, 4$ the sine has the value $1, 0, -1, 0$, respectively, and then repeats these values because of the periodicity. You thus obtain the Fourier sine series

$$f(x) = \frac{4}{\pi}\left(\sin x - \frac{1}{9}\sin 3x + \frac{1}{25}\sin 5x - \frac{1}{49}\sin 7x + - ... \right).$$

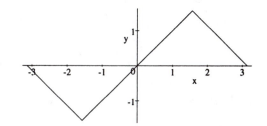

Section 10.4. Problem 13. Given odd periodic function

Sec. 10.5   Complex Fourier Series.   *Optional*

Problem Set 10.5. Page 549

3. **Complex Fourier series** for functions of period 2π have the form (8), which also includes the Euler formulas for the Fourier coefficients c_n. For the given function $f(x) = x$ you obtain

$$c_n = \frac{1}{2\pi}\int_{-\pi}^{\pi} x e^{-inx}\, dx.$$

For $n = 0$ this is simply the integral of x from $-\pi$ to π and gives 0. Hence the series has no constant term. Now determine the further coefficients c_n with $n \neq 0$. Integrate by parts. Since $1/i = -i$, the integral of the exponential function e^{-inx} is

$$\frac{1}{-in}e^{-inx} = \frac{i}{n}e^{-inx}. \tag{A}$$

You thus obtain

$$c_n = \frac{1}{2\pi}\left(\frac{ix}{n}\right)e^{-inx}\Big|_{-\pi}^{\pi} - \frac{i}{2n\pi}\int_{-\pi}^{\pi} e^{-inx}\, dx.$$

The integral-free part (times the factor $1/(2\pi)$) has the value

$$\frac{1}{2\pi}\left(\frac{i\pi}{n}\right)e^{-in\pi} - \frac{1}{2\pi}\left(-\frac{i\pi}{n}\right)e^{in\pi}$$

$$= \frac{i}{2n}(e^{in\pi} + e^{-in\pi}).$$

From (4) you see that this equals

$$\frac{i}{n}\cos n\pi = \frac{i}{n}(-1)^n. \tag{B}$$

Now show that the remaining integral has the value 0 for any integer $n \neq 0$. Indeed, the left side of (A) with $x = \pi$ minus the left side of (A) with $x = -\pi$ gives

$$\frac{1}{-in}(e^{-in\pi} - e^{in\pi}) = \frac{1}{in}(e^{in\pi} - e^{-in\pi}).$$

From (5) you see that this equals $(2/n)\sin n\pi$, which is 0 for any integer n. Hence the Fourier coefficients are 0 (if $n = 0$) and they are given by (B) for $n \neq 0$. This gives the answer on p. A26 in Appendix 2 of the book.

9. **Complex and real Fourier coefficients**. The point is that in complex, the exponential function is closely related to cosine and sine, as shown in (2)-(5). For $n = 0$ the relation between the complex and real Fourier coefficients is immediate because the exponential function then becomes $e^0 = 1$, so that the formulas for a_0 and c_0 are identical. The next formula to be proved in the problem can be best obtained by starting on the right and then using (4). This gives in the case of the periodicity 2π by combining the two integrals into one

$$c_n + c_{-n} = \frac{1}{2\pi} \int_{-\pi}^{\pi} f(x) (e^{-inx} + e^{inx}) \, dx$$

$$= \frac{1}{\pi} \int_{-\pi}^{\pi} f(x) \cos nx \, dx$$

$$= a_n$$

and similarly for arbitrary periods. For b_n use (5) and follow the same idea as for a_n.

Sec. 10.6   Forced Oscillations

Problem Set 10.6. Page 552

3. **Sinusoidal driving force**. $y'' + \omega^2 y = \sin t$ is best solved by undetermined coefficients. Substitute $y = A \sin t$ into the differential equation. Since $y'' = -A \sin t$, this gives

$$(-A + \omega^2 A) \sin t = \sin t, \quad \text{hence} \quad -A + \omega^2 A = 1$$

so that $A = 1/(\omega^2 - 1)$, which becomes the larger in absolute value, the more closely you approach the point of resonance $\omega^2 = 1$. This motivates the values of ω suggested in the problem. Note also that A is negative as long as ω (> 0) is less than 1 and positive for values $\omega > 1$. This illustrates Figs. 57 and 61 (with $c = 0$) in Sec. 2.11.

7. **Forced undamped oscillations**. You need the Fourier series of the driving force $r(t)$, as explained in the text. $r(t)$ is even (sketch it). Hence it is represented by a Fourier cosine series. The period is 2π. Accordingly, the general term of that series is of the form $a_n \cos nt$. As in Example 1, solve the differential equation with a single term on the right, that is,

$$y'' + \omega^2 y = a_n \cos nt.$$

Substitute $y = A \cos nt$ and $y'' = -n^2 A \cos nt$ into the equation. This gives

$$(-n^2 A + \omega^2 A) \cos nt = a_n \cos nt.$$

Solving for A and writing A_n for A, you obtain

$$A_n = a_n/(\omega^2 - n^2). \tag{I}$$

This shows what you have to do next. You must find the Fourier series of $r(t)$. Since $r(t)$ is even, you can use (4*) in Sec. 10.4, with t instead of x, and $r(t)$ instead of $f(x)$. You obtain $a_0 = \pi/2$ (the mean value of $r(t)$ over the interval of periodicity) and, furthermore, for $n = 1, 2, 3, \ldots$, by integration by parts,

$$a_n = \frac{2}{\pi} \int_0^{\pi} (\pi - t) \cos nt \, dt$$

$$= \frac{2}{n\pi} (\pi - t) \sin nt \Big|_0^{\pi} + \frac{2}{n\pi} \int_0^{\pi} \sin nt \, dt.$$

The integral-free part is zero at both limits of integration. Integrating the last integral, you obtain for the last term

$$-\frac{2}{n^2 \pi} \cos nt \Big|_0^{\pi} = -\frac{2}{n^2 \pi} (\cos n\pi - 1)$$

$$= -\frac{2}{n^2 \pi} ((-1)^n - 1).$$

This is 0 for even n, and for odd n you have

$$a_n = \frac{4}{n^2\pi} \quad (n = 1, 3, 5, \dots).$$

Do you see something if you compare these a_n with (3) in Sec. 10.6? These a_n are the same as the coefficients in (3). Why? Well, the present $r(t)$ is obtained from that in (3) by adding the constant $a_0 = \pi/2$. Hence you could have avoided all those calculations, which therefore serve the purpose of filling in the details not given in Example 1 on p. 551. However, your further calculations differ from those on p. 551 because the present differential equation is different. From a_n and (I) you obtain $A_0 = a_0/\omega^2 = \pi/(2\omega^2)$ and

$$A_n = \frac{4}{n^2\pi(\omega^2 - n^2)} \quad (n = 1, 3, 5, \dots).$$

Hence the particular solution obtained has the form

$$\frac{\pi}{2\omega^2} + \frac{4}{\pi}\left(\frac{1}{\omega^2 - 1}\cos t + \frac{1}{9(\omega^2 - 1)}\cos 3t + \dots \right).$$

Adding to it the familiar general solution of the homogeneous equation, you obtain the answer given on p. A26 in Appendix 2 of the book.

13. Forced damped oscillations. The driving force $r(t)$ on the right side of the differential equation

$$y'' + cy' + y = r(t) = \frac{1}{12}t(\pi^2 - t^2) \tag{I}$$

is odd and of period 2π. Hence its Fourier series is a Fourier sine series, with the general term of the form

$$b_n \sin nt \tag{II}$$

and b_n to be determined later. Derive a particular solution of (I) with $r(t)$ replaced by (II), that is,

$$y'' + cy' + y = b_n \sin nt. \tag{III}$$

In contrast to the previous problem you must set (just as in Example 1 on p. 551)

$$y = A \cos nt + B \sin nt,$$

with both a sine and a cosine term because of the presence of the damping term cy'. To keep the formulas simple, we write A and B rather than A_n and B_n. Substitute y and its derivatives

$$y' = -nA \sin nt + nB \cos nt$$

$$y'' = -n^2 A \cos nt - n^2 B \sin nt$$

into (III). The coefficients of the cosine terms must add up to 0 because there is no cosine term on the right side of (III). This gives

$$A + cnB - n^2 A = 0.$$

The sum of the coefficients of the sine terms must equal the coefficient b_n on the right side of (III). This gives

$$B - cnA - n^2 B = b_n.$$

Ordering the terms of these two equations, you obtain

$$(1 - n^2)A + \qquad cnB = 0$$

$$-cnA + (1 - n^2)B = b_n.$$

Solve this for A and B by elimination or by Cramer's rule (Sec. 6.6). If you use Cramer's rule, you need the following determinants. The coefficient determinant is

$$D = (1 - n^2)^2 + c^2 n^2.$$

The determinant in the numerator of the formula for A is

$$\begin{vmatrix} 0 & cn \\ b_n & 1 - n^2 \end{vmatrix} = -cnb_n.$$

The determinant in the numerator of the formula for B is

$$\begin{vmatrix} 1-n^2 & 0 \\ -cn & b_n \end{vmatrix} = (1-n^2)b_n.$$

Hence

$$A = -cnb_n/D, \qquad B = (1-n^2)b_n/D.$$

Finally determine b_n from (6*) in Sec. 10.4 with $r(t)$ instead of $f(x)$. To evaluate the integral you have to apply two successive integrations by parts. The result is

$$(-1)^{n+1}/n^3.$$

Hence

$$A = -cn(-1)^{n+1}/(n^3 D) = (-1)^n c/(n^2 D)$$
$$B = (1-n^2)(-1)^{n+1}/(n^3 D),$$

in agreement with the solution given on p. A26 in Appendix 2 of the book. Note that A and B depend on n, but we did not indicate this in order to keep the formulas simple.

Sec. 10.7   Approximation by Trigonometric Polynomials

Problem Set 10.7. Page 556

1. **Minimum square error**. The minimum square error E^* is given by (6). To compute it, you need the integral of the square of the given function $f(x)$ over the interval of periodicity. In the present problem this is the integral of 1 from $-\pi$ to π, whose value is 2π. You further need the Fourier coefficients of $f(x)$. This function is odd. Hence the a_n are all zero. The b_n can be obtained from (6*) in Sec. 10.4 (or less conveniently from the Euler formulas in Sec. 10.2). This gives

$$b_n = \frac{2}{\pi}\int_0^\pi \sin nx\,dx = -\frac{2}{n\pi}(\cos n\pi - \cos 0)$$

$$= -\frac{2}{n\pi}((-1)^n - 1)$$

$$= 4/n\pi \qquad \text{for } n = 1, 3, 5, \ldots$$

and $b_n = 0$ for even n. (See also Example 1 in Sec. 10.2.) From this and (6) you obtain

$$E^* = 2\pi - \pi(b_1^2 + b_2^2 + b_3^2 + \ldots)$$
$$= 2\pi - \pi(16/\pi^2)(1 + 0 + 1/9 + 0 + 1/25 + \ldots).$$

This gives the numerical values listed on p. A26 in Appendix 2 of the book.

9. **Monotonicity of the minimum square error**. The minimum square error (6) is monotone decreasing, that is, it cannot increase if you add further terms (by choosing a larger N for the approximating polynomial). To prove this, note that the terms in the sum in (6) are squares, hence they are nonnegative. Since the sum is subtracted from the first term in (6) (the integral), the whole expression cannot increase. This is what is meant by 'monotone decreasing', which, by definition, includes the case that an expression remains constant, in our case,

$$E_N^* \leq E_M^* \qquad \text{if } N > M$$

where M and N are upper summation limits in (6). $N = 13$ is found by adding one term after another until $E^* \leq 0.2$.

13. **Parseval's identity** can be used to find the sum of certain series. In this problem you have to find the integral of f^2, that is, of x^2 when $-\pi/2 \leq x \leq \pi/2$, and of $(\pi - x)^2$ when $\pi/2 \leq x \leq 3\pi/2$. Each of these two integrals equals $(2/3)(\pi/2)^3$. Together, $(4/3)(\pi/2)^3$. Division by π (as shown in Parseval's identity) and

simplification gives $\pi^2/6$. You also need the squares of the Fourier coefficients of the given function, which you can find from the Fourier series in the answer to Prob. 13 in Sec. 10.4,

$$\frac{4}{\pi}\left(\sin x - \frac{1}{9}\sin 3x + \frac{1}{25}\sin 5x - +...\right).$$

These squares equal the common factor $16/\pi^2$ times $1, 1/9^2, 1/25^2, 1/49^2$, and so on. Parseval's identity (8) thus gives

$$\frac{\pi^2}{6} = \frac{16}{\pi^2}\left(1 + \frac{1}{9^2} + \frac{1}{25^2} + ...\right).$$

Multiplying by $\pi^2/16$ on both sides gives the desired result.

Sec. 10.8   Fourier Integrals

Problem Set 10.8. Page 563

1. **Fourier integral.** If only the integral were given, the problem would be difficult. The function on the right gives the idea of how to proceed. The function is zero for negative x. For $x = 0$ it is $\pi/2$ (the mean value of the limits from the left and right as x approaches 0). Essential to you is that $f(x) = \pi e^{-x}$ for $x > 0$. Use (4). π cancels, and you have to integrate from 0 to ∞ because $f(x)$ is zero for negative x. Thus

$$A = \int_0^\infty e^{-v} \cos wv\, dv.$$

This integral can be solved by integration by parts, as is shown in calculus. Its value is $A = 1/(1 + w^2)$. Similarly, also from (4),

$$B = \int_0^\infty e^{-v} \sin wv\, dv.$$

This gives $B = w/(1 + w^2)$. Substituting A and B into (5) gives the integral shown in the problem. Integration by parts can be avoided by working in complex. From (4), using $\cos wv + i\sin wv = e^{iwv}$, you obtain

$$A + iB = \int_0^\infty e^{-(v-iwv)}\, dv$$

$$= -\frac{1}{1 - iw}e^{-(1-iw)v}\Big|_0^\infty$$

$$= \frac{1}{1 - iw} = \frac{1 + iw}{1 + w^2},$$

where the last expression is obtained by multiplying numerator and denominator by $1 + iw$. Separation of the real and imaginary parts on both sides gives the integrals for A and B on the left and their values on the right, in agreement with the previous result.

5. **Fourier cosine integral.** The integrand has no sine term. From (10) with $f(x) = e^{-x}$ ($x > 0$) you obtain

$$A = \frac{2}{\pi}\int_0^\infty e^{-v} \cos wv\, dv = \frac{2}{\pi(1 + w^2)}.$$

From this and (11) it follows that

$$f(x) = \frac{2}{\pi}\int_0^\infty \frac{\cos wx}{1 + w^2}\, dw.$$

Multiplication by $\pi/2$ gives the formula in the problem.

17. **Fourier sine integral.** Calculate B by (12), evaluating the integral by integration by parts and recursion, as before. The result can be read from the integral given in the problem. It is more involved than the integrals in the previous problems because of the contribution at $x = 1$, the upper limit of integration in (12)

(because $f(x)$ is zero for $x > 1$). In the previous problems the upper limit infinity gave no contribution.

Sec. 10.9   Fourier Cosine and Sine Transforms

Problem Set 10.9. Page 568

1. Fourier cosine transform. From (2) you obtain (sketch the given function if necessary)

$$\hat{f}_c(w) = \sqrt{\tfrac{2}{\pi}} \left(\int_0^1 \cos wx \, dx + \int_1^2 (-1) \cos wx \, dx \right)$$

$$= \sqrt{\tfrac{2}{\pi}} \left(\frac{\sin wx}{w} \bigg|_0^1 - \frac{\sin wx}{w} \bigg|_1^2 \right)$$

$$= \sqrt{\tfrac{2}{\pi}} \left(\frac{\sin w}{w} - \frac{\sin 2w}{w} + \frac{\sin w}{w} \right)$$

$$= \sqrt{\tfrac{2}{\pi}} \left(2 \frac{\sin w}{w} - \frac{\sin 2w}{w} \right).$$

9. Fourier integral and Fourier cosine transform. These two concepts are related. This is shown in the text and is illustrated by the present problem, as follows. The task is to show that $f(x) = (\sin x)/x$ has the Fourier cosine transform (see (2), Sec. 10.9)

$$\hat{f}_c(w) = \sqrt{\tfrac{2}{\pi}} \int_0^\infty \frac{\sin x \cos wx}{x} \, dx = \sqrt{\tfrac{\pi}{2}} [1 - u(w - 1)]. \qquad (A)$$

This should follow from Example 2 in Sec. 10.8.

Now the integral in (7*), Sec. 10.8, equals $\pi/2$ if $0 < x < 1$ and 0 if $x > 1$. Hence, interchanging the notations w and x in (7*) (and disregarding the value at the single point $x = 1$), you can write the integral in (7*) using a unit step function (see p. 265 of AEM) as

$$\int_0^\infty \frac{\cos wx \sin x}{x} \, dx = \frac{1}{2} \pi (1 - u(w - 1)). \qquad (B)$$

Multiplying (B) by $\sqrt{2/\pi}$ on both sides gives (A) and completes the derivation.

13. Fourier sine transform. The problem amounts to the evaluation of the integral $\int_0^1 x^2 \sin wx \, dx$ (times the factor $\sqrt{2/\pi}$) by two successive integrations by parts. The first of them gives

$$-\frac{x^2}{w} \cos wx \bigg|_0^1 + \frac{1}{w} \int_0^1 2x \cos wx \, dx.$$

Evaluating the integral-free part (which is zero at the lower limit) and another integration by parts gives

$$-\frac{\cos wx}{w} + \frac{2x}{w^2} \sin wx \bigg|_0^1 - \frac{2}{w^2} \int_0^1 \sin wx \, dx.$$

Evaluating the new integral-free part and the remaining integral, you obtain

$$-\frac{\cos wx}{w} + \frac{2}{w^2} \sin w + \frac{2}{w^3} \cos wx \bigg|_0^1.$$

The evaluation of the last expression gives $(2/w^3)(\cos w - 1)$. Taking the common denominator w^3 and multiplying the result by $\sqrt{2/\pi}$ produces the answer given on p. A27 in Appendix 2 of the book.

Sec. 10.10   Fourier Transform

Problem Set 10.10. Page 575

1. **Calculation of Fourier transforms** amounts to evaluating the defining integral (6). For the function in Prob. 1 this simply means integrating a complex exponential function, which is formally the same as in calculus. This integration extends from a to b only because outside this interval the given function is everywhere zero. Thus,

$$\hat{f}(w) = \frac{1}{\sqrt{2\pi}} \int_a^b e^{-iwx}\, dx = \frac{1}{-iw\sqrt{2\pi}}(e^{-iwb} - e^{-iwa}).$$

Now $1/i = i/i^2 = i/(-1) = -i$; keep this in mind, it will also be needed on other occasions. With this you obtain the answer given on p. A27 in Appendix 2 of the book.

13. **Table III in Sec. 10.11** contains formulas of Fourier transforms, some of which are related. For deriving formula 7 from formula 8 start from formula 8 with $-b$ instead of b and set $c = b$. This gives

$$i\left(\frac{e^{-ib(a-w)} - e^{ib(a-w)}}{\sqrt{2\pi}\,(a-w)} \right). \tag{A}$$

The basic formula now to be used is the relation between the sine and the exponential function in complex

$$\sin x = \frac{1}{2i}(e^{ix} - e^{-ix}).$$

This implies

$$e^{ix} - e^{-ix} = 2i \sin x.$$

With $x = b(a - w)$ this gives in (A) (use $i\cdot 2i = -2$)

$$\frac{2 \sin\,(b(a-w))}{\sqrt{2\pi}\,(a-w)} = \sqrt{\frac{2}{\pi}}\left(\frac{\sin\,(b(w-a))}{w-a} \right),$$

the expression given in formula 7.

CHAPTER 11. Partial Differential Equations

Sec. 11.1 Basic Concepts

Problem Set 11.1. Page 584

15. Reduction to ordinary differential equations is possible, for instance, if an equation contains only partial derivatives with respect to x or only with respect to y. The given equation

$$u_y = u$$

with $u = u(x, y)$ can be regarded as $u' = u$ and solved to obtain

$$u = c e^y$$

where c now depends on x, that is, the solution is

$$u(x, y) = c(x) e^y.$$

Check by differentiation that this function with arbitrary $c(x)$ satisfies the given equation.

19. Substitution $u_x = v$. This reduces the given equation $u_{xy} = u_x$ to

$$u_y = v.$$

The general solution is

$$v = c(x) e^y$$

where the arbitrary "constant" of integration c depends on x. Integration with respect to x now gives the answer

$$u = \int v\, dx = e^y c_1(x) + c_2(y),$$

where $c_1(x)$ is the integral of $c(x)$ with respect to x and $c_2(y)$ is the corresponding "constant" of integration. Check your result by differentiation to make sure that it is correct.

23. Boundary value problem. Verify the solution. Observing the chain rule, you obtain by differentiation

$$u_x = \frac{2ax}{x^2 + y^2}$$

and by another differentiation, with the product rule applied to $2ax$ and $1/(x^2 + y^2)$

$$u_{xx} = \frac{2a}{x^2 + y^2} - \frac{4ax^2}{(x^2 + y^2)^2}. \qquad \text{(A)}$$

Similarly, with y instead of x,

$$u_{yy} = \frac{2a}{x^2 + y^2} - \frac{4ay^2}{(x^2 + y^2)^2}. \qquad \text{(B)}$$

Taking the common denominator $(x^2 + y^2)^2$, you obtain in (A) the numerator

$$2a(x^2 + y^2) - 4ax^2 = -2ax^2 + 2ay^2$$

and in (B)

$$2a(x^2 + y^2) - 4ay^2 = -2ay^2 + 2ax^2.$$

Addition of the two expressions on the right gives 0 and completes the verification.

   Now determine a and b in $u(x, y) = a \ln(x^2 + y^2) + b$ from the boundary conditions. For $x^2 + y^2 = 1$ you have $\ln 1 = 0$, so that $b = 0$ from the first boundary condition. From this and the second boundary condition you have $a \ln 4 = 3$. Hence $a = 3/\ln 4$, in agreement with the answer on p. A28 in Appendix 2 of the book.

Sec. 11.3   Separation of Variables. Use of Fourier Series

Problem Set 11.3. Page 594

3. **Vibrating string: simplest solutions.** In (11) a term $\sin px$ is multiplied by $\cos cpt$ because

$$p = \frac{n\pi}{L} \quad \text{in (9) and} \quad \lambda_n = c\frac{n\pi}{L} = cp \quad \text{in (11*)}.$$

p and λ_n differ by a factor c, which is 1 in the problem, by assumption. It follows that for $c = 1$, initial velocity 0, and initial deflection

$$k\left(\sin x - \frac{1}{2}\sin 2x\right)$$

the corresponding solution is

$$k\left(\cos t \sin x - \frac{1}{2}\cos 2t \sin 2x\right).$$

Note that in this problem, $B_1 = k$ and $B_2 = -k/2$ in (11) are given; the initial condition already has the form of a Fourier series, so you need not do anything further.

7. **Use of Fourier series.** Problems 4-9 amount to the determination of the Fourier sine series of the initial deflection. Each term $b_n \sin nx$ is then multiplied by the corresponding $\cos nt$ (since $c = 1$; for arbitrary c it would be $\cos cnt$). The series of the terms thus obtained is the desired solution. For the "triangular" initial deflection in Prob. 7 you obtain the Fourier sine series (see p. 546 with $L = \pi$ and $k = 1/2$)

$$\frac{4}{\pi^2}\left(\sin x - \frac{1}{9}\sin 3x + \frac{1}{25}\sin 5x - + \cdots\right).$$

Multiplying each term $\sin((2n+1)x)$ by the corresponding $\cos((2n+1)t)$, you obtain the answer on p. A28 in Appendix 2 of the book.

17. **Separation of variables** is a relatively general method of solution, particularly when it is used for generating the terms of suitable series developments, as shown in the text.

   The given differential equation is $u_{xy} - u = 0$ or $u_{xy} = u$. To solve it, substitute $u = F(x)\,G(y)$, obtaining

$$\frac{dF}{dx} \cdot \frac{dG}{dy} = FG.$$

Now separate variables. This gives

$$\frac{dF/dx}{F} = \frac{G}{dG/dy}. \tag{A}$$

Both sides must equal a constant, say, k, by the argument used in the text. This gives from the left side of (A), equated to k,

$$\frac{dF}{dx} = kF.$$

The solution is

$$F = c_1 e^{kx}.$$

Equating the right side of (A) to k, you have $G/(dG/dy) = k$; hence

$$\frac{dG}{dy} = \frac{G}{k}.$$

The solution is

$$G = c_2 e^{y/k}.$$

This gives the answer $u = FG = c e^{kx+y/k}$.

Sec. 11.4   D'Alembert's Solution of the Wave Equation

Problem Set 11.4. Page 597

1. Speed. This is a uniform motion (motion with constant speed). Use that in this case, speed is distance divided by time or, equivalently, speed is distance traveled in unit time.

13. Normal form. The given equation

$$u_{xx} + 4u_{xy} + 4u_{yy} = 0$$

is of the form (15) with $A = 1$, $B = 2$, and $C = 4$. Hence $AC - B^2 = 0$, so that the equation is parabolic. Furthermore, the ordinary differential equation in Prob. 11 takes the form

$$y'^2 - 4y' + 4 = 0.$$

You can factor this in the form

$$(y' - 2)^2 = 0. \tag{A}$$

From Prob. 12 you see that for a parabolic equation the new variables leading to the normal form are $v = x$ and $z = \Psi(x, y)$, where $\Psi = const$ is a solution $y = y(x)$ of (A). From (A) you obtain

$$y' - 2 = 0, \quad y - 2x = const, \quad z = 2x - y.$$

($z = y - 2x$ would do it equally well; try it.) Hence the new variables are

$$v = x$$
$$z = 2x - y.$$

The remainder of the work consists of transforming the occurring partial derivatives into partial derivatives with respect to these new variables. This is done by the chain rule. For the first partial derivative with respect to x you obtain

$$u_x = u_v v_x + u_z z_x$$
$$= u_v + 2u_z.$$

The partial derivative of this with respect to x is

$$u_{xx} = (u_v + 2u_z)_v v_x + (u_v + 2u_z)_z z_x$$
$$= u_{vv} + 2u_{zv} + 2(u_{vz} + 2u_{zz}).$$

Assuming the continuity of the partial derivatives involved, you can interchange the order of differentiation and have $u_{zv} = u_{vz}$. Hence u_{xx} becomes

$$u_{xx} = u_{vv} + 4u_{vz} + 4u_{zz}. \tag{B}$$

Now turn to partial differentiation with respect to y. The first partial derivative with respect to y is

$$u_y = u_v v_y + u_z z_y = -u_z$$

because $v_y = 0$ and $z_y = -1$. Take the derivative of this with respect to x, obtaining

$$u_{yx} = u_{xy} = -u_{zv} v_x - u_{zz} z_x \tag{C}$$
$$= -u_{vz} - 2u_{zz}.$$

Finally, taking the partial derivative of u_y with respect to y gives

$$u_{yy} = -u_{zv} v_y - u_{zz} z_y \tag{D}$$
$$= u_{zz}.$$

Substituting the second partial derivatives (B)-(D) into the given equation, you obtain

$$u_{xx} + 4u_{xy} + 4u_{yy} = u_{vv} + 4u_{vz} + 4u_{zz}$$
$$- 4u_{vz} - 8u_{zz}$$
$$+ 4u_{zz}$$
$$= u_{vv} = 0.$$

This is the normal form for parabolic equations.

Sec. 11.5   Heat Equation: Solution by Fourier Series

Problem Set 11.5. Page 608

3. Single sine terms as initial temperatures lead to solutions of the form (9),

$$B_n \sin \frac{n\pi x}{L} e^{-\lambda_n^2 t},$$

that is, single eigenfunctions (which in more complicated cases would not be sufficient for satisfying the inital condition). The initial condition

$$\sin (0.1\,\pi x) = \sin (\pi x/10) = \sin (\pi x/L)$$

(where $L = 10$) shows that $n = 1$, that is, the initial condition is such that the solution is given by the first eigenfunction. From the data for K, σ, and ρ calculate $c^2 = K/(\sigma\rho) = 1.75202$ (see the very first equation in Sec. 11.5). This gives the answer

$$u = \sin (0.1\,\pi x) \exp (-1.75202\pi^2 t/100)$$

where $L^2 = 100$ results from

$$\lambda_1^2 = \frac{c^2\pi^2}{L^2}.$$

17. Heat flow in a plate. The problem corresponds to the situation discussed on p. 606 of the text for a rectangle because the boundary conditions are as in the text, with $f(x) = 20 = const$. Hence you can obtain the solution of the problem from (19) and (20) with $a = b = 24$. You have to begin with (20), which for the problem takes the form

$$A_n^* = \frac{2}{24 \sinh n\pi} \int_0^{24} 20 \sin \frac{n\pi x}{24}\, dx$$

$$= \frac{40}{24 \sinh n\pi} \left(-\frac{24}{n\pi}\right) \cos \frac{n\pi x}{24} \Big|_{x=0}^{24}$$

$$= -\frac{40}{n\pi \sinh n\pi}(\cos n\pi - 1)$$

$$= -\frac{40}{n\pi \sinh n\pi}((-1)^n - 1)$$

$$= +\frac{80}{n\pi \sinh n\pi} \quad \text{if } n = 1, 3, 5, \ldots$$

and 0 for even n. If you substitute this into (19), you obtain the series

$$u = \frac{80}{\pi} \sum \frac{1}{n \sinh n\pi} \sin \frac{n\pi x}{24} \sinh \frac{n\pi y}{24}$$

where you sum over odd n only because $A_n^* = 0$ for even n. If in this answer you write $2m - 1$ instead of n, then you automatically have the desired summation and can drop the condition "over odd n only". This is the form of the answer given on p. A29 in Appendix 2 of the book.

Sec. 11.6   Heat Equation: Solution by Fourier Integrals and Transforms

Problem Set 11.6. Page 615

3. **Solutions in integral form** are obtained by using the Fourier integral (instead of the Fourier series, whose use is restricted to periodic solutions). These solutions are of the form (6). Here, $A(p)$ and $B(p)$ are given by (8), which shows that they are determined by the initial temperature $f(x)$ in the infinitely long (practically: very long) bar or wire. The given initial temperature in Prob. 3 is

$$f(x) = 1 \text{ if } -1 < x < 1 \quad \text{and } f(x) = 0 \text{ otherwise.} \tag{I}$$

This models a situation when a short portion of the bar is heated, whereas the rest of it is kept at temperature 0.

From (8) and (I) you obtain

$$A(p) = \frac{1}{\pi} \int_{-1}^{1} \cos pv \, dv = \frac{1}{\pi p} \sin pv \Big|_{v=-1}^{1} \tag{II}$$

$$= \frac{1}{\pi p}(\sin p - \sin(-p)) = \frac{2 \sin p}{\pi p}.$$

Furthermore, from (8) and (I),

$$B(p) = \frac{1}{\pi} \int_{-1}^{1} \sin pv \, dv = 0; \tag{III}$$

this follows immediately without calculation by noting that $f(x)$ is an even function, so that $f(x) \sin px$ is odd, and you integrate from -1 to 1, so that the area under the curve from -1 to 0 is minus the area under the curve from 0 to 1. Substituting (II) and (III) into (6) in the text gives the answer shown on p. A29 in Appendix 2 of the book.

5. **Use of Sec. 10.8.** That section and its problem set contain integrals that can be used for the present purpose. The initial temperature in Prob. 5 is "wavy" with a relatively fast decreasing maximum amplitude and alternatingly positive and negative temperatures given by $f(x)$.

$\sin x$ is odd. Hence $(\sin x)/x$ is even, so that $B(p)$ in (8) is zero. For $A(p)$ you obtain from (8)

$$A(p) = \frac{1}{\pi} \int_{-1}^{1} \frac{\sin v}{v} \cos pv \, dv$$

$$= \frac{2}{\pi} \int_{0}^{1} \frac{\sin v}{v} \cos pv \, dv. \tag{IV}$$

This is precisely the integral in Prob. 2 of Problem Set 10.8, except for notation. Its value is $\pi/2$ if $0 < x < 1$. Multiplication by $2/\pi$ (the factor in (IV)) gives the values $A(p) = 1$ if $0 < x < 1$ and $A(p) = 0$ if $x > 1$.

(The value at $x = 1$ is $(\pi/4)(2/\pi) = 1/2$; this is of no interest here because we are concerned with an integral, (6).) Substitution of $A(p)$ into (6) gives the integral from 0 to 1 shown on p. A29 in Appendix 2.

Sec. 11.8   Rectangular Membrane. Use of Double Fourier Series

Problem Set 11.8. Page 626

7. **Coefficients of a double Fourier series** can be obtained following the idea in the text. For $f(x, y) = y$ in the square $0 < x < \pi, 0 < y < \pi$ the calculations are as follows. (Here we use the formula numbers of the text.) The desired series is obtained from (18) with $a = b = \pi$ in the form

$$f(x, y) = y = \sum \left(\sum B_{mn} \sin mx \sin ny \right) \tag{18}$$

$$= \sum K_m(y) \sin mx \quad (\text{sum over } m)$$

where the notation

$$K_m(y) = \sum B_{mn} \sin ny \quad (\text{sum over } n) \tag{19}$$

was used. Now fix y. Then the second line of (18) is the Fourier sine series of $f(x,y) = y$ considered as a function of x (hence as a constant, but this is not essential). Thus, by (6) in Sec. 10.4 its Fourier coefficients are

$$b_m = K_m(y) = \frac{2}{\pi} \int_0^\pi y \sin mx \, dx. \tag{20}$$

You can pull out y from under the integral sign (since you integrate with respect to x) and integrate, obtaining

$$K_m(y) = \frac{2y}{m\pi}(-\cos m\pi + 1)$$

$$= \frac{2y}{m\pi}(-(-1)^m + 1)$$

$$= \frac{4y}{m\pi} \quad \text{if } m \text{ is odd and 0 for even } m$$

(because $(-1)^m = 1$ for even m). By (6) in Sec. 10.4 (with y instead of x and $L = \pi$) the coefficients of the Fourier series of the function $K_m(y)$ just obtained are

$$B_{mn} = \frac{2}{\pi} \int_0^\pi K_m(y) \sin ny \, dy$$

$$= \frac{2}{\pi} \int_0^\pi \frac{4y \sin ny}{m\pi} \, dy$$

$$= \frac{8}{m\pi^2} \int_0^\pi y \sin ny \, dy.$$

Integration by parts gives

$$B_{mn} = \frac{8}{nm\pi^2} \left(-y \cos ny \Big|_{y=0}^\pi + \int_0^\pi \cos ny \, dy \right).$$

The integral gives a sine, which is zero at $y = 0$ and $y = n\pi$. The integral-free part is zero at the lower limit. At the upper limit it gives

$$\frac{8}{nm\pi^2}(-\pi(-1)^n) = \frac{(-1)^{n+1}8}{nm\pi}.$$

Remember that this is the expression when m is odd, whereas for even m these coefficients are zero.

Sec. 11.9   Laplacian in Polar Coordinates

Problem Set 11.9. Page 628

3. **Laplacian in polar coordinates**. The alternative form is sometimes more practical than (4), which is obtained from it simply by performing the indicated product differentiation.

5. **Solution depending only on r.** If r is the only independent variable involved, you have to write the usual derivatives instead of partial derivatives. Hence (4), with the last term absent, now gives

$$u'' + u'/r = 0.$$

This can be solved for u' by separating variables and subsequent exponentiation, giving $u' = a/r$, where a is an arbitrary constant. Now integrate.

9. **Temperature in a circular disk**. The disk is bounded by the circle $x^2 + y^2 = r^2 = 1$. The boundary temperature means that the left half of the circle is kept at temperature 0. At $x = 0$, $y = -1$ (the lowest point of the circle) the temperature jumps from 0 to $-\pi/2$. Then it increases steadily to 0 (this temperature is reached where the circle crosses the x-axis) and on to $\pi/2$ (reached at the uppermost point $x = 0$, $y = 1$ of the circle), where it jumps down to 0.

   The temperature inside the disk is obtained from (6) (see Team Project 6). For this you need the Fourier series of the boundary temperature (sketch it over the θ-axis from $-\pi$ to in π the usual fashion). This is an odd function; hence that series is a Fourier sine series. Because it has period 2π, the Fourier coefficients are obtained from (6*) in Sec. 10.4 (or from the original Euler formulas in Sec. 10.2). To evaluate the integral, use integration by parts. The result is

$$u = \frac{2}{\pi} \sin \theta + \frac{1}{2} \sin 2\theta - \frac{2}{9\pi} \sin 3\theta - \frac{1}{4} \sin 4\theta + \cdots,$$

in agreement with the answer to Prob. 15 in Problem Set 10.2. From this the temperature in the disk is obtained by multiplying each term by a power of r, namely, $\sin n\theta$ is multiplied by r^n; thus

$$u = \frac{2r}{\pi} \sin \theta + \frac{r^2}{2} \sin 2\theta - \frac{2r^3}{9\pi} \sin 3\theta - \frac{r^4}{4} \sin 4\theta + \cdots.$$

Try to sketch the isotherms by noting the following. You know that the left half of the boundary circle is at temperature 0. So is the x-axis because it corresponds to $\theta = 0$ and $\theta = \pi$, for which the sine is 0. All positive isotherms must begin at the corresponding boundary points of the circle in the first quadrant and must end at the jump at $y = 1$ on the y-axis, and they must remain in the upper half of the disk. Similarly in the lower half of the disk, for the negative isotherms between $-\pi/2$ and 0. Since the boundary temperature is symmetric with respect to the x-axis, so are the isotherms, that is, a negative isotherm is obtained by reflecting the corresponding positive isotherm in the x-axis. At the upper jump, isotherms, such as $u = 0$, $\pi/8$, $\pi/4$, $3\pi/8$, must make equal angles with one another ($\pi/4$ for these four isotherms). Why?

   This discussion is typical. It shows that in many cases, one can obtain substantial qualitative information without actual computation.

Sec. 11.10   Circular Membrane. Use of Fourier-Bessel Series

Problem Set 11.10. Page 634

7. **Nonzero initial velocity**. Differentiating (12) with respect to t, you obtain

$$u_t(r, t) = \sum_{m=1}^{\infty} \left(-a_m \lambda_m \sin \lambda_m t + b_m \lambda_m \cos \lambda_m t \right) J_0\left(\frac{\alpha_m r}{R} \right).$$

For $t = 0$ this must equal $g(r)$, the given inital velocity in (4). Since $\sin 0 = 0$ and $\cos 0 = 1$, you have

$$u_t(r, 0) = \sum_{m=1}^{\infty} b_m \lambda_m J_0\left(\frac{\alpha_m r}{R} \right) = g(r).$$

Hence the $b_m \lambda_m$ are the coefficients of the Fourier-Bessel series of $g(r)$. Accordingly, you obtain their form from (14) with a_m replaced by $b_m \lambda_m$ on the left and $f(r)$ replaced by $g(r)$ under the integral sign. Division by $\lambda_m = c k_m = c \alpha_m/R$ (see before) gives (15) because a factor R cancels.

13. **Periodicity**. This follows from the fact that θ is an angular variable (an angular coordinate), which is determined only up to integer multiples of 2π, so that its increase or decrease by 2π, 4π, etc. does not change the point to which it belongs.

   To express (20) in terms of the usual Bessel equation, set $kr = s$, thus $r = s/k$, $d/dr = k d/ds$, so that by the chain rule,

$$(s^2/k^2)k^2\ddot{W} + (s/k)k\dot{W} + (s^2 - n^2)W = 0, \qquad s^2\ddot{W} + s\dot{W} + (s^2 - n^2)W = 0,$$

where dots denote derivatives with respect to s.

15. Membrane vibrations also depending on angle. These solutions are given by (22). This formula is a consequence of the steps of separation in Probs. 11-14. To understand it completely, it is worth thinking about how this case relates to that of independence of angle. In the latter case, the Bessel equation for the radial coordinate r is (7*), which is Bessel's equation with $v = 0$. The latter is obtained from Bessel's equation (7) involving a constant k (not to be confused with the parameter v; k slips into the argument, whereas v specifies the whole function and distinguishes it from a Bessel function with another value of v). That arbitrary k is later determined by the boundary condition that the membrane be fixed on the boundary. In this case of independence of the angle you are concerned exclusively with the Bessel function J_0. And there is no differential equation for the angle–of course not because the angle does not appear. In the case of angular dependence you do obtain Bessel functions with v different from 0 (hence, greater than zero; Bessel's equation involves the square v^2, so you can always assume v to be positive or 0). And you have (19), the equation for the angle, which for reasons of periodicity leads to the integer values $v = n = 1, 2, 3,...$. Instead of the zeros α_m of J_0 resulting from fixing the membrane along the boundary circle you now have the zeros α_{mn} of J_n resulting for the very same reason.

For $n = 0$ the solutions u_{mn} in the first line of (22) reduce to those in (11), and the solutions u_{mn}^* in the second line of (22) are no longer present when $n = 0$ because then $\sin n\theta$ is identically zero.

Sec. 11.11   Laplace's Equation in Cylindrical and Spherical Coordinates. Potential

Problem Set 11.11. Page 641

7. Potential between two spheres. Since the region is bounded by two concentric spheres, and since these spheres are kept at constant potentials, the potential between the spheres will be spherically symmetric, that is, the equipotential surfaces will be concentric spheres. Now a spherically symmetric solution of the three-dimensional Laplace equation is

$$u(r) = c/r + k$$

(see Prob. 6). The constants c and k can be determined from the two boundary conditions $u(2) = 220$ and $u(4) = 140$; thus,

$$u(2) = c/2 + k = 220 \qquad (A)$$

and $u(4) = c/4 + k = 140$ or, multiplied by 2,

$$2u(4) = c/2 + 2k = 280. \qquad (B)$$

(B) minus (A) gives $k = 60$. From this and (A) you have $c/2 = 160$, hence $c = 320$. Check the result.

13. Special Fourier-Legendre series. These series were introduced in Example 2 of Sec. 4.8. They are of the form

$$a_0 P_0 + a_1 P_1 + a_2 P_2 + ...$$

Since x is one of our coordinates in space, you must choose another notation. Choose w or use ϕ obtained by setting $w = \cos \phi$. The Legendre polynomials $P_n(w)$ involve powers of w; thus $P_n(\cos \phi)$ involves powers of $\cos \phi$. Accordingly, you have to transform $\cos 2\phi$ into powers of $\cos \phi$. This gives

$$\Phi = 2\cos^2 \phi - 1 = 2w^2 - 1.$$

This must now be expressed in terms of Legendre polynomials. From the occurring powers you see that you need P_2 and P_0. The result is

$$2w^2 - 1 = \frac{4}{3} P_2(w) - \frac{1}{3} P_0(w). \qquad (C)$$

Check this by using the definitions of P_2 and P_0 in Sec. 4.3.

From (C) and (16) you finally see that the answer is

$$u(r, \phi) = -\frac{1}{3} P_0(\cos \phi) + \frac{4}{3} P_2(\cos \phi).$$

Note that in the present case the coefficient formulas (18) or (18*) were not needed because the boundary condition was so simple that the coefficients were already known to us. Note further that $P_0 = 1 = const$, but our notations $P_0(w)$ and $P_0(\cos\phi)$ are correct because a constant is a special case of a function of any variable.

21. Transformation of the Laplacian. Set $r = 1/s$, hence $s = 1/r$. You have to transform the partial derivatives occurring in the Laplacian in polar coordinates. By the chain rule you obtain for the first partial derivative

$$u_r = u_s s_r = u_s \left(-\frac{1}{r^2} \right) = -s^2 u_s.$$

From $u_r = u_s s_r$ you obtain for the second partial derivative

$$u_{rr} = u_{ss} s_r^2 + u_s s_{rr} = u_{ss} \left(-\frac{1}{r^2} \right)^2 + u_s \frac{2}{r^3} = s^4 u_{ss} + 2 s^3 u_s.$$

Substituting these expressions into the Laplacian gives

$$u_{rr} + (1/r)u_r + (1/r^2)u_{\theta\theta} = s^4 u_{ss} + 2 s^3 u_s + s(-s^2 u_s) + s^2 u_{\theta\theta}.$$

On the right side take the second and third terms together. Then you see that this is s^4 times the Laplacian in $s = 1/r$ and θ. Hence if $u(r,\theta)$ satisfies Laplace's equation, so does $v(r,\theta) = u(1/r,\theta)$, as claimed.

Sec. 11.12   Solution by Laplace Transforms

Problem Set 11.12. Page 646

5. First-order differential equation. The boundary conditions mean that $u(x,t)$ vanishes on the positive parts of the coordinate axes in the xt-plane. Let U be the Laplace transform of $u(x,t)$ considered as a function of t; write $U = U(x, s)$. The derivative u_t has the transform sU because $u(x,0) = 0$. The transform of t on the right is $1/s^2$. Hence you first have

$$x U_x + s U = \frac{x}{s^2}.$$

This is a first-order linear differential equation with the independent variable x. Division by x gives

$$U_x + \frac{s U}{x} = \frac{1}{s^2}.$$

Its solution is given by the integral formula (4) in Sec. 1.6. Using the notation in that section, you obtain

$$p = s/x, \quad h = \int p\,dx = s \ln x, \quad e^h = x^s, \quad e^{-h} = 1/x^s.$$

Hence (4) in section 1.6, with the "constant" of integration depending on s, gives, since $1/s^2$ does not depend on x,

$$U(x, s) = \frac{1}{x^s} \left(\int \frac{x^s}{s^2} dx + c(s) \right) = \frac{c(s)}{x^s} + \frac{x}{s^2 (s + 1)}$$

(note that x^s cancels in the second term, leaving the factor x). Here you must have $c(s) = 0$ for $U(x, s)$ to be finite at $x = 0$. Then

$$U(x,s) = \frac{x}{s^2 (s + 1)}.$$

Now

$$\frac{1}{s^2 (s + 1)} = \frac{1}{s^2} - \frac{1}{s} + \frac{1}{s + 1}.$$

This has the inverse Laplace transform $t - 1 + e^{-t}$ and gives the solution $u(x, t) = x(t - 1 + e^{-t})$.

PART D.   COMPLEX ANALYSIS

CHAPTER 12.   Complex Numbers and Functions. Conformal Mapping

Sec. 12.1   Complex Numbers. Complex Plane

Example 2. The check is $z z_2 = (\frac{66}{85} + \frac{43}{85} i)(9 - 2i) = 8 + 3i$.

Problem Set 12.1. Page 656

1. **Powers of i.** $i^2 = -1$ and $\frac{1}{i} = -i$ will be used quite frequently. A formal derivation of $i^2 = -1$ from the multiplication formula is shown in the text. $1/i = -i/i(-i) = -i/1 = -i$ follows from (7).

3. **Multiplication.** For $z_1 = 4 + 3i$ and $z_2 = 2 - 5i$ the recipe on p. 654 at the top [which results from (3)] gives

$$
\begin{aligned}
z_1 z_2 &= (4 + 3i)(2 - 5i) \\
&= 4 \cdot 2 - 4 \cdot 5i + 3i \cdot 2 + i^2 \cdot 3 \cdot (-5) \\
&= 8 - 20i + 6i + 15 = 23 - 14i.
\end{aligned}
$$

With a little training you can go faster and write down first the two terms of the real part and then the two terms of the imaginary part; thus

$$
\begin{aligned}
(4 + 3i)(2 - 5i) &= 8 - (-15) + i(-20 + 6) \\
&= 23 - 14i.
\end{aligned}
$$

5. **Division.** Given $z_1 = 4 + 3i$, find $1/z_1$.

   This is a simple special case of (7), which gives

$$
\frac{1}{z_1} = \frac{1}{4 + 3i} = \frac{4 - 3i}{(4 + 3i)(4 - 3i)} = \frac{4 - 3i}{16 + 9} = 0.16 - 0.12 i.
$$

17. **Real part. Complex conjugate.** Let $z = x + iy$. Find $\text{Re}(z^2/\bar{z})$.

    First determine $z^2/\bar{z}$. According to (7) multiply numerator and denominator by the conjugate of the denominator, which is z, and use that $z\bar{z} = x^2 + y^2$. This gives

$$
\begin{aligned}
z^2/\bar{z} = z^3/z\bar{z} &= \frac{(x + iy)^3}{x^2 + y^2} \\
&= \frac{x^3 + 3ix^2 y - 3xy^2 - iy^3}{x^2 + y^2},
\end{aligned}
$$

where the minus signs come from i^2. The real part of this is obtained by omitting the two terms that have an i. This gives the answer

$$
\text{Re}(z^2/\bar{z}) = \frac{x^3 - 3xy^2}{x^2 + y^2}.
$$

Sec. 12.2   Polar Form of Complex Numbers. Powers and Roots

Generalized triangle inequality (6). Drawing the complex numbers as little arrows and letting each tail coincide with the preceding head, you get a zigzag line of n parts. The left side of (6) equals the distance from the tail of z_1 to the head of z_n. Can you "*see*" it? Now take your zigzag line and pull it taut, then you have the right side as the length of the zigzag line straightened out.

This inequality is very important and we shall use it on various occasions. In almost all cases it will not matter whether the right side is much larger than the left or not. It will be essential that we have such an upper bound for the absolute value of the sum on the left.

Problem Set 12.2. Page 662

1. **Polar form.** Sketch $z = 1 + i$, to understand what is going on. z is the point $(1, 1)$ in the complex plane. From this you see that the distance of z from the origin is $|z| = \sqrt{2}$. This is the absolute value of z. Furthermore, z lies on the bisecting line of the first quadrant, so that its argument (the angle between the positive ray of the x-axis and the segment from 0 to z) is 45 degrees or $\pi/4$.

Now show how the results follow from (3) and (4). In the notation of (3) and (4) you have $z = x + iy = 1 + i$. Hence the real part of z is $x = 1$ and the imaginary part of z is $y = 1$. From (3) you thus obtain

$$|z| = \sqrt{1^2 + 1^2} = \sqrt{2},$$

as before. From (4) you obtain

$$\tan \theta = y/x = 1, \qquad \theta = 45 \text{ degrees or } \pi/4.$$

Hence the polar form (2) is

$$z = \left(\sqrt{2} \cos \frac{\pi}{4} + i \sin \frac{\pi}{4} \right).$$

7. **Polar form.** It is probably best to first perform the division and then square the result. From (7) in Sec. 12.1 you obtain

$$\frac{6 + 8i}{4 - 3i} = \frac{(6 + 8i)(4 + 3i)}{(4 - 3i)(4 + 3i)}$$

$$= \frac{24 - 24 + 18i + 32i}{16 + 9} = 2i.$$

The square of this is $(2i)^2 = 2^2 i^2 = -4$. Denoting the given expression by z, you thus have $|z| = 4$ and $\theta = \pi$,

$$z = 4(\cos \pi + i \sin \pi).$$

Check this by noting that $\cos \pi = -1$ and $\sin \pi = 0$.

Furthermore, if a somewhat lengthy calculation leads to a simple result (namely, $2i$), suspect that there may be a simpler way. There is. Write

$$\frac{6 + 8i}{4 - 3i} = \frac{2i(-3i + 4)}{4 - 3i}$$

and you can see the result $(2i)^2 = -4$ immediately.

13. **Principal value of the argument.** $3 + 4i$ lies in the first quadrant. Any number $z \neq 0$ in this quadrant has principal value of the argument between 0 and $\pi/2$. Indeed,

$$\theta = \text{Arg}(3 + 4i) = \arctan \frac{4}{3} = 0.927295.$$

Since $3 - 4i$ is the conjugate of $3 + 4i$, the principal value of its argument is the negative of the previous one, that is,

$$\theta = \text{Arg}(3 - 4i) = \arctan \left(-\frac{4}{3} \right) = -\arctan \frac{4}{3} = -0.927295.$$

21. **Cube root.** You have to find the cube root of $1 + i$. You can use (15) with $n = 3$ and $k = 0, 1, 2$. In (15) you need

$$r = \sqrt{1^2 + 1^2} = \sqrt{2},$$

which implies

$$\sqrt[3]{r} = \sqrt[6]{2},\tag{a}$$

and

$$\theta = \text{Arg}\,(1 + i) = \arctan\,(1/1) = \arctan 1 = \frac{\pi}{4}.\tag{b}$$

Equation (b) gives the three arguments of the cosine and sine in (15) with $n = 3$ in the form

$$\frac{1}{3}(\pi/4) \qquad\qquad\qquad \text{when } k = 0\tag{c}$$

$$\frac{1}{3}(\pi/4 + 2\pi) = 9\pi/12 = 3\pi/4 \quad \text{when } k = 1\tag{d}$$

$$\frac{1}{3}(\pi/4 + 4\pi) = 17\pi/12 \qquad\quad \text{when } k = 2.\tag{e}$$

For $k = 3$ you would obtain

$$\frac{1}{3}\left(\frac{\pi}{4} + 6\pi\right) = \frac{1}{3}\frac{\pi}{4} + 2\pi,$$

which gives the same complex number as in (c). Indeed, a cube root of a number $z \neq 0$ has 3 values, not 4, and they form the vertices of an equilateral triangle. Figure 297 illustrates this for a special case, and for other complex numbers the figure looks quite similar, with a different isosceles triangle that has its vertices at some other points on a circle with center at the origin.

From (a)-(e) and (15) you obtain the answer given on p. A31 in Appendix 2 of the book.

Sec. 12.3   Derivative. Analytic Function

Problem Set 12.3. Page 668

1. **Closed circular disk**. The given inequality represents a closed circular disk, as explained near the beginning of the section. The inequality can be written

$$|z - (-2 - 5i)| \le 1/2.$$

From this you see that the center is $-2 - 5i$ (not $2 + 5i$). The corresponding equation

$$|z - (-2 - 5i)| = 1/2$$

represents the boundary, the circle of radius $1/2$ and center $-2 - 5i$.

7. **Region bounded by a hyperbola**. You have $z^2 = x^2 + 2ixy - y^2$. The real part is

$$\text{Re}\,(z^2) = x^2 - y^2.$$

Hence the given region, call it R, is

$$\text{Re}\,(z^2) = x^2 - y^2 \le 1.\tag{A}$$

Hence R is bounded by the hyperbola

$$x^2 - y^2 = 1$$

familiar from calculus; call it H and sketch it. Now H has two branches, that is, it consists of two curves without common points. It is the boundary of the region between the two branches as well as of the two regions to the left of the left branch and to the right of the right branch. You can easily find out which of these regions is R. Indeed, if you set $x = 0$ and $y = 0$ in (A) you obtain $0 \le 1$, which is true. This shows that the origin $(0, 0)$ must lie in R. Hence R is the region *between* the two branches of H. If the inequality had come out false, $0 \ge 1$, then R would have consisted of those other two regions.

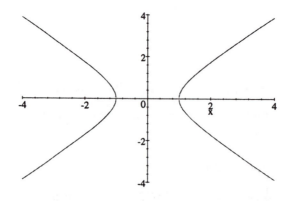

Section 12.3. Problem 7. Hyperbola H bounding the region R between the two branches

11. Function values. $z = 7 + 2i$ gives $1 - z = -6 - 2i$, hence

$$\frac{1}{1-z} = \frac{-6+2i}{(-6-2i)(-6+2i)} = \frac{-6+2i}{36+4} = -\frac{3}{20} + \frac{i}{20}.$$

Hence Re $(f) = -3/20$ and Im $(f) = 1/20$.

13. Continuity. First of all, the only point where $f(z) = (\text{Im } z)/|z|$ could be discontinuous is the origin because Im $z = y$ is continuous everywhere. Now

$$f(z) = (\text{Im } z)/|z| = y/\sqrt{x^2 + y^2}.\tag{A}$$

Continuity at 0 would mean that $f(z)$ approaches 0 as z approaches 0 from any direction. But from (A) you see that on the x-axis ($y = 0$) the function has the value 0 for any $x \neq 0$, whereas on the y-axis ($x = 0$) you obtain the value $y/\sqrt{y^2} = y/|y|$, which is $+1$ for positive y and -1 for negative y. Hence $f(z)$ is discontinuous at $z = 0$.

   More simply, if you use polar coordinates, you have

$$\text{Im } z = r \sin \theta, \quad |z| = r, \quad f(z) = (r \sin \theta)/r = \sin \theta.$$

From the last expression you get 0 on the positive ray of the x-axis ($\theta = 0$), 1 on the positive ray of the y-axis ($\theta = \pi/2$), 0 on the negative ray of the x-axis ($\theta = \pi$), and -1 on the negative ray of the y-axis ($\theta = 3\pi/2$ or $-\pi/2$, etc.). This agrees with the results obtained just before.

19. Derivative. The differentiation rules are the same as in calculus. Hence $f(z) = (5 + 3i)/z^3$ has the derivative

$$f'(z) = -\frac{3(5+3i)}{z^4}.\tag{A}$$

For $z = 2 + i$ you obtain by the usual division rule (Sec. 12.1)

$$\frac{1}{z^4} = \frac{1}{(2+i)^4} = \frac{(2-i)^4}{(2+i)^4(2-i)^4} = \frac{((2-i)^2)^2}{(4+1)^4} = \frac{(3-4i)^2}{625} = \frac{9-16-24i}{625}.$$

Multiplying this by the factor $-3(5 + 3i)$ in (A) finally gives

$$\frac{(-15-9i)(-7-24i)}{625} = \frac{-111+423i}{625}.$$

If you had difficulties with this problem, review a few calculations from Sec. 12.1 and the corresponding problem set because the differentiation as such was done as in calculus.

Sec.12.4   Cauchy-Riemann Equations. Laplace's Equation

Problem Set 12.4. Page 673

1. **Check of analyticity**. The form of the given function, $f(z) = z^6$, shows that in the present case, (7) will be simpler. Indeed, in polar coordinates you simply have

$$f(z) = r^6 (\cos 6\theta + i \sin 6\theta).$$

Hence

$$u = r^6 \cos 6\theta, \quad v = r^6 \sin 6\theta.$$

The expressions needed in (7) are obtained by straightforward differentiation. In the first Cauchy-Riemann equation in (7) you need

$$u_r = 6r^5 \cos 6\theta \quad \text{and} \quad v_\theta = 6r^6 \cos 6\theta,$$

with the factor 6 in v_θ resulting from the chain rule. From this you see that this first equation $u_r = v_\theta/r$ is satisfied. In the second Cauchy-Riemann equation you need

$$v_r = 6r^5 \sin 6\theta \quad \text{and} \quad u_\theta = -6r^6 \sin 6\theta.$$

If you divide u_θ by $-r$, you obtain v_r. Hence the second Cauchy-Riemann equation is also satisfied, and you can conclude that z^6 is analytic for all $z \neq 0$.

z^6 is also analytic at $z = 0$. This does not follow from (7), but you have to use (1), which involves more work. (Of course, this will make your work on (7) superfluous.) You can get u and v by using the binomial theorem, obtaining

$$(x + iy)^6 = x^6 + 6x^5(iy) + 15x^4(iy)^2 + 20x^3(iy)^3 + 15x^2(iy)^4 + 6x(iy)^5 + (iy)^6$$
$$= x^6 + 6ix^5y - 15x^4y^2 - 20ix^3y^3 + 15x^2y^4 + 6ixy^5 - y^6.$$

The terms without an i give the real part

$$u = x^6 - 15x^4y^2 + 15x^2y^4 - y^6.$$

The terms containing i give the imaginary part

$$v = 6x^5y - 20x^3y^3 + 6xy^5.$$

In the first Cauchy-Riemann equation you need the partial derivatives

$$u_x = 6x^5 - 60x^3y^2 + 30xy^4$$

and

$$v_y = 6x^5 - 60x^3y^2 + 30xy^4.$$

Hence the first Cauchy-Riemann equation is satisfied. The second one involves

$$v_x = 30x^4y - 60x^2y^3 + 6y^5$$

and

$$u_y = -30x^4y + 60x^2y^3 - 6y^5.$$

You see that $v_x = -u_y$, so that the second Cauchy-Riemann equation is satisfied, too. This proves analyticity of z^6 for all z.

3. **Cauchy-Riemann equations. Analyticity**. The function

$$f(z) = u + iv = e^x (\cos y + i \sin y) \tag{A}$$

has the real part $u = e^x \cos y$ and the imaginary part $v = e^x \sin y$. The familiar rules for differentiating the exponential function and cosine and sine show that the Cauchy-Riemann equations (1) are satisfied for all $z = x + iy$. Indeed,

$$u_x = e^x \cos y = v_y$$
$$u_y = -e^x \sin y = -v_x.$$

You will see in Sec. 12.6 that (A) defines the complex exponential function. When $y = 0$, so that $z = x$ is real, then $\cos y = \cos 0 = 1$, $\sin y = \sin 0 = 0$, and $f(z)$ becomes e^x, the exponential function known from calculus. More on this follows in Sec. 12.6.

17. Harmonic functions appear as real and imaginary parts of analytic functions. If you remember that the given function u is the real part of $1/z$, you are done; indeed, by the division rule,

$$\frac{1}{z} = \frac{1}{x + iy} = \frac{x - iy}{x^2 + y^2}.$$

This also shows that a conjugate harmonic of u is $-y/(x^2 + y^2)$.

If you don't remember that, you have to work systematically by differentiation, beginning with proving that the Laplace equation (8) is satisfied. Such somewhat lengthy differentiations (as well as other calculations) can often be simplified (and made more reliable) by introducing suitable shorter notations for certain expressions. In the present case you can write

$$u = \frac{x}{F}, \quad \text{where} \quad F = x^2 + y^2. \quad \text{Then} \quad F_x = 2x, \quad F_y = 2y. \tag{A}$$

By applying the product rule of differentiation (and the chain rule), not the quotient rule, you obtain the first partial derivative

$$u_x = \frac{1}{F} - \frac{x(2x)}{F^2}. \tag{B}$$

By differentiating this again, using the product and chain rules you obtain the second partial derivative

$$u_{xx} = -\frac{2x}{F^2} - \frac{4x}{F^2} + \frac{8x^3}{F^3}. \tag{C}$$

Similarly, the partial derivative of u with respect to y is obtained from (A) in the form

$$u_y = -\frac{2xy}{F^2}. \tag{D}$$

The partial derivative of this with respect to y is

$$u_{yy} = -\frac{2x}{F^2} + \frac{8xy^2}{F^3}. \tag{E}$$

Adding (C) and (E) and remembering that $F = x^2 + y^2$ give

$$u_{xx} + u_{yy} = -\frac{8x}{F^2} + \frac{8x(x^2 + y^2)}{F^3} = -\frac{8x}{F^2} + \frac{8x}{F^2} = 0.$$

This shows that $u = x/F = x/(x^2 + y^2)$ is harmonic.

Now determine a conjugate harmonic. From (D) and the second Cauchy-Riemann equation you obtain

$$u_y = -\frac{2xy}{F^2} = -v_x.$$

Integration of $2x/F^2$ with respect to x gives $-1/F$, so that integration of v_x with respect to x gives

$$v = -\frac{y}{F} = -\frac{y}{x^2 + y^2} + h(y).$$

Now show that $h(y)$ must be a constant (which you can choose to be 0). By differentiation with respect to y and taking the common denominator F^2 you obtain

$$v_y = -\frac{1}{F} + \frac{2y^2}{F^2} = \frac{-x^2 + y^2}{F^2} + h'(y).$$

On the other hand, you have from (B)

$$u_x = \frac{1}{F} - \frac{2x^2}{F^2} = \frac{y^2 - x^2}{F^2}.$$

By the first Cauchy-Riemann equation, $v_y = u_x$, so that $h'(y) = 0$ and $h(y) = const$, as claimed.

Sec. 12.5   Geometry of Analytic Functions: Conformal Mapping

Problem Set 12.5. Page 678

3. **Mapping** $w = 1/z$. Taking absolute values, you obtain
$$|w| = |1/z| = 1/|z|.$$

This shows that the concentric circles $|z| = 1/3, 1/2, 1, 2, 3$ are mapped onto the concentric circles $|w| = 3, 2, 1, 1/2, 1/3$, respectively, in the w-plane. In particular, the unit circle is mapped onto the unit circle. For this reason this mapping is often called a *reflection in the unit circle*. The points $z = 1$ and $z = -1$ are "mapped onto itself"; this means, they are mapped onto $w = 1$ and $w = -1$, respectively. Other points on the unit circle are not mapped onto itself; for example, $z = i$ is mapped onto $w = 1/i = -i$. Circles inside the unit circle are mapped onto circles outside the unit circle, and conversely.

For arguments you have
$$\arg w = \arg\left(\frac{1}{z}\right) = -\arg z,$$

as follows from (11) in Sec. 12.2 with $z_1 = 1$ and $z_2 = z$. Hence Arg $z = 0$ (the positive ray of the real axis in the z-plane) maps onto Arg $w = 0$. Furthermore, Arg $z = \pi/4$ (the bisecting line of the first quadrant) is mapped onto Arg $w = -\pi/4$ (the bisecting line of the fourth quadrant in the w-plane). And so on.

9. **Mapping of a sector.** The given region in the z-plane is the sector bounded by the positive ray of the y-axis and the bisecting line $y = -x$ of the second quadrant. Since $w = z^2$ doubles angles at the origin, the image of the sector is the sector bounded by the negative ray of the u-axis and the negative ray of the v-axis, where $w = u + iv$, as usual. Thus this image is the third quadrant in the w-plane. More formally,
$$\text{Arg } w = \text{Arg } z^2 = 2\,\text{Arg } z = 2\frac{\pi}{2} \quad \text{and} \quad 2\frac{3\pi}{4},$$

respectively.

17. **Parametric representations of curves** are of great importance in our further work, not only in connection with mappings, but also in integration methods in the complex plane, as we shall see in the next chapter. A circle of radius r with center at $z = 0$ is given by
$$x^2 + y^2 = r^2. \tag{A}$$
Parametrically represented, you have
$$x = r\cos t, \quad y = r\sin t. \tag{B}$$
Indeed, substituting (B) into (A), you see from $\cos^2 t + \sin^2 t = 1$ that (A) is satisfied. In complex form this can be written
$$z = x + iy = r\cos t + ir\sin t = r(\cos t + i\sin t).$$
For a circle of radius r with center at (a, b), thus at $z = a + ib$ in complex notation, you can write instead of (B)
$$x - a = r\cos t, \quad y - b = r\sin t. \tag{C}$$
Then you obtain instead of (A) more generally the familiar nonparametric representation
$$(x - a)^2 + (y - b)^2 = r^2.$$
From (C) you now have
$$x = a + r\cos t, \quad y = b + r\sin t.$$
or in complex form
$$z = x + iy = a + r\cos t + i(b + r\sin t).$$
In the problem, $a = 3$, $b = -1$, $r^2 = 4$, so that $r = 2$ and the answer is

$$z = 3 + 2\cos t + i(-1 + 2\sin t).$$

Sec. 12.6   Exponential Function

Problem Set 12.6. Page 682

1. **Function values.** From (1) you obtain
$$e^{2+3\pi i} = e^2(\cos 3\pi + i\sin 3\pi) = e^2(-1 + i\cdot 0) = -e^2 = -7.389.$$
   From (10) you obtain the absolute value
$$|e^z| = e^x = e^2 = 7.389.$$

7. **Polar form (6).** $z = 4 + 3i$ has the absolute value
$$|z| = \sqrt{4^2 + 3^2} = 5$$
   and the argument
$$\text{Arg } z = \arctan(3/4).$$
   Hence (6) gives the polar form
$$z = 5\,e^{i\arctan(3/4)}$$
$$= 5\,e^{0.643501\,i}$$
   You can check this by calculating
$$z = 5(\cos 0.643501 + i\sin 0.643501)$$
$$= 5(0.8 + 0.6i)$$
$$= 4 + 3i.$$

13. **Equation.** Since $e^x > 0$, the given equation $e^z = -3$ has no real solution. Taking absolute values and using (10) gives
$$|e^z| = e^x = 3, \quad \text{hence} \quad x = \ln 3.$$
   From this and (1) you obtain
$$e^z = 3(\cos y + i\sin y) = -3,$$
   hence $\cos y = -1$ and $\sin y = 0$. A solution is $y = \pi$. Further solutions are $\pi \pm 2n\pi$, where n is a positive integer. Together,
$$z = x + iy = \ln 3 + i(\pi \pm 2n\pi).$$

17. **Conformal mapping.** The given domain in the z-plane is a horizontal strip of width $\pi/2$ bounded by the x-axis and the horizontal straight line $y = \pi/2$. From (10) you see that y is the argument of $w = e^z$. Hence that domain is mapped onto the open first quadrant Q, because Q is precisely the domain consisting of all complex w whose principal argument lies between 0 and $\pi/2$.

Sec. 12.7   Trigonometric Functions. Hyperbolic Functions

Problem Set 12.7. Page 686

1. **Real and imaginary parts of cosh z.** Use the definition (11), multiply it by 2 (in order not to carry 1/2 along), and set $z = x + iy$ as usual. Because of the definition of the exponential function in Sec. 12.6 this gives

$$2 \cosh z = e^z + e^{-z}$$

$$= e^x (\cos y + i \sin y) + e^{-x} (\cos y - i \sin y).$$

Next collect cosine and sine terms, obtaining

$$2 \cosh z = (e^x + e^{-x}) \cos y + i (e^x - e^{-x}) \sin y.$$

The expressions in the parentheses are the real hyperbolic functions $2 \cosh x$ and $2 \sinh x$, respectively. Division by 2 now gives the expected result

$$\cosh z = \cosh x \cos y + i \sinh x \sin y.$$

The other formula follows by a similar straightforward calculation.

7. **Function values.** Your CAS may be able to give function values of the complex trigonometric or hyperbolic functions directly. On a calculator you may use the formulas (6) and those in Prob. 1 to calculate those values from values of the real exponential function, cosine, and sine. In the problem,

$$\cosh (-3 - 6i) = \cosh (-3) \cos (-6) + i \sinh (-3) \sin (-6)$$

$$= \cosh 3 \cos 6 + i \sinh 3 \sin 6$$

because in the last term you have a product of two minus signs. Expressing cosh and sinh in terms of exponential functions and evaluating them, you obtain from the second line

$$\cosh (-3 - 6i) = \frac{1}{2} (20.0855 + 0.0498) \cdot 0.960170 + \frac{1}{2} i (20.0855 - 0.0498) \cdot (-0.279415)$$

$$= 9.66667 - 2.79915 i.$$

11. **Equation.** $\cosh z = 1/2$ has no real solution because $\cosh x \geq 1$ for any real x. Use the formula in Prob. 1. For the real parts you have

$$\cosh x \cos y = 1/2 \qquad\qquad\qquad\text{(A)}$$

and for the imaginary parts

$$\sinh x \sin y = 0. \qquad\qquad\qquad\text{(B)}$$

From (B) you have $x = 0$ or $y = n\pi$, where n is any integer, positive, zero, or negative.

For $x = 0$ you get from (A) the equation $\cos y = 1/2$. Hence

$$y = \frac{\pi}{3} + 2n\pi \quad \text{or} \quad y = \frac{5\pi}{3} + 2n\pi,$$

where n is any integer. This agrees with the answer on p. A32 in Appendix 2 of the book (which is merely slightly differently written; note that $5\pi/3 - 2\pi = -\pi/3$; this explains it).

For $y = n\pi$ you have in (A)

$$\cosh x \cos n\pi = (-1)^n \cosh x.$$

This is either at least equal to $+1$ (if n is even), or at most equal to -1 (if n is odd). Hence in none of these two cases it can be equal to $1/2$, so that you get no further solutions.

17. **Mapping** $w = \cos z$.

$$w = u + iv = \cos z = \cos x \cosh y - i \sin x \sinh y. \qquad\qquad\text{(A)}$$

For the x-axis ($y = 0$) this becomes $\cos x$ and varies from 1 to -1 as x varies from 0 to π (the lower edge of the rectangle in the z-plane to be mapped). For the y-axis ($x = 0$) the equation (A) gives $\cosh y$, which varies from 1 to $\cosh 1$ as y varies from 0 to 1.

For the vertical line $x = \pi$ the equation (A) gives $-\cosh y$; this varies from -1 to $-\cosh 1$ as y varies from y to 1 along the right boundary of the given rectangle. Hence three edges of the rectangle are mapped into the real axis (the u-axis) of the w-plane.

Finally, for the upper edge $y = 1$ you have in (A)

$$u = \cos x \cosh 1, \qquad v = -\sin x \sinh 1. \qquad\qquad\text{(B)}$$

Using $\cos^2 x + \sin^2 x = 1$, you obtain from this

$$\frac{u^2}{\cosh^2 1} + \frac{v^2}{\sinh^2 1} = 1.$$

This represents an ellispse with semiaxes $\cosh 1 = 1.54308$ and $\sinh 1 = 1.17520$. Since part of the u-axis is part of the boundary of the image, the image must be the upper or lower half of the interior of this ellipse. To find out, calculate the image of the midpoint of the upper edge of the rectangle, which has the coordinates $x = \pi/2$, $y = 1$. As the image of this point you obtain from (A)

$$\cos(\pi/2 + i) = 0 - i \sin \pi/2 \sinh 1 = -i \sinh 1.$$

Since this is a point in the lower half-plane, the image of the rectangle must lie in the lower half-plane, not in the upper. More simply: $\sin x$ in (B) is positive (except at $x = 0$ and $x = \pi$), hence v in (B) is negative.

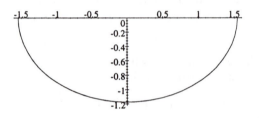

Section 12.7. Problem 17. Image of the given rectangle under $w = \cos z$

Sec. 12.8   Logarithm. General Power

Problem Set 12.8. Page 691

3. **Analyticity.** Use $\text{Ln}\, z = \ln |z| + i\,\text{Arg}\,(z) = \ln r + i\theta$. Then show that (7) in Sec. 12.4 is satisfied everywhere except at $z = 0$ and on the negative ray of the x-axis.

5. **Principal value.** Note that the real logarithm of a negative number is undefined. The principal value $\text{Ln}\, z$ of $\ln z$ is defined by (2), where $\text{Arg}\, z$ is the principal value of $\arg z$. Now recall from Sec. 12.2 that the principal value of the argument is defined by

$$-\pi < \text{Arg}\,\theta \leq \pi.$$

In particular, for a negative real number you always have $\text{Arg}\,\theta = +\pi$, as you should keep in mind. From this and (2) you obtain the answer

$$\text{Ln}\,(-5) = \ln 5 + i\pi.$$

13. **All values of a complex logarithm.** You need the absolute value and the argument of $-e^{-i}$ because by (1) and (2),

$$\ln(-e^{-i}) = \ln |-e^{-i}| + i \arg(-e^{-i}) = \ln |-e^{-i}| + i\,\text{Arg}\,(-e^{-i}) \pm 2n\pi i.$$

Now the absolute value of the exponential function e^z with a pure imaginary exponent always equals 1, as you should memorize.; the derivation is

$$|e^{iy}| = |\cos y + i \sin y| = \sqrt{\cos^2 y + \sin^2 y} = 1.$$

(Can you see where this calculation would break down if y were not real?) In our case,

$$|-e^{-i}| = 1, \quad \text{hence} \quad \ln |-e^{-i}| = 0. \tag{A}$$

The argument of $-e^{-i}$ is obtained from (10) in Sec. 12.6, that is,

$$\arg(e^z) = \text{Arg}(e^z) \pm 2n\pi = y \pm 2n\pi.$$

In Prob. 13 you have $z = -i$, hence $y = -1$, and, therefore,

$$\arg(e^{-i}) = -1 \pm 2n\pi. \tag{B}$$

Finally, by (9) in Sec. 12.2, the argument of a product is the sum of the arguments of the factors, up to integer multiples of 2π. Hence multiplying e^{-i} by -1 corresponds to adding π in the argument. From (B) you thus obtain

$$\arg(-e^{-i}) = \pi - 1 \pm 2n\pi. \tag{C}$$

From (A) and (C) you obtain the answer

$$\ln(-e^{-i}) = (\pi - 1)i \pm 2n\pi i.$$

21. General power. You first have

$$3^{4-i} = 3^4 \, 3^{-i} = 81 \cdot 3^{-i}. \tag{A}$$

To the last factor apply (8) with $a = 3$ and $z = -i$. This gives

$$3^{-i} = e^{-i \ln 3}.$$

On the right now use the definition of the exponential function (Sec. 12.6). This gives

$$3^{-i} = \cos(\ln 3) - i \sin(\ln 3).$$

Substituting this into (A) gives the answer

$$3^{4-i} = 81 \left(\cos(\ln 3) - i \sin(\ln 3) \right)$$
$$= 36.841 - 72.137i.$$

Sec. 12.9   Linear Fractional Transformations.   *Optional*

Problem Set 12.9. Page 698

1. Inverse. Write (1) as $(cz + d)w = az + b$ and and take the z-terms to the left and the other terms to the right,

$$z(cw - a) = b - dw. \tag{A}$$

Now divide. Note that the result is determined only up to a common factor in the numerator and the denominator. For instance, to obtain (4) from (A), multiply (A) by -1 on both sides.

3. Occurrence of infinity. For the given data the left side of (6) takes the form

$$\frac{w - (-1)}{w - 1} \cdot \frac{-i - 1}{-i - (-1)}. \tag{B}$$

The rule of complex division in Sec. 12.1 shows that the second quotient has the value $-i$. Hence (B) reduces to

$$\frac{-i(w + 1)}{w - 1}. \tag{C}$$

Infinity occurs on the right side of (6), which for the gives data becomes

$$\frac{z - 0}{z - \infty} \cdot \frac{1 - \infty}{1 - 0} \tag{D}$$

By Theorem 2 you have to replace the quotient of $1 - \infty$ divided by $z - \infty$ by the value 1. Hence the whole expression (D) reduces to z. From this and (C) you have by multiplying by $w - 1$

$$-i(w + 1) = z(w - 1).$$

Reshuffling terms gives

$$w(-i - z) = -z + i$$

Division by $-i - z$ and multiplying both the numerator and the denominator of the result by -1 you obtain

$$w = \frac{-z + i}{-i - z} = \frac{z - i}{z + i},$$

in agreement with Example 2.

13. **Determination of a linear fractional transformation.** Problems 7-14 can be solved by a straightforward use of (6), a formula not to be remembered, but to be looked up when needed. For the data in Prob. 13 the left side of (6) is

$$\frac{w - 0}{w - (-1)} \cdot \frac{\infty - (-1)}{\infty - 0}.$$

Replacing the quotient containing the infinities by 1, there remains

$$\frac{w}{w + 1}.$$

The right side of (6) is

$$\frac{z - i}{z - 0} \cdot \frac{-i - 0}{-i - i} = \frac{z - i}{2z}.$$

Equating the two results and multiplying through by the two denominators, you obtain

$$w \cdot 2z = (w + 1)(z - i).$$

Collecting the w-terms on the left and the others on the right, you have

$$w(2z - (z - i)) = z - i.$$

Simplification and division finally gives

$$w = \frac{z - i}{z + i}.$$

You may also get the result by using the given data one after another, as follows, starting from (1). $z = i$ maps onto $w = 0$. Hence in the numerator of (2) you have $ai + b = 0$, $b = -ia$. $z = -i$ maps onto $w = \infty$. Hence in the denominator of (2) you must have $c(-i) + d = 0$. Hence, so far you have

$$w = \frac{az - ia}{cz + ic}. \tag{E}$$

$z = 0$ maps onto $w = -1$. This gives $-1 = -ia/(ic)$, hence $c = a$, so that the quotient in (E) becomes $(az - ia)/(az + ia)$. Now divide by a.

17. **Matrices.** It is clear that there must be a condition on the coefficients a, b, c, d in (1) and (4), as stated in the problem, because these coefficients are determined only up to a constant factor.

Sec. 12.10   Riemann Surfaces.   *Optional*

Problem Set 12.10. Page 700

1. **Square root.** If z moves on the unit circle, it has absolute value $|z| = 1$. The image under $w = \sqrt{z}$ also has absolute value $|w| = 1$, as you can see from

$$z = re^{i\theta}, \qquad w = \sqrt{r}\, e^{i\theta/2}. \tag{A}$$

Hence the image point also moves around the unit circle (in the w-plane). Since (A) shows that

$$\text{Arg}\, w = \frac{1}{2} \arg z,$$

it follows that as z begins its motion and moves once around the unit circle, w will move from $w = 1$ to

$w = -1$ in the upper half-plane. When z continues and moves once more around the unit circle, w will move from -1 to 1 in the lower half-plane.

5. **Logarithm.** Use Fig. 318 in Sec. 12.8 as a guide to visualizing the answer on p. A32 in Appendix 2. Note that on the unit circle $z = 1$ you have $\ln |z| = \ln 1 = 0$, which gives the indicated motion of w on the imaginary axis (the v-axis).

9. **Branch points and sheets.** The radicand $2z + i$ is zero at $z = -i/2$. This is the location of a branch point, the only one. The Riemann surface of a cube root has 3 sheets and looks as shown in Fig 323b. In the figure the branch point is at 0, whereas in the present problem it is at $-i/2$. The corresponding function value is $w = 3$, whereas in that figure it is $w = 0$.

CHAPTER 13. Complex Integration

Sec. 13.1 Line Integral in the Complex Plane

Problem Set 13.1. Page 711

1. Straight-line segment. The endpoints of the segment are $z_0 = 0$ and $z_1 = 4 - 7i$. Sketch it. If you set

$$z(t) = x(t) + iy(t) = (4 - 7i)t \qquad \text{(A)}$$

you see that for variable real t this represents a straight line because

$$\frac{y(t)}{x(t)} = -\frac{7t}{4t} = -\frac{7}{4} = const;$$

this is the slope of the straight line. Furthermore, from (A) with $t = 0$ and $t = 1$ you obtain

$$z(0) = 0 = z_0, \quad z(1) = 4 - 7i = z_1$$

respectively. This shows that this line passes through the given points. For reasons of continuity, the points on the segment between z_0 and z_1 must correspond to values of t between 0 and 1. Hence the answer is

$$z(t) = (4 - 7i)t, \quad \text{where} \ \ 0 \leq t \leq 1.$$

7. Circle. The given equation

$$|z - (-3 + i)| = 5$$

represents the circle of radius 5 with center at $-3 + i$. Since $z = x + iy$, you can write this as

$$|x + iy + 3 - i| = |x + 3 + i(y - 1)| = \sqrt{(x + 3)^2 + (y - 1)^2} = 5$$

or, squaring this equation, as

$$(x + 3)^2 + (y - 1)^2 = 25.$$

This is the usual nonparametric equation for a circle known from calculus.

Now for a circle of radius 5 with center at 0 you can write

$$z(t) = x(t) + iy(t) = 5e^{it} = 5(\cos t + i \sin t). \qquad \text{(B)}$$

Note that as t increases the circle is traced in the counterclockwise sense. All you now have to do is moving the center from 0 to $-3 + i$. This can be done simply by adding $-3 + i$ in (B). Using again the notations $z(t)$, $x(t)$, and $y(t)$, for simplicity, you have

$$z(t) = x(t) + iy(t) = -3 + i + 5e^{it}$$
$$= -3 + 5\cos t + i(1 + 5\sin t),$$

where $0 \leq t \leq 2\pi$. This is the desired representation of the given circle.

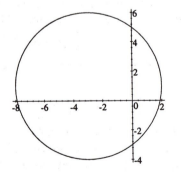

Section 13.1. Problem 7. Given circle to be represented parametrically

15. Integration by the use of the path. The integrand

$$w = u + iv = f(z) = \text{Re } z = x$$

is not analytic. Indeed, the first Cauchy-Riemann equation $u_x = v_y$ (Sec 12.4) is not satisfied, $u_x = 1$ but $v = 0$, hence $v_y = 0$. (The second Cauchy-Riemann equation is satisfied, but, of course, this is not enough for analyticity.) Hence you cannot apply the first method (which would be more convenient), but must use the second.

The shortest path from $z_0 = 1 + i$ to $z_1 = 3 + 2i$ is the straight-line segment with these points as endpoints. Sketch the path. The difference of these points is

$$z_1 - z_0 = 3 + 2i - (1 + i) = 2 + i. \tag{I}$$

Set

$$z(t) = z_0 + (z_1 - z_0)t. \tag{II}$$

Then by taking the values $t = 0$ and $t = 1$ you have

$$z(0) = z_0 \quad \text{and} \quad z(1) = z_1$$

(because z_0 cancels when $t = 1$). Hence (II) is a general representation of a segment with given endpoints z_0 and z_1, and t ranges from 0 to 1.

In Prob. 15, Eq. (II) is

$$z(t) = x(t) + iy(t) = 1 + i + (2 + i)t \tag{III}$$
$$= 1 + 2t + i(1 + t).$$

Integrate by using (10) on p. 708. In (10) you need

$$f(z(t)) = x(t) = 1 + 2t,$$

as well as the derivative of $z(t)$ with respect to t, that is,

$$\dot{z}(t) = 2 + i.$$

Both of these expressions are obtained from (III).

You are now ready to integrate. From (10) you obtain

$$\int_C f(z)\, dz = \int_0^1 (1 + 2t)(2 + i)\, dt$$
$$= (2 + i) \int_0^1 (1 + 2t)\, dt$$
$$= (2 + i)(t + t^2)\big|_{t=0}^1$$
$$= 4 + 2i.$$

17. Indefinite integration and substitution of limits. The integrand $f(z) = \sin^2 z$ is analytic. Hence you can apply the first method in the text. This integration is as in calculus. If you have difficulties with this method, review calculus, in particular, integration. The standard trick in this integral is to use the formula

$$\sin^2 z = \frac{1}{2}(1 - \cos 2z).$$

Incidentally, this is easy to remember if you recall what the $\sin^2$-curve looks like. The indefinite integral of the function on the right is

$$\frac{1}{2}\left(z - \frac{1}{2}\sin 2z\right) = \frac{1}{2}z - \frac{1}{4}\sin 2z \tag{IV}$$

because $(\sin 2z)' = 2\cos 2z$, where we used the chain rule.

Now comes the second step, the evaluation at the given limits of integration. The given path is of no interest to you; just its endpoints are essential. You have to take the value of (IV) at the terminal point $z = \pi i$ minus the value of (IV) at the initial point $z = -\pi i$. This gives

$$\tfrac{1}{2}\pi i - \tfrac{1}{4}\sin(2\pi i) - \left(\tfrac{1}{2}(-\pi i) - \tfrac{1}{4}\sin(-2\pi i)\right) \qquad \text{(V)}$$
$$= \tfrac{1}{2}\pi i + \tfrac{1}{2}\pi i - \tfrac{1}{4}\sin(2\pi i) - \tfrac{1}{4}\sin(2\pi i)$$
$$= \pi i - \tfrac{1}{2}\sin(2\pi i).$$

You can evaluate the sine either by expressing it in terms of exponential functions (see (1) in Sec. 12.7) or by using (6b) in Sec. 12,7, that is,

$$\sin z = \sin x \cosh y + i \cos x \sinh y,$$

which for $x = 0$ and $y = 2\pi$ gives simply $i \sinh 2\pi$. Your answer thus obtained is $\pi i - \tfrac{1}{2}i\sinh 2\pi$ and agrees with that on p. A33 in Appendix 2 of the book.

Sec. 13.2   Cauchy's Integral Theorem

Problem Set 13.2. Page 720

7. **Cauchy's integral theorem** applies to the given function, which is an entire function (can you still remember from Sec. 12.6 what this means?). Hence the integral of e^{-z^2} around the unit circle or, as a matter of fact, around any closed path of integration is zero.

11. **Cauchy's integral theorem not applicable. Deformation of path.** You see that $2z - 1 = 0$ at $z = 1/2$. Hence at this point the function $f(z) = 1/(2z - 1)$ is not analytic. Since $z = 1/2$ lies inside the contour of integration (the unit circle), Cauchy's theorem is not applicable. Hence you have to integrate by the use of path. However, you can choose a most convenient path by applying the principle of deformation of path. You can move the unit circle to the right by 1/2; that is, you can choose the path C given by

$$z(t) = \tfrac{1}{2} + e^{it} \quad (0 \le t \le 2\pi).$$

Note that C is traversed counterclockwise as t increases from 0 to 2π. This is required in the problem. Then

$$f(z(t)) = \frac{1}{2z(t) - 1} = \frac{1}{2e^{it}}$$

and

$$\dot{z}(t) = i e^{it}.$$

With these functions, (10) in Sec. 13.1 gives the desired integral

$$\int_C f(z)\,dz = \int_0^{2\pi} \frac{ie^{it}}{2e^{it}}\,dt.$$

e^{it} cancels, and integration of the remaining $i/2$ from 0 to 2π gives

$$(i/2)2\pi = \pi i.$$

Note that the answer also follows directly from (6) with $m = -1$ and $z_0 = \tfrac{1}{2}$ without any further calculation.

21. **Use of partial fractions.** $z^2 - 1 = (z + 1)(z - 1)$ shows that the given function is not analytic at -1 and $+1$. You can write the integrand $f(z) = 1/(z^2 - 1)$ in terms of two partial fractions, namely,

$$\frac{1}{z^2 - 1} = \frac{1}{2}\left(\frac{1}{z - 1} - \frac{1}{z + 1}\right).$$

The integration of the first fraction (together with the factor 1/2) over the right loop gives πi by (6) with $m = -1$ and $z_0 = 1$; and over the left loop it gives 0 by Cauchy's integral theorem because the integrand is analytic inside and on that loop, in particular, at $z = -1$. Note that over the right loop you integrate counterclockwise.

For evaluating the integral of the second fraction (together with the factor 1/2) the idea is the same. Over the left loop you now obtain $-\pi i$ because you integrate clockwise; together with the minus sign in front of the fraction you get $+\pi i$. Over the right loop you now get 0 because the fraction is analytic everywhere on and inside the loop, including at $z = 1$. Adding your two results, you obtain the answer $2\pi i$.

Sec. 13.3   Cauchy's Integral Formula

Problem Set 13.3. Page 724

1. **Cauchy's integral formula (1).** The given function to be integrated is

$$g(z) = \frac{z^2}{z^4 - 1}.$$

Your first task is to find out where $g(z)$ is not analytic. These are the points where $z^4 = 1$. The solutions of this equation are the four values of the fourth root of 1, namely $1, i, -1, -i$, shown in Fig. 298 in Sec. 12.2.

Your next task is to find out which of those four values lie inside the contour and to make sure that none of them lies on the contour (a case we would not yet be able to handle). The contour is the circle (sketch it!)

$$|z + 1| = 1.$$

Its center is $z_0 = -1$. Hence for this contour you have to set in (1)

$$g(z) = \frac{z^2}{z^4 - 1} = \frac{f(z)}{z - z_0} = \frac{f(z)}{z + 1} \qquad (A)$$

Since

$$z^4 - 1 = (z^2 + 1)(z^2 - 1)$$
$$= (z^2 + 1)(z + 1)(z - 1)$$

you see that (A) gives

$$f(z) = \frac{z^2}{(z^2 + 1)(z - 1)}. \qquad (B)$$

Alternatively, you can obtain the denominator of $f(z)$ by the division

$$(z^4 - 1)/(z + 1) = z^3 - z^2 + z - 1.$$

(Verify!) From (1) and (B) you thus obtain the answer

$$2\pi i f(z_0) = 2\pi i f(-1) = 2\pi i (-1)^2/(-4) = -\pi i/2.$$

7. **Another application of (1).** The given function $g(z) = (\cosh 3z)/(2z)$ is not analytic at $z = z_0 = 0$. Hence in (1) you have $g(z) = f(z)/z$

$$f(z) = zg(z) = \frac{z \cosh 3z}{2z} = \frac{\cosh 3z}{2}.$$

This calculation was simpler than that in Prob. 1. Do you see why? Since $\cosh 0 = 1$, the answer is

$$2\pi i f(z_0) = 2\pi i/2 = \pi i.$$

19. **Partial fractions** were used in Prob. 21 of the previous section because the integrand had two points at which it was not analytic. In the present problem the situation is similar as long as $z_1 \neq z_2$

$$\frac{1}{(z - z_1)(z - z_2)} = K\left(\frac{1}{z - z_1} - \frac{1}{z - z_2}\right),$$

where

$$K = 1/(z_1 - z_2).$$

Now apply (6) in Sec. 13.2 with $m = -1$ to each of the two fractions separately. For the first fraction [with z_1 instead of z_0 in (6)] this gives $K2\pi i$. For the second fraction [with z_2 instead of z_0 in (6)] this gives $-K2\pi i$. Hence the answer is 0, as claimed.

If $z_1 = z_2$, the integrand is $1/(z - z_2)^2$. In this case the result also follows from (6) in Sec. 13.2, but this time you have to use (6) with $m = -2$, one of the infinitely many cases in which (6) gives 0.

Sec. 13.4   Derivatives of Analytic Functions

Example 1. Evaluate $\sin(\pi i)$ by (1) or (6b) in Sec. 12.7.

Example 2. The factor $\pi i = 2\pi i/2$ results from (1) since $n = 2$ (second derivative).

Problem Set 13.4. Page 729

1. **Use of a third derivative.** The given function $(\sinh 2z)/z^4$ is of the form of the integrand in (1) with $z_0 = 0, n + 1 = 4$, hence $n = 3$, and $f(z) = \sinh 2z$. For these data you conclude from (1) that the integral around the unit circle in the counterclockwise sense equals $2\pi i/3!$ times the third derivative of $f(z) = \sinh 2z$, which is $8 \cosh 2z$, taken at $z = 0$; this gives 8; here the factor 8 comes from the chain rule of differentiation. Together this gives the answer $2\pi i8/6 = 8\pi i/3$.

5. **Use of a second derivative.** $(\tan z)/(z - \frac{\pi}{4})^3$ requires the evaluation of the second derivative of $\tan z$ at $z_0 = \pi/4$, as can be seen from (1). You obtain

$$(\tan z)'' = (1/\cos^2 z)' = (-2/\cos^3 z)(-\sin z). \tag{A}$$

Since $\cos(\pi/4) = \sin(\pi/4) = 1/\sqrt{2}$, the value of (A) at $\pi/4$ is 4. Hence, by (1) the integral equals $2\pi i \cdot 4/2! = 4\pi i$.

13. **First derivative. Logarithm.** The given integrand is $\operatorname{Ln}(z)/(z - 2)^2$. The contour of integration is the circle of radius 2 with center at 3. At 0 and at the points on the negative ray of the real axis the function $\operatorname{Ln} z$ is not analytic, and it is essential that these points lie outside that circle. Otherwise, that is, if that ray intersected or touched the contour, we would not be able to integrate.

Furthermore, the integrand is not analytic at $z = z_0 = 2$, which lies inside the contour. Thus, according to (1) with $n + 1 = 2$, hence $n = 1$, and $z_0 = 2$, the integral equals $2\pi i$ times the value of the first derivative of $\operatorname{Ln} z$, which is $1/z$, at $z = z_0 = 2$; this gives a factor 1/2. Hence the answer is πi.

Section 13.4. Problem 13. Behavior of the integrand and path of integration

CHAPTER 14.   Power Series, Taylor Series

Sec. 14.1   Sequences, Series, Convergence Tests

Problem Set 14.1. Page 740

1. **Uniqueness of limit.** A formal proof is given on p. A34 in Appendix 2. A standard idea for many uniqueness proofs is to proceed indirectly, that is, one assumes that there are two objects of the kind considered and shows that they are identical. In the present problem one assumes the existence of two limits and shows that they are identical. The idea of doing this is that one draws two circles, one around each of the two limits and so small that they do not intersect. Then, by the definition of a limit. the first of these circles must contain all the terms of the sequence in its interior, except for at most finitely many of them. But the same must also be true for the second circle, again because of the definition of a limit; that is, it must also contain all the terms in its interior, except for finitely many of them, But this is impossible because the two circles lie outside of each other, their interiors have no points in common.

   How comes that nothing can happen if the two limits are "very close to each other"? Well, they are distinct points and they are kept fixed. Hence they have a positive distance d from each other (which may be extremely small but not zero–otherwise the two points would be identical). And if you choose circles of radius, say, $d/3$ or $d/4$, you obtain disjoint circular disks, as needed.

11. **Boundedness.** Let $\{z_n\}$ be bounded, say, $|z_n| < K$ for some K and all n. Set $z_n = x_n + iy_n$ as in the text. Then boundedness of the sequences $\{x_n\}$ and $\{y_n\}$ can be seen from

    $$|x_n| \le |z_n| < K, \qquad |y_n| \le |z_n| < K.$$

    Here it was used that for $z = x + iy$ you always have

    $$x^2 \le x^2 + y^2 = |z|^2 \quad \text{hence} \quad |x| \le |z|$$

    and similarly for the imaginary part y, namely, $|y| \le |z|$.

       Conversely, let $\{x_n\}$ and $\{y_n\}$ be bounded, say,

    $$|x_n| < K, \qquad |y_n| < K.$$

    Then $x_n^2 < K^2$, $y_n^2 < K^2$, so that

    $$|z_n|^2 = x_n^2 + y_n^2 < 2K^2$$

    By taking square roots this gives

    $$|z_n| < k \qquad (k = K\sqrt{2}).$$

    Hence $\{z_n\}$ is bounded.

       Can you see that this proof is very similar to that of Theorem 1? Just set $c = a + ib = 0$, and write K for ϵ. Then you see that the idea is practically the same in both proofs.

13. **Convergence test.** Apply the ratio test. For this you need

    $$a_n = n^2 i^n / 2^n \quad \text{hence} \quad |a_n| = n^2 / 2^n$$

    where it was used that $|i^n| = 1$ for all n, and

    $$a_{n+1} = (n+1)^2 i^{n+1} / 2^{n+1}$$

    hence

    $$|a_{n+1}| = (n+1)^2 / 2^{n+1}.$$

    The quotient is

    $$\left| \frac{a_{n+1}}{a_n} \right| = \frac{|a_{n+1}|}{|a_n|} = \frac{(n+1)^2 / 2^{n+1}}{n^2 / 2^n} = \frac{1}{2} \left(\frac{n+1}{n} \right)^2.$$

    Obviously, it approaches $1/2$ as n approaches infinity. This shows that the series converges.

The intuitive qualitative reason for this result is the fact that the exponential factor 2^n in the denominator increases eventually much more rapidly than the factor n^2 in the numerator.

15. Failure of the ratio test. Divergence by comparison. The ratio

$$\frac{a_{n+1}}{a_n} = \frac{1/\sqrt{n+1}}{1/\sqrt{n}} = \sqrt{\frac{n}{n+1}}$$

approaches 1 as n approaches infinity. Hence no conclusion can be drawn from the ratio test. However, the answer is obtained by comparing with the harmonic series. You have

$$\sqrt{2} < 2 \quad \text{hence} \quad 1/\sqrt{2} > 1/2 \tag{A}$$

$$\sqrt{3} < 3 \quad \text{hence} \quad 1/\sqrt{3} > 1/3, \text{ etc.}$$

Now the harmonic series diverges. Hence its partial sums must eventually become greater than any (fixed) bound, no matter how large. But because of the infinitely many inequalities (A) each partial sum of the given series (except for the first, which equals 1) must be larger than the corresponding partial sum of the harmonic series; hence these partial sums must also eventually become larger than any given bound. This means that the given series also diverges.

Sec. 14.2   Power Series

Problem Set 14.2. Page 745

1. Radius of convergence. You can immediately see that the center is $-i\sqrt{2}$. The radius of convergence equals 1 because $a_n = n$, $a_{n+1} = n + 1$, and the Cauchy-Hadamard formula (6) gives

$$R = \lim_{n\to\infty} \left| \frac{a_n}{a_{n+1}} \right| = \lim_{n\to\infty} \frac{n}{n+1} = 1.$$

11. Cauchy-Hadamard formula. The center is 0. The radius of convergence can be determined by the Cauchy-Hadamard formula (6). For this you need

$$a_n = \frac{(3n)!}{2^n (n!)^3} \quad \text{and} \quad a_{n+1} = \frac{(3(n+1))!}{2^{n+1} ((n+1)!)^3}.$$

Now

$$(3(n+1))! = (3n+3)! = (3n+3)(3n+2)(3n+1)(3n)!$$

and $(3n)!$ will cancel when you form the quotient a_n/a_{n+1}. Similarly,

$$((n+1)!)^3 = (n+1)^3 (n!)^3$$

and $(n!)^3$ will cancel. Finally, $2^{n+1}/2^n = 2$. Together,

$$\frac{a_n}{a_{n+1}} = \frac{2(n+1)^3}{(3n+3)(3n+2)(3n+1)}.$$

The limit of this quotient as n approaches infinity equals the quotient of the highest power of n, which is n^3 in both the numerator and the denominator; thus,

$$2n^3/(27n^3) = 2/27.$$

This is the radius of convergence. It is relatively small. The reason is that $(3n)!$ in the numerator of the general coefficient grows much faster than $(n!)^3$ in the denominator, about 27 times as fast, the first few values of the quotient being

$$1, \ 6, \ 90, \ 1680, \ 34650, \ 756756, \ 17153136, \ 399072960.$$

17. Extension of Theorem 2. The given series

$$3^2 z^2 + z^3 + 3^4 z^4 + z^5 + 3^6 z^6 \tag{A}$$

consists of the geometric series $z^3 + z^5 + z^7 + ...$, which has radius of convergence 1, and the geometric series $(3z)^2 + (3z)^4 + (3z)^6 + ...$, which converges for $|3z|^2 < 1$, hence $|3z| < 1$, thus $|z| < 1/3$. It follows that the given series has radius of convergence 1/3.

In principle, the series (A) is similar to that in Example 6. It can be written

$$\sum_{n=2}^{\infty} a_n z^n$$

where

$$a_n = \frac{1}{2}(1 + (-1)^n) 3^n + \frac{1}{2}(1 + (-1)^{n+1});$$

indeed, the first summand in a_n equals 3^n if n is even and 0 if n is odd; the second summand in a_n equals 0 if n is even and 1 if n is odd. The sequence of the nth roots $|a_n|^{1/n}$ of $|a_n|$ has the two limit points 3 and 1, and the reciprocal 1/3 of the greatest limit point is the radius of convergence, as in Example 6 in the text.

Sec. 14.3   Functions Given by Power Series

Problem Set 14.3. Page 750

3. Radius of convergence by differentiation (Theorem 3). The geometric series

$$\sum_{n=0}^{\infty} \left(\frac{z}{5}\right)^n$$

converges for $|z/5| < 1$, thus for $|z| < 5$. By Theorem 3, the same holds for the derived series

$$\sum_{n=1}^{\infty} \frac{n z^{n-1}}{5^n} \tag{A}$$

(where you can sum from $n = 1$ because the term for $n = 0$ is 0) and for the derived series of (A)

$$\sum_{n=2}^{\infty} \frac{n(n-1) z^{n-2}}{5^n}.$$

Hence the same is true for

$$z^2 f''(z) = \sum_{n=2}^{\infty} n(n-1) \left(\frac{z}{5}\right)^n.$$

This is the given series and proves that it has the radius of convergence 5.

7. Radius of convergence by integration (Theorem 4). The factors $n + 2$ and $n + 1$ in the denominator of the coefficients suggests determining the radius of convergence by using two successive integrations. Since $(-4)^n/2^n = (-2)^n$, you may start from

$$\sum_{n=0}^{\infty} (-1)^n 2^n z^{2n} = \sum_{n=0}^{\infty} (-1)^n (2z^2)^n.$$

This geometric series converges for $|2z^2| < 1$, hence $|z^2| < 1/2$ or $|z| < 1/\sqrt{2}$. The same is true for this series multiplied by z, that is,

$$\sum_{n=0}^{\infty} (-1)^n 2^n z^{2n+1}. \tag{B}$$

Integration and cancellation of a factor 2 in the numerator and denominator give

$$\sum_{n=0}^{\infty} \frac{(-1)^n 2^n z^{2n+2}}{2n+2} = \sum_{n=0}^{\infty} \frac{(-1)^n 2^{n-1} z^{2n+2}}{n+1}.$$

This series has the same radius of convergence $1/\sqrt{2}$ as (B). The same is true for this series multiplied by z, that is,

$$\sum_{n=0}^{\infty} \frac{(-1)^n 2^{n-1} z^{2n+3}}{n+1}.$$

Another integration and cancellation of 2 give

$$\sum_{n=0}^{\infty} \frac{(-1)^n 2^{n-1} z^{2n+4}}{(2n+4)(n+1)} = \sum_{n=0}^{\infty} \frac{(-1)^n 2^{n-2} z^{2n+4}}{(n+2)(n+1)}.$$

By Theorem 4 this series also has the radius of convergence $1/\sqrt{2}$. Multiplication by $4/z^4$ yields the given series, which thus has the radius of convergence $1/\sqrt{2}$.

15. Cauchy product. The observation that

$$\frac{1}{(1-z)^2} = \frac{1}{1-z} \cdot \frac{1}{1-z}$$

suggests trying the geometric series

$$(1 + z + z^2 + z^3 + \dots)(1 + z + z^2 + z^3 + \dots) = \sum_{n=0}^{\infty} a_n z^n.$$

Now you obtain the power z^n on the left as the sum of the products

$$1 \cdot z^n + z \cdot z^{n-1} + z^2 \cdot z^{n-2} + \dots + z^{n-1} \cdot z + z^n \cdot 1.$$

These are $n + 1$ terms. Hence $a_n = n + 1$, as claimed.

A more natural approach seems differentiation of the geometric series and of its sum, obtaining

$$\frac{1}{(1-z)^2} = \left(\frac{1}{1-z}\right)' = \sum_{n=1}^{\infty} n z^{(n-1)} = \sum_{s=0}^{\infty} (s+1) z^s,$$

where $n = s + 1$, hence $s = n - 1$, so that the summation over s starts with 0.

Sec. 14.4   Taylor Series and Maclaurin Series

Example 2 shows the Maclaurin series of the exponential function. Using it for defining e^z would have forced us to introduce series rather early. I tried this out several times, but found the approach chosen in this book didactically superior.

Problem Set 14.4. Page 757

1. Cosine. Use the familiar series for $\cos s$ and set $s = 2z^2$.

5. Geometric series. The denominator of

$$f(z) = (z+2)/(1-z^2)$$

suggests starting from the geometric series (with z^2 instead of z), that is,

$$1/(1-z^2) = 1 + z^2 + z^4 + z^6 + \dots.$$

Multiplication by $z + 2$ gives the result

$$f(z) = (2+z)(1 + z^2 + z^4 + z^6 + \dots) = 2 + z + 2z^2 + z^3 + 2z^4 + z^5 + 2z^6 + \dots. \qquad \text{(A)}$$

The radius of convergence is 1 because the multiplication by $2 + z$ does not change it. In terms of summation signs the calculation is

$$f(z) = (2 + z) \sum_{n=0}^{\infty} z^{2n} = \sum_{n=0}^{\infty} (2z^{2n} + z^{2n+1}) = \sum_{n=0}^{\infty} \frac{(3 + (-1)^n) z^n}{2}.$$

Indeed, $(-1)^n = +1$ if n is even, so that $(3 + (-1)^n)/2 = 4/2 = 2$; this is the coefficient of every even power of z in (A). And $(-1)^n = -1$ if n is odd, so that then $(3 + (-1)^n)/2 = 2/2 = 1$; this is the coefficient of every odd power in (A).

11. Fresnel integral. Start from the Maclaurin series of $\sin x$. Set $x = t^2$. Perform termwise integration, obtaining

$$\int_0^z \sum_{n=0}^{\infty} \frac{(-1)^n t^{(4n+2)}}{(2n+1)!} \, dt = \sum_{n=0}^{\infty} \frac{(-1)^n t^{(4n+3)}}{(2n+1)!(4n+3)} \Bigg|_{t=0}^{z}.$$

Now set $t = z$; this is the contribution from the upper limit of integration. The lower limit of integration gives 0, so that you obtain the answer by setting $t = z$ in the series on the right.

17. Taylor series. Use the method explained in Example 7, based on the geometric series, as follows.

$$\frac{1}{z} = \frac{1}{[z-2] + 2} = \frac{1}{2[1 + \frac{z-2}{2}]} = \frac{1}{2} \sum_{n=0}^{\infty} (-1)^n \left(\frac{z-2}{2} \right)^n = \sum_{n=0}^{\infty} (-1)^n 2^{-n-1} (z-2)^n. \qquad \text{(B)}$$

This series converges for $|(z-2)/2| < 1$, thus $|z - 2| < 2$. Hence the radius of convergence is $R = 2$.

In the present case the use of the coefficient formula in (1) would also be quite simple and straightforward. Indeed, by differentiation,

$$f(z) = 1/z, \quad f'(z) = -1/z^2, \quad f''(z) = +2/z^3, \quad f'''(z) = -3!/z^3$$

and in general,

$$f^{(n)}(z) = (-1)^n n!/z^{n+1}.$$

This implies for the center $z = 2$

$$f^{(n)}(2) = (-1)^n n!/2^{n+1}.$$

Division by $n!$ gives the coefficient

$$a_n = f^{(n)}(2)/n! = (-1)^n/2^{n+1},$$

in agreement with (B).

Sec. 14.5   Uniform Convergence.   *Optional*

Problem Set 14.5. Page 766

1. Power series. This follows from Theorem 1 because the series has radius of convergence $R = 1$, so that it converges for $|z - i| < 1$.

7. Power series. By Theorem 1, a power series in powers of $z - z_0$ converges uniformly in the closed disk $|z - z_0| \leq r$, where $r < R$ and R is the radius of convergence of the series. Hence solving Probs. 7 – 12 amounts to determining the radius of convergence.

In Prob. 7 you have a power series in powers of

$$Z = (z + i)^2 \qquad \text{(A)}$$

of the form

$$\sum_{n=0}^{\infty} a_n Z^n \qquad \text{(B)}$$

with coefficients $a_n = 1/5^n$. Hence the Cauchy-Hadamard formula in Sec. 14.2 gives the radius of convergence R^* of this series in Z in the form

$$\frac{a_n}{a_{n+1}} = \frac{5^{-n}}{5^{-(n+1)}} = 5.$$

Hence the series (B) converges uniformly in every closed disk $|Z| \leqq r^* < R^* = 5$. Substituting (A) and taking square roots, you see that this means uniform convergence of the given power series in powers of $z + i$ in every closed disk

$$|z + i| \leqq r < R = \sqrt{5}. \tag{C}$$

You can also write this differently by setting

$$\delta = R - r. \tag{D}$$

Then from $R > r$ by subtracting r on both sides you have $\delta = R - r > r - r = 0$, thus $\delta > 0$. Furthermore, from (D) you have $r = R - \delta$. Together,

$$|z + i| \leqq R - \delta = \sqrt{5} - \delta \qquad (\delta > 0).$$

This is the form in which the answer is given in Appendix 2 of the book.

CHAPTER 15. Laurent Series. Residue Integration

Sec. 15.1   Laurent Series

Problem Set 15.1. Page 775

1. **Laurent series near a singularity**. Examples 4 and 5 in the text illustrate that a function may have different Laurent series in different annuli with the same center. However, practically most important of these is the Laurent series that converges directly near the center at which the given function has a singularity. (In Example 4 this is $z = 0$.) In each of Probs. 1-8 that Laurent series is obtained by using a familiar Maclaurin series or (in Probs. 5 and 7) a series in powers of $1/z$. Thus, in Prob. 1 you consider the Maclaurin series

$$\cos z = 1 - \frac{z^2}{2!} + \frac{z^4}{4!} - \frac{z^6}{6!} + - \dots .$$

Division by z^4 gives the Laurent series

$$z^{-4} \cos z = \frac{1}{z^4} - \frac{1}{2z^2} + \frac{1}{24} - \frac{z^2}{720} + - \dots .$$

The principal part consists of the first two terms on the right. The series converges for all $z \neq 0$.

7. **Infinite principal part**. Use the familiar Maclaurin series of the exponential funcion,

$$e^t = \sum_{n=0}^{\infty} \frac{t^n}{n!} = 1 + t + \frac{t^2}{2} + \frac{t^3}{6} + \frac{t^4}{24} + \dots .$$

Substituting $t = -1/z^2$, you obtain

$$\exp(-1/z^2) = \sum \frac{(-1)^n}{z^{2n} n!} = 1 - \frac{1}{z^2} + \frac{1}{2z^4} - \frac{1}{6z^6} + - \dots .$$

Now divide by z^2. You see that the series consists of an infinite principal part and there are no nonnegative powers. The series converges for all $z \neq 0$.

13. **Use of the binomial theorem**. Develop the numerator z^4 of the given function

$$f(z) = z^4/(z + 2i)^4$$

in terms of powers of $z + 2i$ by means of the binomial theorem and then divide by $(z + 2i)^4$. (If you have forgotten that theorem, you will find it on p. 1069 of the book.) Since for $(a + b)^4$ the binomial coefficients needed are 1, 4, 6, 4, 1, you obtain

$$z^4 = ([z + 2i] - 2i)^4$$
$$= (z + 2i)^4 - 4(2i)(z + 2i)^3 + 6(2i)^2(z + 2i)^2 - 4(2i)^3(z + 2i) + (2i)^4$$
$$= (z + 2i)^4 - 8i(z + 2i)^3 - 24(z + 2i)^2 + 32i(z + 2i) + 16.$$

Division by $(z + 2i)^4$ gives the Laurent series

$$\frac{z^4}{(z + 2i)^4} = 1 - \frac{8i}{z + 2i} - \frac{24}{(z + 2i)^2} + \frac{32i}{(z + 2i)^3} + \frac{16}{(z + 2i)^4}.$$

Instead of the binomial theorem you may use the Taylor series (1) in Sec. 14.4, which in the present case reduces to a polynomial because z^4 and its derivatives are $z^4, 4z^3, 12z^2, 24z, 24, 0, 0, \dots$. At $z = 2i$ the values of these expressions are

$$(-2i)^4 = 16, \quad 4(-2i)^3 = 32i, \quad 12(-2i)^2 = -48, \quad 24(-2i) = -48i, \quad 24.$$

Division by $0!, 1!, 2!, \dots$ gives the Taylor coefficients

$$16, \quad 32i, \quad -24, \quad -8i, \quad 1, \quad 0, \quad \dots .$$

Hence the development is the same as before, with the terms being in reverse order,

$$z^4 = 16 + 32i(z+2i) - 24(z+2i)^2 - 8i(z+2i)^3 + (z+2i)^4$$

The amount of work was not much more than before, because it would not have been necessary to write down all the intermediate expressions.

Sec. 15.2 Singularities and Zeros. Infinity

Problem Set 15.2. Page 780

1. Zeros. Since tan z is periodic with period π, it follows that tan πz is periodic with period 1. Since tan $0 = 0$, you see that tan πz has zeros at $0, \pm 1, \pm 2, \ldots$. Determine the order. The derivative is (by the chain rule!)

$$(\tan \pi z)' = \pi / \cos^2 \pi z.$$

Since the cosine is not zero at $z = 0$, the zero of tan πz at 0 is simple. Because of periodicity all those other zeros are simple, too.

Show that tan πz has no further zeros. To have simpler formulas, write $\pi z = s + it$. Then (6b) in Sec. 12.7 becomes

$$\sin \pi z = \sin s \cosh t + i \cos s \sinh t.$$

This is zero if and only if the real part is zero,

$$\sin s \cosh t = 0, \quad \text{hence} \quad \sin s = 0$$

(since cosh $t \neq 0$, note that s and t are *real*) and the imaginary part is zero,

$$\cos s \sinh t = 0, \quad \text{hence} \quad \sinh t = 0$$

because cos $s \neq 0$ where sin $s = 0$ (sin and cos have no zeros in common). Now sin $s = 0$ gives exactly the zeros at $s = \pi x = 0, \pm \pi, \pm 2\pi, \ldots$, that is, at $z = 0, \pm 1, \pm 2, \ldots$; these are the zeros discovered before. Furthermore, sinh $t = 0$ only at $t = 0$ (note again that t is real!); this gives no additional zeros.

19. Pole, essential singularity. Since sinh z is an entire function, the only singularity the given function

$$f(z) = (z - \pi i)^{-2} \sinh z$$

can have in the finite complex plane (see p. 693) is at $z = \pi i$. It seems to be a pole of second order, but you must be cautious because sinh z may perhaps be zero at that point. Now, indeed, by the definition of sin and sinh in Sec. 12.7 you obtain

$$\sinh \pi i = (e^{\pi i} - e^{-\pi i})/2 = i(e^{\pi i} - e^{-\pi i})/(2i) = i \sin \pi = 0.$$

Fortunately, this zero is simple because the derivative is cosh z, and at πi,

$$\cosh \pi i = (e^{\pi i} + e^{-\pi i})/2 = \cos \pi \neq 0.$$

Hence the given function still has a pole, albeit a simple one, due to the occurrence of that zero. Indeed, for sinh z to have a simple zero at $z = \pi i$, the Taylor series of sinh z with center πi must be of the form

$$a_1(z - \pi i) + a_2(z - \pi i)^2 + \ldots,$$

so that

$$\frac{\sinh z}{(z - \pi i)^2} = \frac{a_1}{z - \pi i} + a_2 + a_3(z - \pi i) + \ldots .$$

By definition this is the Laurent series near a simple pole. (If sinh z had a double zero or a zero of still higher order at πi, the given function would be analytic at πi.)

Furthermore, $f(z)$ has an essential singularity at infinity, for reasons given in Example 5 in the text. To see this directly, consider

$$g(w) = f(1/w) = \frac{\sinh (1/w)}{(1/w - \pi i)^2}$$

at $w = 0$. The function

$$\frac{1}{(1/w - \pi i)^2} = \frac{w^2}{(1 - \pi i w)^2}$$

is analytic at $w = 0$. The other factor of $g(w)$, the function sinh $(1/w)$ has near $w = 0$ the Laurent series

$$\sinh \left(\frac{1}{w}\right) = \frac{1}{w} + \frac{1}{3! w^3} + \frac{1}{5! w^5} + ...,$$

as obtained from the familiar Maclaurin series of sinh s by setting $s = 1/w$. Since this series has an infinite series as its principal part, $w = 0$ is an essential singularity of $g(w)$, by definition. Again by definition, this means that $f(z)$ has an essential singularity at infinity.

Sec. 15.3 Residue Integration Method

Problem Set 15.3. Page 786

1. **Simple poles.** $1 + z^2 = 0$ at $z^2 = -1$, $z = i$ and $-i$. Hence the given function $f(z) = 4/(1 + z^2)$ has simple poles at $z = i$ and $-i$. For $z = z_0 = i$, using $(1 + z^2)' = 2z$ and $1/i = -i$, you obtain from (4)

$$\operatorname*{Res}_{z=i} f(z) = 4/2i = -2i. \tag{A}$$

Similarily, for $z = -i$ you obtain

$$\operatorname*{Res}_{z=-i} f(z) = 4/(-2i) = 2i.$$

Formula (3) gives the same answers. In (3) you need for $z = z_0 = i$

$$(z - i) \cdot 4/(z^2 + 1) = 4/(z + i).$$

At $z = i$ this has the value $4/(2i) = -2i$, in agreement with (A). Similarly for the pole at $z = -i$.

3. **Use of the Laurent series.** In Prob. 1 you have used formulas that gave the residue directly, without reference to the whole Laurent series. For the function $f(z) = (\sin 2z)/z^6$ you may use the familiar Maclaurin series of the sine function and find the coefficient a_5 of the power z^5 because $a_5 z^5/z^6 = a_5/z$, which shows that a_5 is the residue of $f(z)$ at $z = 0$, where $f(z)$ has a pole of fifth order (not sixth because $\sin 2z$ has a simple zero at $z = 0$). You obtain

$$\frac{\sin 2z}{z^6} = \frac{1}{z^6}\left(2z - \frac{(2z)^3}{3!} + \frac{(2z)^5}{5!} - +...\right).$$

Hence $a_5 = 2^5/5! = 32/120 = 4/15$.

13. **Residue theorem.** $\tan \pi z = (\sin \pi z)/(\cos \pi z)$ is singular where $\cos \pi z = 0$, that is, at $\pi z = \pm\pi/2, \pm 3\pi/2, ...,$ hence at $z = \pm 1/2, \pm 3/2, ...$ These are simple zeros of $\cos \pi z$, hence simple poles of $\tan \pi z$, as follows from Theorem 4 in Sec. 15.2. Here you have used that $\sin \pi z \neq 0$ at points where $\cos \pi z = 0$. Hence $\tan \pi z$ has infinitely many simple poles. But only those at $z = 1/2$ and $z = -1/2$ lie inside the contour of integration, which is the unit circle $|z| = 1$.

You can apply (4) to

$$\tan \pi z = p(z)/q(z) = (\sin \pi z)/(\cos \pi z).$$

In (4) you need

$$p(z)/q'(z) = (\sin \pi z)/(\cos \pi z)' = (\sin \pi z)/(-\pi \sin \pi z) = -1/\pi$$

where the factor π results from the chain rule. From this and the residue theorem (Theorem 1) you obtain the answer $2\pi i (-1/\pi - 1/\pi) = -4i$.

17. Residue integration.

The given function

$$\frac{z+1}{z^4 - 2z^3} = \frac{z+1}{z^3(z-2)}$$

has a pole of third order at 0 and a simple pole at 2. The latter lies outside the circle of integration $|z| = 1/2$ and is of no interest to us. You need the residue of $f(z)$ at $z = 0$. You can use (5) with $m = 3$. This involves the second derivative of

$$g(z) = z^3 f(z) = \frac{z+1}{z-2} = \frac{z-2+3}{z-2} = 1 + \frac{3}{z-2}.$$

Differentiation gives

$$g'(z) = -3/(z-2)^2, \quad g''(z) = 6/(z-2)^3, \quad g''(0) = -3/4.$$

Hence the residue is $-3/8$ because of the factor $1/(m-1)! = 1/2$ in (5). This gives the answer $2\pi i(-3/8) = -3\pi i/4$.

Confirm this by developing $f(z)$ in a Laurent series near 0, that is,

$$\frac{z+1}{z^3(z-2)} = \frac{z+1}{(-2z^3)(1-\frac{z}{2})} = \frac{z+1}{-2z^3}\left(1 + \frac{z}{2} + \frac{z^2}{4} + \dots\right)$$

$$= -\frac{1}{2z^3}\left(1 + \frac{3}{2}z + \frac{3}{4}z^2 + \dots\right)$$

where the dots denote terms in z^3 and higher powers of z. Performing the indicated multiplication, you see that $1/z$ has the coefficient $(-1/2)(3/4) = -3/8$. This is the residue of $f(z)$ at $z = 0$, in agreement with the result obtained before.

Sec. 15.4   Evaluation of Real Integrals

Problem Set 15.4. Page 793

1. **An integral involving cos θ.** Use (2), that is, $k + \cos\theta = k + (z + 1/z)/2$ and $d\theta = dz/iz$ [see the book after (2)]. Hence the transformed integrand equals

$$\frac{1}{iz\left(k + \frac{1}{2}\left(z + \frac{1}{z}\right)\right)} = \frac{2}{i(z^2 + 2kz + 1)}. \tag{A}$$

The quadratic equation $z^2 + 2kz + 1 = 0$ has the roots

$$z_1 = -k + \sqrt{k^2 - 1} \quad \text{and} \quad z_2 = -k - \sqrt{k^2 - 1}.$$

You can simplify the next formulas a little by abbreviating the root by r, writing

$$z_1 = -k + r, \quad z_2 = -k - r, \quad \text{where} \quad r = \sqrt{k^2 - 1}. \tag{B}$$

Since $k > 1$ by assumption, these roots are real, and $z_2 < -k < -1$. Hence z_2 lies outside the unit circle. The constant term 1 in (B) is the product of the two roots; hence $|z_1| < 1$ because $|z_2| > 1$, so that z_1 lies inside the unit circle. This gives a simple pole of (A) at z_1.

In the given integral the limits of integration are 0 and π, but you must have the limits 0 and 2π. Now since the integrand is an even function of θ, the integral equals 1/2 times the integral from 0 to 2π; this solves that difficulty. (Integration from 0 to π would not give a closed path, as it is needed in residue integration.)

The residue of (A) at the simple pole at z_1 is probably best obtained from (4) in Sec. 15.3. Differentiating the denominator of (A) and setting $z = z_1$, you obtain

$$\frac{2}{i(2z_1 + 2k)} = \frac{1}{i(z_1 + k)}.$$

From (B) you see that $z_1 + k = r$. Hence the residue of (A) at z_1 is $1/(ir)$. This times $2\pi i$ times 1/2 (the factor that occurred from the extension of the limit of integration from π to 2π) gives the value of the integral, namely,

$$2\pi i\,(1/ir)(1/2) = \frac{\pi}{r} = \pi/\sqrt{k^2-1}\,.$$

9. Infinite interval of integration. Use (7), where, for the given integral,

$$f(x) = 1/(1+x^2)^2, \qquad \text{hence} \quad f(z) = 1/(1+z^2)^2.$$

Now $1+z^2 = 0$ has the solutions i and $-i$. In the derivation of (7) it was explained why and how one uses a path that includes all the singularities (poles) in the upper half-plane (see Fig. 360 in the book). In the present problem there is only the pole at $z = i$ in the upper half-plane. It is of second order because of the square outside the parentheses.

The residue of $f(z)$ at $z = i$ can be obtained from (5*) in Sec. 15.3. Since $z^2 + 1 = (z+i)(z-i)$, you obtain in (5*)

$$[(z-i)^2 f(z)]' = [(z-i)^2 (1/((z+i)^2 (z-i)^2))]'$$
$$= [(z+i)^{-2}]' = -2(z+i)^{-3},$$

At $z = i$ this equals $-2/(2i)^3 = -i/4$. This is the residue needed. Hence the integral equals $2\pi i\,(-i/4) = \pi/2$.

21. Singularities on the path of integration. Formula (14) indicates that residues of poles in the upper half-plane are multiplied by $2\pi i$, whereas those on the path of integration (on the x-axis) are multiplied by πi. In Prob. 21 you have $f(x) = 1/(x^2 - ix)$, hence $f(z) = 1/(z^2 - iz)$. Now

$$z^2 - iz = z(z-i) = 0$$

shows that there is a simple pole at i and another simple pole at 0. The residues are obtained from (4) in Sec. 15.3 with the numerator 1 and the denominator $(z^2 - iz)' = 2z - i$ evaluated at i and 0, respectively. This gives at i the residue

$$1/(2i - i) = 1/i = -i \tag{C}$$

and at 0 the residue

$$1/(-i) = i. \tag{D}$$

To obtain the answer, you must multiply (C) by $2\pi i$ and (D) by πi, as indicated in (14), and add. This gives $2\pi i(-i) + i\pi i = 2\pi - \pi = \pi$.

CHAPTER 16.   Complex Analysis Applied to Potential Theory

Sec. 16.1   Electrostatic Fields

Problem Set 16.1. Page 802

3. **Potential between sloping plates.** $y = x$ and $y = x + 1$ are conducting plates, hence they must be equipotential lines (equipotential planes in xyz-space, with the potential not depending on z; this is commonly called a *two-dimensional problem*). Setting $\Phi_0 = y - x$, the first condition is satisfied. You see that then on the second plate $y = x + 1$ the potential is $\Phi_0 = y - x = 1$. Hence you can satisfy the second condition by multiplying Φ_0 by 110; thus,

$$\Phi = 110(y - x). \tag{A}$$

The first condition is still satisfied; hence this gives the answer to the problem as far as the real potential is concerned. The corresponding complex potential $F = \Phi + i\Psi$ can be obtained by the Cauchy-Riemann equations

$$\Phi_x = -110 = \Psi_y, \qquad \text{hence} \qquad \Psi = -110y + h(x),$$

$$\Phi_y = 110 = -\Psi_x = -h'(x), \qquad \text{hence} \qquad h(x) = -110x.$$

Together,

$$F = \Phi + i\Psi = 110(y - x) + i(-110y - 110x) = -110(-y + x + iy + ix).$$

In the parentheses, $x + iy = z$ and $ix - y = iz$. Hence

$$F = -110(z + iz) = -110(1 + i)z.$$

More simply, $-x$ in (A) is the real part of $-z$, and y is the real part of $-iz = -i(x + iy) = -ix + y$. Adding this and multiplying by 110 gives

$$\Phi = 110(\text{Re}(-iz) + \text{Re}(-z)), \qquad \text{hence} \qquad F = -110(iz + z) = F = -110(1 + i)\,z.$$

11. **Two source lines.** The equipotential lines in Example 7 are

$$|(z - c)/(z + c)| = k = const \qquad (k \text{ and } c \text{ real}).$$

Hence $|z - c| = k|z + c|$. Squaring gives

$$|z - c|^2 = K|z + c|^2 \qquad (K = k^2).$$

Writing this in terms of the real and imaginary parts and taking all the terms to the left, you obtain

$$(x - c)^2 + y^2 - K((x + c)^2 + y^2) = 0.$$

Writing out the squares gives

$$x^2 - 2cx + c^2 + y^2 - K(x^2 + 2cx + c^2 + y^2) = 0. \tag{B}$$

For $k = 1$, hence $K = 1$, most terms cancel, and you are left with $-4cx = 0$, hence $x = 0$ (because $c \neq 0$). This is the y-axis. Then

$$|z - c|^2 = |z + c|^2 = y^2 + c^2, \qquad |z - c|/|z + c| = 1, \qquad \text{Ln } 1 = 0.$$

Hence this shows that the y-axis has potential 0. You can now continue with (B), assuming that $K \neq 1$. Collecting terms in (B), you have

$$(1 - K)(x^2 + y^2 + c^2) - 2cx(1 + K) = 0.$$

Division by $1 - K$ ($\neq 0$ because $K \neq 1$) gives

$$x^2 + y^2 + c^2 - 2Lx = 0 \qquad (L = c(1 + K)/(1 - K)).$$

Completing the square in x, you finally obtain

$$(x - L)^2 + y^2 = L^2 - c^2.$$

This is a circle with center at L on the real axis and radius $\sqrt{L^2 - c^2}$. If you insert L and simplify, you see that the radius equals $2kc/(1 - k^2)$.

15. Potential in a sector. $z^2 = x^2 - y^2 + 2ixy$ gives the potential in sectors of opening $\pi/2$ bounded by the bisecting straight lines of the quadrants because $x^2 - y^2 = 0$ when $y = \pm x$. Similarly, higher powers of z give potentials in sectors of smaller openings on whose boundaries the potential is zero. For $z^3 = x^3 + 3ix^2y - 3xy^2 - iy^3$ the real potential is

$$\Phi_0 = \operatorname{Re} z^3 = x^3 - 3xy^2 = x(x^2 - 3y^2)$$

and $\Phi = 0$ when $y = \pm x/\sqrt{3}$; these are the boundaries given in the problem, the opening of the sector being $\pi/3$, that is, 60 degrees. To satisfy the other boundary condition, multiply Φ_0 by 220.

Sec. 16.2   Use of Conformal Mapping

Problem Set 16.2. Page 807

3. Quarter of an elliptical disk. The given potential in the image domain D^* is $\Phi^*(u, v) = u^2 - v^2$. This is the real part of w^2, where $w = u + iv$. The domain D^* is obtained as the image of the rectangle $D : 0 \leqq x \leqq \pi/2, 0 \leqq y \leqq 1$ under the mapping

$$w = u + iv = \sin z = \sin x \cosh y + i \cos x \sinh y. \tag{A}$$

Figure 316 in Sec. 12.7 shows that D^* lies in the first quadrant. Its boundary consists of the segment of the u-axis from 0 to 1 (this is the image of the segment of the x-axis from 0 to $\pi/2$) and from 1 on to $\cosh 1 = 1.5431$ (this is the image of the right vertical edge of D), then along the second largest ellipse in the figure up to the point $\sinh 1 = 1.1752$ on the v-axis (the image of the upper edge of D) and along the v-axis down to the origin from which we started (this is the image of the segment from 1 to 0 on the y-axis) The potential Φ in D is obtained by substituting u and v from (A) into the potential Φ^* in D^*, that is,

$$\Phi(x, y) = \Phi^*(u(x, y), v(x, y)) = u^2(x, y) - v^2(x, y) = \sin^2 x \cosh^2 y - \cos^2 x \sinh^2 y. \tag{B}$$

The boundary values are obtained from (B) as follows. On the lower edge ($y = 0$) you have $\Phi = \sin^2 x$, which increases from 0 to 1. On the right edge ($x = \pi/2$) you have $\Phi = \cosh^2 y$, which increases from 1 to $\cosh^2 1 = 2.3811$. On the upper edge ($y = 1$) you obtain

$$\Phi = \sin^2 x \cosh^2 1 - \cos^2 x \sinh^2 1 = 2.3811 \sin^2 x - 1.3811 \cos^2 x,$$

which decreases from 2.3811 to -1.3811. Finally, on the left vertical edge ($x = 0$), you have from (B) the potential $\Phi = -\sinh^2 y$, which increases from -1.3811 to 0.

9. Angular region. The potential in the w-plane is $\Phi^*(w) = (6/\pi) \operatorname{Arg} w$. The suggested mapping function is $w = z^2$. Hence $\operatorname{Arg} w = 2 \operatorname{Arg} z$. Substitution gives the potential in the angular region in the form

$$\Phi(z) = \Phi^*(w(z)) = (12/\pi) \operatorname{Arg} z.$$

Indeed, for $\operatorname{Arg} z = \pm\pi/4$ (the boundary of the angular region) you obtain ± 3, as required.

13. Linear fractional transformation. Z here plays the role of z in (7), Sec. 12.9, and z plays the role of w. Thus, $z = (Z - i/2)/(-iZ/2 - 1)$. Now multiply both the numerator and the denominator by 2. This gives the answer on p. A36 in Appendix 2.

Sec. 16.3   Heat Problems

Problem Set 16.3. Page 811

3. Mixed problem. A potential in a sector (angular region) whose sides are at constant temperatures is, once and for all, of the form

$$T = a + b \, \text{Arg } z. \tag{A}$$

Here you use the fact that Arg $z = \theta = \text{Im (Ln } z)$ is a harmonic function. The two constants a and b can be determined from the given values on the two sides Arg $z = 0$ and Arg $z = \pi/2$. Namely, for Arg $z = 0$ (the x-axis) we have $T = a = 100$. Then for Arg $z = \pi/2$ you have

$$T = 100 + b \cdot \pi/2 = -40.$$

Solving for b gives $b = -280/\pi$. Hence a potential giving the required values on the two sides is

$$T = \frac{280}{\pi} \text{Arg } z. \tag{B}$$

Now comes an important observation. The curved portion of the boundary (a circular arc) is insulated. Hence on this arc the normal derivative of the temperature T must be zero. But the normal direction is the radial direction; so the partial derivative with respect to r must vanish. Now formula (B) shows that T is independent of r, that is, the condition under discussion is automatically satisfied. (If this were not the case, the whole solution would not be valid.) Finally derive the complex potential F. From Sec. 12.8 recall that

$$\text{Ln } z = \ln |z| + i \, \text{Arg } z. \tag{C}$$

Hence for Arg z to become the real part (as it must be the case because $F = T + i\Psi$), you must multiply (C) by $-i$. Indeed, then

$$-i \, \text{Ln } z = -i \ln |z| + \text{Arg } z.$$

Hence from this and (B) you see that the complex potential is

$$F = 100 - \frac{280}{\pi}(-i \, \text{Ln } z) = 100 + \frac{280}{\pi} i \, \text{Ln } z.$$

13. Another use of Arg z. You can proceed similarly as in Prob. 3, starting from

$$T = a + b \, \text{Arg } z.$$

For Arg $z = 0$ (the positive ray of the x-axis) you must have

$$T = a = T_0.$$

For Arg $z = \pi/2$ (the positive ray of the y-axis) you must have

$$T = T_0 + b(\pi/2) = T_1.$$

Solving for b gives $b = (2/\pi)(T_1 - T_0)$. Hence the answer is

$$T = T_0 + \frac{2}{\pi}(T_1 - T_0) \text{Arg } z.$$

Sec. 16.4   Fluid Flow

Example 2. For another interesting application of $w = z + 1/z$ see Sec. 12.5.

Problem Set 16.4. Page 817

1. Parallel flow. A flow is completely determined by its complex potential

$$F(z) = \Phi(x, y) + i\Psi(x, y).$$

The stream function Ψ gives the streamlines $\Psi = const$ and is generally more important than the velocity potential Φ, which gives the equipotential lines $\Phi = const$. The flow can best be visualized in terms of the velocity vector V, which is obtained from the complex potential in the form

$$V = V_1 + iV_2 = \bar{F}'(z).$$

(Here we need no special vector notation because a complex function V can always be regarded as a vector function with components V_1 and V_2. Hence for the given

$$F(z) = Kz = K(x + iy) \tag{A}$$

with positive real K you have $\bar{F} = F' = K$; thus,

$$V = V_1 = k.$$

It is essential that K is real (this makes $V_2 = 0$, the velocity vector is parallel to the x-axis) and is positive, so that $V = V_1$ points to the right (in the positive x-direction). Hence you are dealing with a uniform flow (a flow of constant velocity) that is parallel (the streamlines are straight lines parallel to the x-axis) and is flowing to the right (because K is positive). From (A) you see that the equipotential lines are vertical parallel straight lines; indeed,

$$\Phi(x,y) = \operatorname{Re} F(z) = kx = const, \qquad \text{hence} \quad x = const.$$

9. Flow around a cylinder. Since a cylinder of radius r_0 is obtained from a cylinder of radius 1 by a dilatation (a uniform stretch in all directions in the complex plane), it is natural to replace z by az with a real constant a because this corresponds to such a stretch. That is, replace the complex potential

$$z + 1/z \tag{A}$$

in Example 2 by

$$F(z) = \Phi(r,\theta) + i\Psi(r,\theta) = az + \frac{1}{az} = are^{i\theta} + \frac{1}{ar}e^{-i\theta}. \tag{B}$$

The stream function Ψ is the imaginary part of F. Since by Euler's formula, $e^{\pm i\theta} = \cos\theta \pm i\sin\theta$, you obtain

$$\Psi(r,\theta) = \left(ar - \frac{1}{ar}\right)\sin\theta.$$

The streamlines are the curves $\theta = const$. As in Example 2 of the text, the streamline $\Psi = 0$ consists of the x-axis ($\theta = 0$ and π), where $\sin\theta = 0$, and of the locus where the other factor of Ψ is zero, that is,

$$ar - \frac{1}{ar} = 0, \qquad \text{thus} \quad (ar)^2 = 1 \quad \text{or} \quad a = 1/r.$$

Since the cylinder has radius $r = r_0$, you must have $a = 1/r_0$. With this, the answer is

$$F(z) = az + \frac{1}{az} = \frac{z}{r_0} + \frac{r_0}{z}.$$

Sec.16.5   Poisson's Integral Formula

Problem Set 16.5. Page 822

7. Sinusoidal boundary values lead to a series (7) that reduces to finitely many terms (a "trigonometric polynomial"). In Prob. 7 the given boundary function $\Phi(1,\theta) = 4\sin^3\theta$ is not immediately one of the terms in (7), but can be expressed in terms of sine functions of multiple angles. The corresponding formula is

$$\sin^3\theta = \frac{3}{4}\sin\theta - \frac{1}{4}\sin 3\theta.$$

Hence the boundary value can be written

$$\Phi(1,\theta) = 3\sin\theta - \sin 3\theta.$$

From (7) you now see immedately that the potential in the unit disk satisfying the given boundary conditionis

$$\Phi(r,\theta) = 3r\sin\theta - r^3\sin 3\theta.$$

11. Piecewise linear boundary values lead to a series (7) whose coefficients can be found from Chap. 11. For instance, the Fourier series of the present boundary values is a special case of the odd periodic extension in

Example 3 of Sec. 10.4 with $k = \pi/2$ and $L = \pi$, that is,

$$f(x) = \frac{4}{\pi}\left(\sin x - \frac{1}{9}\sin 3x + \frac{1}{25}\sin 5x - +...\right).$$

From this with $x = \theta$ you obtain the potential (7) in the disk in the form

$$\Phi(r, \theta) = \frac{4}{\pi}\left(r \sin \theta - \frac{1}{9}r^3 \sin 3\theta + \frac{1}{25}r^5 \sin 5\theta - +...\right). \qquad (A)$$

The figure shows the given boundary potential, an approximation of it (the sum of the first three terms of the series (A)), which is rather good, and an approximation of the potential on the circle of radius $r = 1/2$ (the sum of those three terms for $r = 1/2$)). The latter is practically a sine curve because the terms in (A) with $r = 1/2$ have coefficients $4/\pi$ times $1/2$, $1/72$, $1/800$, etc, which decrease very fast. This illustrates that the partial sums of the series (A) give good approximations of the potential in the disk.

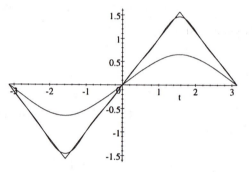

Section 16.5. Problem 11. Boundary potential and approximations for $r = 1$ and $r = 1/2$

13. Terms of (7). Use the Laplacian in polar coordinates, given by (5) in Sec. 11.11 (without the z-term) .

Sec. 16.6   General Properties of Harmonic Functions

Problem Set 16.6. Page 825

1. Mean value of an analytic function. Verify Theorem 1 for

$$F(z) = (z - 1)^2, \qquad z_0 = 1/2, \qquad |z - 1/2| = 1.$$

The last of these formulas arises from the requirement that you integrate around a circle of radius 1 and that the center of the circle be z_0. Since $F(z_0) = F(1/2) = (-1/2)^2 = 1/4$, your task is to verify that the integral in (2) in the proof of Theorem 1 has the value $1/4$. Use (2) with $z_0 = 1/2$ and $r = 1$ (since the circle of integration has radius 1). The path of integration is

$$z = z_0 + e^{i\alpha} = \frac{1}{2} + e^{i\alpha}.$$

Hence on this path the integrand is

$$F(z_0 + e^{i\alpha}) = \left(\frac{1}{2} + e^{i\alpha} - 1\right)^2 = \left(-\frac{1}{2} + e^{i\alpha}\right)^2.$$

Squaring as indicated, you obtain

$$F(z_0 + e^{i\alpha}) = \frac{1}{4} + e^{i\alpha} + e^{2i\alpha}.$$

Indefinite integration over α gives

$$\frac{\alpha}{4} - \frac{1}{i}e^{i\alpha} + \frac{1}{2i}e^{2i\alpha}.$$

(the factor $1/2\pi$ in front of the integral in (2) will not be carried along, but introduced at the end of the calculation). Now insert the limits of integration 0 and 2π. Then the first term gives

$$\frac{1}{4}2\pi = \frac{1}{2}\pi. \tag{A}$$

Since $e^{2\pi i} = 1$, the next term gives $1 - 1 = 0$. Similarly for the last term. Multiplying (A) by the factor $1/2\pi$ (which you did not carry along in (2), in order to have simpler formulas), you obtain $(\pi/2)/(2\pi) = 1/4$. This completes your verification.

7. **Mean values of harmonic functions (Theorem 2).** The two formulas in the proof of Theorem 2 give the mean values to be calculated in order to verify Theorem 2. The given function, point (x_0, y_0), and circle are

$$\Phi(x, y) = (x - 1)(y - 1), \qquad (x_0, y_0) = (3, -3), \qquad z = 3 - 3i + e^{i\alpha}. \tag{B}$$

Φ is harmonic; verify this by differentiation. Note that $z_0 = x_0 + iy_0 = 3 - 3i$ is the center of the circle in (B). In terms of the real and imaginary parts it is (by Euler's formula (5) in Sec. 12.7)

$$x = 3 + \cos\alpha, \qquad y = -3 + \sin\alpha. \tag{C}$$

This is the representation you need, since Φ is a real function of the two real variables x and y. You see that

$$\Phi(z_0, y_0) = \Phi(3, -3) = (x_0 - 1)(y_0 - 1) = (3 - 1)(-3 - 1) = -8.$$

Hence you have to show that each of the two mean values equals -8. Substituting (C) into (B) (which is a completely schematical process) gives

$$\Phi(3 + \cos\alpha, -3 + \sin\alpha) = (3 + \cos\alpha - 1)(-3 + \sin\alpha - 1) \tag{D}$$

$$= (2 + \cos\alpha)(-4 + \sin\alpha)$$

$$= -8 + 2\sin\alpha - 4\cos\alpha + \cos\alpha\sin\alpha.$$

Consider the mean value over the circle. Integrate each of the four terms in the last line of (D) over α from 0 to 2π. The first of them gives -16π. The second term gives 0, and so does the third. The fourth term equals $(1/2)\sin 2\alpha$, and its integral is 0, too. Multiplication by $1/(2\pi)$ (the factor in front of the first integral in the proof of Theorem 2) gives $-16\pi/(2\pi) = -8$. This is the mean value of the given harmonic function over the circle considered and completes the verification of the first part of the theorem for our given data. Now calculate the mean value over the disk of radius 1 and center $(3, -3)$. The integrand of the double integral (second formula in the proof of Theorem 2) is similar to that in (D), but in (D) you had $r = 1$ (the circle over which you integrated), whereas you now have r variable and integrate over it from 0 to 1, and you also have a factor r resulting from the element of area in polar coordinates, which is $r\,dr\,d\theta$. Hence instead of $(2 + \cos\alpha)(-4 + \sin\alpha)$ in (D) you now have

$$(2 + r\cos\alpha)(-4 + r\sin\alpha)r = -8r + 2r^2\sin\alpha - 4r^2\cos\alpha + r^3\cos\alpha\sin\alpha.$$

The factors of r have no influence on the integration over α from 0 to 2π. Accordingly, the present four terms on the right give upon integration over α the values $-8r \cdot 2\pi = -16r\pi$, 0, 0, and 0. Integration of r from 0 to 1 gives $1/2$, so that the double integral equals -8π. In front of the double integral you have the factor $1/(\pi r_0^2) = 1/\pi$, because the circle of integration has radius 1. Hence your second result is $-8\pi/\pi = -8$. This completes the verification.

15. **Location of maxima.** Look for a counterexample, as simple as possible. $x = \operatorname{Re} z$ and $y = \operatorname{Im} z$ are harmonic in any region, say, to have a simple situation, in the square $0 \le x \le 1$, $0 \le y \le 1$. Then you have max $x = 1$ at all points on the right boundary and max $y = 1$ at all points of the upper boundary. Hence there is a point, $(1, 1)$, that is, $z = 1 + i$, where both functions have a maximum. But this should give you the idea: omit the point $(1, 1)$ from the region. Or take a rectangle, a triangle, a square with vertices ± 1, $\pm i$, and so on.

PART E.   NUMERICAL METHODS

CHAPTER 17.   Numerical Methods in General

Sec. 17.1   Introduction

Problem Set 17.1. Page 836

7. **Quadratic equation.** Given $x^2 - 30x + 1 = 0$. First use (6), where $a = 1$, $b = -30$, and $c = 1$. Calculating with 4S, you obtain $\sqrt{(-30)^2 - 4} = \sqrt{896} = 29.93$. Hence

$$x_1 = (30 + 29.93)/2 = 59.93/2 = 29.96$$

and

$$x_2 = (30 - 29.93)/2 = 0.07/2 = 0.04.$$

Now use (7). The root x_1 equals 29.96, as before. For x_2 you now obtain

$$x_2 = \frac{c}{ax_1} = 1/29.96 = 0.03338.$$

With 2S the calculations are as follows. You have to calculate the square root of

$$90 \cdot 10^1 - 4 = 90 \cdot 10^1$$

(remember that on the right you may retain only two significant digits) or, differently written,

$$0.90 \cdot 10^3 - 0.40 \cdot 10^1 = 0.90 \cdot 10^3.$$

With 2S, this gives 30. Hence by (6),

$$x_1 = (30 + 30)/2 = 60/2 = 30$$

and

$$x_2 = (30 - 30)/2 = 0.$$

In contrast, from (7) you obtain better results for the second root. You have $x_1 = 30$, as before, and

$$x_2 = 1/x_1 = 1/30 = 0.033.$$

The point of this and similar examples and problems is not to show that calculations with fewer significant digits generally give inferior results (this is fairly plain, although not always the case). The point is to show in terms of simple numbers what will happen in principle, regardless of the number of digits used in a calculation. Here, formula (6) illustrates the loss of significant digits, easily recognizable when we work with pencil (or calculator) and paper, but difficult to spot in a long calculation in which only a few intermediate results are printed out. This explains the necessity of developing programs that are virtually free of possible cancellation effects.

9. **Change of formula.** Given

$$\sqrt{9 + x^2} - 3, \tag{A}$$

where $|x|$ is small. Multiplication and division by

$$\sqrt{9 + x^2} + 3 \tag{B}$$

gives the numerator

$$\sqrt{9 + x^2}^{\,2} - 9 = 9 + x^2 - 9 = x^2$$

and the denominator (B), thus

$$x^2 / \left(\sqrt{9 + x^2} + 3 \right). \tag{C}$$

For instance, if $x = 0.1$ and you use 4S, you obtain from (A)

$$\sqrt{9.01} - 3 = 3.002 - 3.000 = 0.002.$$

The improved formula (C) gives

$$0.01000/(3.002 + 3.000) = 0.01000/6.002 = 0.001666.$$

The 10S-value is 0.001666 203961.

17. Rounding and adding. For instance, in rounding to, say, 1D, the given numbers $a_1 = 1.03$ and $a_2 = 0.24$ you get $\tilde{a}_1 = 1.0$ and $\tilde{a}_2 = 0.2$, hence the sum 1.2. But if you add first, you obtain 1.27. Rounded to 1D this gives 1.3, which is a more accurate approximation of the true value 1.27 than the approximation 1.2 obtained before. In terms of general formulas you have

$$\tilde{a}_1 = a_1 - \epsilon_1$$
$$\tilde{a}_2 = a_2 - \epsilon_2,$$

where ϵ_1 and ϵ_2 are the errors due to rounding, hence they are less than or equal to 1/2 unit of the last decimal in absolute value. If you round first and add then, you add the rounded numbers $\tilde{a}_1$ and $\tilde{a}_2$, that is,

$$\tilde{a}_1 + \tilde{a}_2 = a_1 + a_2 - (\epsilon_1 + \epsilon_2).$$

You see that in this case the error $\epsilon_1 + \epsilon_2$ is a number between 0 and 1 unit of the last decimal in absolute value. But if you add first, the sum is $a_1 + a_2$, and in rounding it you make an error between 0 and 1/2 unit of the last decimal in absolute value. Similarly for n numbers, where the sum of the rounded numbers is a number with an error between 0 and $n/2$ units of the last decimal in absolute value, whereas in adding and then rounding the error is between 0 and 1/2 unit of the last decimal in absolute value, as before in the case of two numbers.

Sec. 17.2 Solution of Equations by Iteration

Problem Set 17.2. Page 847

1. Nonmonotonicity (as in Example 2) occurs if $g(x)$ is monotone decreasing, that is,

$$g(x_1) \leq g(x_2) \quad \text{if} \quad x_1 > x_2. \tag{A}$$

(Make a sketch to better understand the reasoning.) Then

$$g(x) \geq g(s) \quad \text{if and only if} \quad x \leq s \tag{B}$$

and

$$g(x) \leq g(s) \quad \text{if and only if} \quad x \geq s. \tag{C}$$

Start from an $x_1 > s$. Then $g(x_1) \leq g(s)$ by (C). If $g(x_1) = g(s)$ (which could happen if $g(x)$ is constant between s and x_1), then x_1 is a solution of $f(x) = 0$, and you are done. If $g(x_1) < g(s)$, then by the definition of x_2 (formula (3) in the text) and since s is a fixed point ($s = g(s)$), you obtain

$$x_2 = g(x_1) < g(s) = s \quad \text{so that} \quad x_2 < s.$$

Hence by (B),

$$g(x_2) \geq g(s).$$

The equality sign would give a solution, as before. Strict inequality and the use of (3) in the text give

$$x_3 = g(x_2) > g(s) = s, \quad \text{so that} \quad x_3 > s,$$

and so on. This gives a sequence of values that are alternatingly larger and smaller than s, as illustrated in Fig. 395 of the text.

11. Newton's method. The derivation of this and similar formulas is schematical. Denote the quantity to be computed by x, that is,

$$x = \sqrt[3]{7}.$$

Then try to find an equation for x, in many cases an equation by which x is (explicitly or implicitly)

defined. In the present problem, using the definition of a cube root, you have
$$x^3 = 7.$$
The equation obtained is written as $f(x) = 0$, simply by collecting all the terms of the equation on the left side. In our case, $f(x) = x^3 - 7 = 0$. You also need $f'(x) = 3x^2$. With this, you can now set up the basic relation of Newton's method. This is equation (5) in the algorithm in Sec. 17.2,
$$x_{n+1} = x_n - \frac{f(x_n)}{f'(x_n)} = x_n - \frac{x_n^3 - 7}{3x_n^2} = \frac{2}{3}x_n + \frac{7}{3x_n^2}.$$
Some computational operations are avoided by pulling out the factor 1/3,
$$x_{n+1} = \frac{1}{3}\left(2x_n + \frac{7}{x_n^2}\right).$$

21. Secant method. You have $f(x) = \cos x \cosh x - 1$. Hence formula (10) gives
$$x_{n+1} = x_n - (\cos x_n \cosh x_n - 1)\frac{x_n - x_{n-1}}{\cos x_n \cosh x_n - \cos x_{n-1} \cosh x_{n-1}}.$$
In the answer on p. A38 in Appendix 2 the first value listed is the suggested $x_1 = 5$. From $x_0 = 4$ and $x_1 = 5$ you obtain $x_2 = 4.48457$, and so on. The convergence is slower than in Prob. 17 for Newton's method. The sequence of approximate values is not monotone, in contrast to that in Prob. 17 (but these properties are not typical, they depend on the kind of curve you are dealing with).

Sec. 17.3   Interpolation

Problem Set 17.3. Page 860

7. Extrapolation. In the case of extrapolation the various factors tend to be larger than in the case of interpolation because in the latter case the point of interpolation lies more "in the middle" between the nodes. In general, interpolation will give better results than extrapolation far enough away from the nodes. However, our simple figures illustrate that we cannot make statements that are always true. In Fig. A, interpolation gives better results than extrapolation at points much smaller than 5 or much larger than 15. In Fig. B, extrapolation is more accurate than interpolation near $x = 0$. Of course, these naive examples should merely make you aware of similar possibilities in more complicated cases in which you cannot see immediately what is going on.

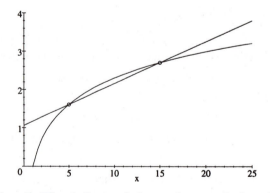

Section 17.3. Problem 7. Fig. A. Interpolation and extrapolation. (Logarithmic curve)

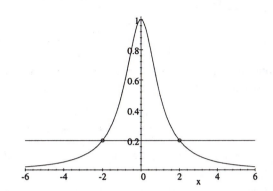

Section 17.3. Problem 7. Fig. B. Interpolation and Extrapolation ($y = 1/(1 + x^2)$).

9. Lagrange polynomial for the error function. From (3) and the given data you obtain the Lagrange polynomial

$$p_2(x) = \frac{(x - 0.5)(x - 1.0)}{-0.25(-0.75)}0.27633 + \frac{(x - 0.25)(x - 1.0)}{0.25(-0.5)}0.52050 + \frac{(x - 0.25)(x - 0.5)}{0.75 \cdot 0.5}0.84270.$$

Expanding and simplifying, you obtain the answer given on p. A38 of Appendix 2. The approximate value $p_2(0.75) = 0.70929$ is not very accurate. The exact 5D-value is erf $(0.75) = 0.71116$.

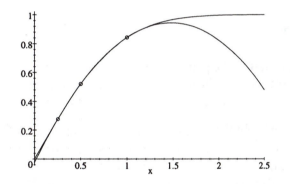

Section 17.3. Problem 9. erf (x) and Lagrange polynomial $p_2(x)$ (lower curve)

11. Newton's forward difference formula (14) applies to the given data since these are equally spaced, with $h = 0.02$. Set up a difference table as in Example 5, but containing one column less because you have only three given x-values $x_0 = 1.00$, $x_1 = 1.02$, $x_2 = 1.04$ and corresponding function values of the gamma function rounded to 4D. In (14) you need $\Gamma(1.00) = 1$, $\Delta^1 = -0.0112$, and $\Delta^2 = 0.0008$. With this you can read $p_2(x)$ in the answer on p. A38 of Appendix 2 in terms of r directly from (14). Then calculate $r = (x - x_0)/h = (x - 1.00)/0.02 = 50(x - 1)$. In this r you can substitute $x = 1.01, 1.03, 1.05$ and then calculate p_2 by using the corresponding $r = 0.5, 1.5, 2.5$, respectively. Or you can convert p_2 from r to x (which amounts to expanding p_2 in powers of x, as shown in the answer) and then substitute the x-values into this polynomial in x. The 4D-values in the answer are correct, also the last one (obtained by extrapolation).

15. Newton's divided difference formula (10) is less frequently used in practice than Newton's formulas for equally spaced data, which occur more often. Example 4 illustrates that the difference table contains the divided differences needed in calculating those that appear in (10); the latter are circled. In Prob. 15 use erf $(0.25) = 0.27633$. Then calculate the two first divided differences. The first of them is

$$f[0.25, 0.50] = \text{erf } (0.5) - \text{erf } (0.25)/(0.50 - 0.25)$$

$$= 4(0.52050 - 0.27633)$$

$$= 0.97668$$

and appears in (10). The second of them is

$$f[0.50, 1.00] = \text{erf } (1.0) - \text{erf } (0.5)/(1.0 - 0.5)$$

$$= 2(0.84270 - 0.52050)$$

$$= 0.6444$$

and is needed for calculating the second divided difference (you have only one because you have only three nodes, three function values). You obtain

$$f[0.25, 0.50, 1.00] = (f[0.50, 1.00] - f[0.25, 0.50])/(1.00 - 0.25)$$

$$= (0.6444 - 0.97668)/0.75 = -0.44304.$$

This is the last coefficient needed in (10). From this and (10) you obtain the expression for $p_2(x)$ given in the answer. Developing it in powers of x, you obtain the same polynomial in x as in Prob. 9 obtained by Lagrange's method. This illustrates the fact mentioned in the text, that the Lagrange's and Newton's formulas merely give different forms of the same interpolation polynomial, which is uniquely determined by the given data.

Sec. 17.4   Splines

Problem Set 17.4. Page 867

3. **Derivation of (7) and (8) from (6).** The point of the problem is that you minimize a chance of errors by introducing suitable short notations. For instance, for the expressions involving x you may set

$$X_j = x - x_j, \qquad X_{j+1} = x - x_{j+1},$$

and for the occurring constant quantities in (6) you may choose the short notations

$$A = f(x_j) c_j^2, \quad B = 2c_j, \quad C = f(x_{j+1}) c_j^2, \quad D = k_j c_j^2, \quad E = k_{j+1} c_j^2.$$

Then formula (6) becomes simply

$$p_j(x) = A X_{j+1}^2 (1 + BX_j) + C X_j^2 (1 - BX_{j+1}) + D X_j X_{j+1}^2 + E X_j^2 X_{j+1}.$$

Differentiate this twice with respect to x, applying the product rule for the second derivative, that is,

$$(uv)'' = u''v + 2u'v' + uv'',$$

and noting that the first derivative of X_j is simply 1, and so is that of X_{j+1}. (Of course, you may do the differentiations in two steps if you want.) You obtain

$$p_j''(x) = A (2(1 + BX_j) + 4X_{j+1}B + 0) + C(2(1 - BX_{j+1}) + 4X_j(-B) + 0) \qquad \text{(I)}$$

$$+ D(0 + 4X_{j+1} + 2X_j) + E(2X_{j+1} + 4X_j + 0),$$

where $4 = 2 \cdot 2$ with one 2 resulting from the product rule and the other from differentiating a square. And the zeros arise from factors whose second derivative is zero. Now calculate p_j'' at $x = x_j$. Since $X_j = x - x_j$, you see that $X_j = 0$ at $x = x_j$. Hence in each line the term containing X_j disappears. This gives

$$p_j''(x_j) = A (2 + 4BX_{j+1}) + C(2 - 2BX_{j+1}) + 4D X_{j+1} + 2E X_{j+1}.$$

Also, when $x = x_j$, then $X_{j+1} = x_j - x_{j+1} = -1/c_j$ (see the formula without number between (4) and (5), which defines c_j). Inserting this as well as the expressions for $A, B, ..., E$, you obtain (7). Indeed,

$$p_j''(x_j) = f(x_j) c_j^2 \left(2 + 2 \cdot \frac{4c_j}{-c_j} \right) + f(x_{j+1}) c_j^2 \left(2 - 2 \cdot \frac{2c_j}{-c_j} \right) + \frac{4 k_j c_j^2}{-c_j} + \frac{2 k_{j+1} c_j^2}{-c_j}$$

and cancellation of some of the factors c_j gives

$$p_j''(x_j) = -6f(x_j)c_j^2 + 6f(x_{j+1})c_j^2 - 4k_j c_j - 2k_{j+1} c_j.$$

The derivation of (8) is similar. For $x = x_{j+1}$ you have $X_{j+1} = x_{j+1} - x_{j+1} = 0$, so that (I) simplifies to

$$p_j''(x_{j+1}) = A(2 + 2BX_j) + C(2 - 4BX_j) + 2DX_j + 4EX_j.$$

Furthermore, for $x = x_{j+1}$ you have $X_j = x_{j+1} - x_j = 1/c_j$, and by substituting $A, ..., E$ into the last equation you obtain

$$p_j''(x_{j+1}) = f(X_j)c_j^2\left(2 + \frac{4c_j}{c_j}\right) + f(x_{j+1})c_j^2\left(2 - \frac{8c_j}{c_j}\right) + \frac{2k_j c_j^2}{c_j} + \frac{4k_{j+1} c_j^2}{c_j}.$$

Cancellation of some factors c_j and simplification fianlly gives (8), that is,

$$p_j''(x_{j+1}) = 6c_j^2 f(x_j) - 6c_j^2 f(x_{j+1}) + 2c_j k_j + 4c_j k_{j+1}.$$

11. Determination of a spline. Proceed as in Example 1. Arrange the given data in a table for easier work.

j	x_j	$f(x_j)$	k_j
0	-1	0	0
1	0	4	
2	1	0	0

Since there are three nodes, the spline will consist of two polynomials, $p_0(x)$ and $p_1(x)$. The polynomial $p_0(x)$ gives the spline for x from -1 to 0, and $p_1(x)$ gives the spline for x from 0 to 1.

Step 1. Since $n = 2$, you have just one equation in (12), from which you can determine k_1. The equation is obtained by taking $j = 1$ and noting that $h = 1$; thus

$$k_0 + 4k_1 = \frac{3}{1}(f_2 - f_0) = 0.$$

Hence $k_1 = 0$. Geometrically this means that at $x = 0$ the spline will have a horizontal tangent.

Step 2 for $p_0(x)$. Determine the coefficients of the spline from (14). You see that in general, $j = 0, ..., n - 1$, so that in the present case you have $j = 0$ (this will give the spline from -1 to 0) and $j = 1$ (which will give the other half of the spline, from 0 to 1). Take $j = 0$. Then (14) gives

$$a_{00} = p_0(p_0) = f_0 = 0$$

$$a_{01} = p_0'(x_0) = k_0 = 0$$

$$a_{02} = \frac{1}{2}p_0''(x_0) = \frac{3}{1^2}(f_1 - f_0) - \frac{1}{1}(k_1 - 2k_0) = 3 \cdot 4 - 0 = 12$$

$$a_{03} = \frac{1}{6}p_0'''(x_0) = \frac{2}{1^3}(f_0 - f_1) + \frac{1}{1^2}(k_1 + k_0) = 2 \cdot (-4) + 0 = -8.$$

With these Taylor coefficients you obtain from (13) the first half of the spline in the form

$$\begin{aligned}p_0(x) &= a_{00} + a_{01}(x - x_0) + a_{02}(x - x_0)^2 + a_{03}(x - x_0)^3 \\ &= 0 + 0 + 12(x - (-1))^2 - 8(x - (-1))^3 \\ &= 12x^2 + 24x + 12 - 8(x^3 + 3x^2 + 3x + 1) = 4 - 12x^2 - 8x^3.\end{aligned}$$

Step 2 for $p_1(x)$. This is slightly simpler because $x_j = x_1 = 0$, so that (13) will give powers of x directly. From the given data and (14) with $j = 1$ you obtain the Taylor coefficients

$$a_{10} = p_1(x_1) = f_1 = 4$$

$$a_{11} = p_1'(x_1) = k_1 = 0$$

$$a_{12} = \frac{1}{2}p_1''(x_1) = \frac{3}{1^2}(f_2 - f_1) - \frac{1}{1}(k_2 + 2k_1) = 3 \cdot (-4) - 0 = -12$$

$$a_{13} = \frac{1}{6}p_1'''(x_1) = \frac{2}{1^3}(f_2 - f_1) + \frac{1}{1^2}(k_2 + k_1) = 2 \cdot 4 + 0 = 8.$$

With these coefficients and $x_1 = 0$ you obtain from (13) with $j = 1$ the polynomial

$$p_1(x) = 4 - 12x^2 + 8x^3,$$

giving the spline on the interval from 0 to 1. As a check of the answer, you should verify that the spline gives the function values $f(x_j)$ and the values k_j of the derivatives in the table at the beginning. Also make sure that the first and second derivatives of the spline at 0 are continuous by verifying that

$$p_0'(0) = p_1'(0) = 0 \quad \text{and} \quad p_0''(0) = p_1''(0) = -24.$$

The third derivative is no longer continuous,

$$p_0'''(0) = -48 \quad \text{but} \quad p_1'''(0) = 48.$$

(Otherwise the spline would consist of a single cubic polynomial for the whole x-interval from -1 to 1.)

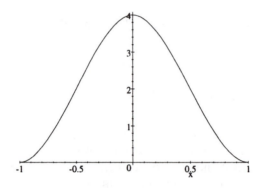

Section 17.4. Problem 11. Spline

Sec. 17.5   Numerical Integration and Differentiation

Problem Set 17.5. Page 880

5. **Error estimate (5) for the trapezoidal rule (2).** In (5) you need two approximate values. Since you calculate the integral

$$J = \int_0^1 \sin\left(\frac{\pi x}{2}\right) dx = -\cos\left(\frac{\pi x}{2}\right) \Big/ \frac{\pi}{2} \Big|_0^1 = \frac{2}{\pi} = 0.63662 \tag{A}$$

by (2) for three choices of h, namely, for $h = 1, 1/2, 1/4$, you can make two error estimates (5). Sketch the integrand to see what is going on. Now apply the trapezoidal rule (2). By using the exact 5D-value 0.63662 in (A) you can immediately determine the actual error, which we write after each result obtained from (2). The trapezoidal rule (2) with $h = 1$ gives

$$J_{1.0} = 1.0(0 + (1/2)\cdot 1) = 0.50000. \qquad \text{Error} \quad 0.13662. \tag{B}$$

With $h = 0.5$ you have the x-values $0, 1/2, 1$, for which the integrand has the values $0, 1/\sqrt{2} = 0.70711, 1$, respectively, so that (2) gives

$$J_{0.5} = 0.5(0 + 0.70711 + (1/2)\cdot 1) = 0.60355. \qquad \text{Error} \quad 0.03307. \tag{C}$$

With $h = 0.25$ you have the cosine values just used plus 0.38268 at $x = 1/4$ and 0.92388 at $x = 3/4$, so that (2) gives

$$J_{0.25} = 0.25(0 + 0.38268 + 0.70711 + 0.92388 + 0.50000) = 0.62842. \qquad \text{Error} \quad 0.00820. \tag{D}$$

Note that the error (3) contains the factor h^2. Hence in halving you can expect the error to be multiplied by about $(1/2)^2 = 1/4$. This property is nicely reflected by the numerical values in (B)-(D). Now turn to error estimating by (5). You obtain

$$\epsilon_{0.5} \approx \frac{1}{3}(J_{0.5} - J_{1.0}) = \frac{1}{3}(0.60355 - 0.50000) = 0.03452$$

$$\epsilon_{0.25} \approx \frac{1}{3}(J_{0.25} - J_{0.5}) = \frac{1}{3}(0.62842 - 0.60355) = 0.00829.$$

The agreement of these estimates with the actual value of the errors is very good, Although in other cases the difference between estimate and actual value may be larger, estimation will still serve its purpose, namely, to give an impression of the order of magnitude of the error.

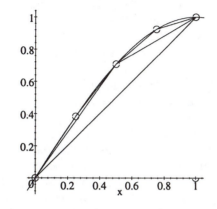

Section 17.5 Problem 5. Given sine curve and approximating polygons in the three trapezoidal rules used

21. **Three-eights rule**. For the present problem, this rule is very practical because the values of the integrand needed are simple,

$$\cos 30^o = \cos \frac{1}{6}\pi = \frac{1}{2}\sqrt{3},$$

$$\cos 60^o = \cos \frac{1}{3}\pi = \frac{1}{2}.$$

Also, the fourth derivative in the error term is simply $\cos \hat{t}$. You thus obtain

$$J = \int_0^{\pi/2} \cos x \, dx \approx \frac{3}{8}\cdot\frac{\pi}{6}\left(1 + 3\cdot\frac{\sqrt{3}}{2} + 3\cdot\frac{1}{2} + 0\right) - \frac{\pi/2}{80}\left(\frac{\pi}{6}\right)^4 \cos \hat{t} \qquad (E)$$

$$= \frac{\pi}{16}\cdot 5.098076 - 0.001476 \cos \hat{t} = 1.001005 - 0.001476 \cos \hat{t}.$$

Note that this approximation 1.001005 is much inferior to that in Prob. 23 obtained by Gauss integration with almost as little work as in the present problem. Error bounds are now readily obtained from (E) by noting that in the interval of integration, $\cos \tilde{t}$ varies between 0 and 1. Hence $\cos \pi/2 = 0$ gives the upper bound 0 for the error, and $\cos 0 = 1$ gives the lower bound $-0.001476\cdot 1 = -0.001476$ for the error. From this and (E) you have $1.001005 - 0.001476 = 0.999529$. Hence bounds for the approximate value $\hat{J} = 1.001005$ of $J = 1$ given by (E) are

$$1.001005 - 0.001476\cdot 1 = 0.999529 \leqq \hat{J} \leqq 1.001005.$$

23. **Gauss integration**. The answer on p. A39 shows that the transformation of a given integral to the standard interval $-1 \leqq x \leqq 1$ can often be avoided. This gives an additional reduction of the amount of work involved in this integration. You see that you obtain almost 7D accuracy with very little work. This result is much more accurate than that in Prob. 21 just considered.

CHAPTER 18. Numerical Methods in Linear Algebra

Sec. 18.1 Linear Systems: Gauss Elimination

Problem Set 18.1. Page 893

5. **System without solution.** The left side of the second equation equals minus three times the left side of the first equation. Hence for a solution to exist the right sides should be related in the same fashion; they should equal, for instance, 16 and −48 (instead of 48). Of course, for most systems with more than two equations, one cannot immediately see whether there will be solutions, but the Gauss elimination (with partial pivoting) will work in each case, giving the solution(s) or indicating that there is none.

7. **System with a unique solution. Pivoting.** Worked-out examples are given in the text. They show all the details. Review those first because there is little we can do for a better understanding and we shall have to restrict ourselves to a more detailed discussion of Table 18.1, which contains the algorithm for the Gauss elimination, and the addition of a few remarks. Consider Table 18.1. To follow the discussion, control it for Prob. 7 in terms of matrices with paper and pencil. In each case, write down all three rows of a matrix, not just one or two rows, as was done below to save some space and to avoid copying the same numbers several times. At the beginning, $k = 1$. Since $a_{11} = 0$, you must pivot. Line 2 in Table 18.1 requests to look for the absolutely greatest a_{j1}. This is a_{31}. According to the algorithm, you have to interchange Equations 1 and 3, that is, Rows 1 and 3 of the *augmented* matrix . This gives

$$\begin{bmatrix} 13 & -8 & 0 & | & 79 \\ 6 & 0 & -8 & | & -38 \\ 0 & 6 & 13 & | & 61 \end{bmatrix}. \tag{A}$$

Don't forget to interchange the entries on the right side (that is, in the last column of the augmented matrix). In line 2 of Table 18.1, the phrase 'the smallest' $j \geq k$ is necessary since there may be several entries of the same absolute value (or even of the same size), and the computer needs unique instructions what to do in each operation. To get 0 as the first entry of Row 2, subtract 6/13 times Row 1 from Row 2. The new Row 2 is

$$\begin{bmatrix} 0 & 3.692308 & -8 & | & -74.461538 \end{bmatrix}. \tag{B}$$

This was $k = 1$ and $j = 2$ in lines 3 and 4 in the table.

Now comes $k = 1$ and $j = n = 3$ in line 3. The calculation is $m_{31} = a_{31}/a_{11} = 0/13 = 0$. Hence the operations in line 4 simply have no effect, they merely reproduce Row 3 of the matrix in (A). This was $k = 1$.

Now comes $k = 2$. Look at line 2 in the table. Since $6 > 3.692308$, interchange Row 2 in (B) and Row 3 in (A). This gives the matrix

$$\begin{bmatrix} 13 & -8 & 0 & | & 79 \\ 0 & 6 & 13 & | & 61 \\ 0 & 3.692308 & -8 & | & -74.461538 \end{bmatrix}. \tag{C}$$

In line 3 of the table with $k = 2$ and $j = k + 1 = 3$ calculate

$$m_{32} = a_{32}/a_{22} = 3.692308/6 = 0.615385.$$

Performing the operations in line 4 of the table for $p = 3, 4$, you obtain the new Row 3

$$\begin{bmatrix} 0 & 0 & -16 & | & -112 \end{bmatrix}.$$

The system and its matrix have now reached triangular form, and back substitution begins with line 6 of the table,

$$x_3 = a_{34}/a_{33} = -112/(-16) = 7.$$

(Remember that in the table the right sides b_1, b_2, b_3 are denoted by a_{14}, a_{24}, a_{34}, respectively.) Line 7 of the table with $i = 2, 1$ gives

$$x_2 = \frac{1}{6}(61 - 13 \cdot 7) = -5 \qquad (i = 2)$$

and

$$x_1 = \frac{1}{13}(79 - (-8 \cdot (-5) + 0 \cdot 7)) = 3 \qquad (i = 1).$$

Depending on the number of digits you use in your calculation, your values may be slightly affected by round-off.

11. **System with more than one solution.** Solutions exist if and only if the coefficient matrix and the augmented matrix have the same rank (see Sec. 6.5). If these matrices have equal rank $r < n$ (n the number of unknowns), there exists more than one solution and, in fact, infinitely many solutions. In this case, to one or more suitable unknowns there can be assigned arbitrary values. In the present problem, $n = 3$ and the system is nonhomogeneous. For such a system you may have $r = 3$ (a unique solution), $r = 2$ (one (suitable) unknown remains arbitrary), $r = 1$ (two (suitable) variables remain arbitrary). $r = 0$ is impossible because then the matrices would be zero matrices. In most cases you have choices which of the variables you want to leave arbitrary; the present result will show this. To avoid misunderstandings: you need not determine those ranks, but the Gauss elimination will automatically give all solutions. *Your CAS may give only some solutions* (for example, those obtained by equating arbitrary unknowns to zero); so be careful. Following line 2 in Table 18.1, exchange Rows 1 and 2, so that the augmented matrix is

$$\begin{bmatrix} -5 & 7 & 2 & | & -4 \\ 2 & 5 & 7 & | & 25 \\ 1 & 22 & 23 & | & 71 \end{bmatrix}.$$

For $k = 1$ the operations in lines 3 and 4 of the table with $j = 2$ and 3 give

$$\begin{bmatrix} -5 & 7 & 2 & | & -4 \\ 0 & 7.8 & 7.8 & | & 23.4 \\ 0 & 23.4 & 23.4 & | & 70.2 \end{bmatrix} \begin{matrix} \\ \text{Row 2 + 0.4 Row 1} \\ \text{Row 3 + 0.2 Row 1.} \end{matrix} \qquad (D)$$

For $k = 2$ the operations in lines 3 and 4 of the table with $j = 3$ give the new Row 3 as a row of zeros,

$$\begin{bmatrix} 0 & 0 & 0 & | & 0 \end{bmatrix} \text{ Row 3 - 3 Row 2.}$$

This was the elimination. Now begins the back substitution. From Row 2 in (D) you obtain

$$x_2 = \frac{1}{7.8}(23.4 - 7.8x_3) = 3 - x_3. \qquad (E)$$

With this, Row 1 in (D) gives

$$x_1 = \frac{1}{-5}(-4 - 7x_2 - 2x_3) = \frac{1}{5}(4 + 7(3 - x_3) + 2x_3) = \frac{1}{5}(25 - 5x_3) = 5 - x_3. \qquad (F)$$

You see that you have no condition on x_3; hence x_3 is arbitrary. Solving (E) for x_3, you have

$$x_3 = 3 - x_2. \qquad (G)$$

Substituting (G) into (F), you obtain

$$x_1 = 5 - (3 - x_2) = 2 + x_2. \qquad (H)$$

This shows that you can leave x_2 arbitrary; then x_1 and x_3 are uniquely determined in terms of x_2. Equations (G) and (H) give the form of the solution shown on p. A39 in Appendix 2 of the book.

Sec. 18.2   Linear Systems: LU-Factorization, Matrix Inversion

Example 1. Doolittle's method (p.895). In the calculation of the entries of L and U (or L^T in Cholesky's method) in the factorization $A = LU$ with given A you employ the usual matrix multiplication

Row times Column.

In all three methods in this section, the point is that the calculation can proceed in an order such that you solve only one equation at a time. This is possible because you are dealing with triangular matrices, so that the sums of $n = 3$ products often reduce to sums of 2 products or even to a single product, as you will see. This will be a discussion of the steps of the calculation on p. 895 in terms of the matrix equation $A = LU$, written out (see the result on p, 896 at the top)

$$A = \begin{bmatrix} 3 & 5 & 2 \\ 0 & 8 & 2 \\ 6 & 2 & 8 \end{bmatrix} = LU = \begin{bmatrix} 1 & 0 & 0 \\ m_{21} & 1 & 0 \\ m_{31} & m_{32} & 1 \end{bmatrix} \begin{bmatrix} u_{11} & u_{12} & u_{13} \\ 0 & u_{22} & u_{23} \\ 0 & 0 & u_{33} \end{bmatrix}.$$

Remember that in Doolittle's method the main diagonal of L is $1,1,1$. Also, the notation m_{jk} suggests *multiplier*, because in Doolittle's method the matrix L is the matrix of the multipliers in the Gauss elimination. Begin with Row 1 of A. The entry $a_{11} = 3$ is the dot product of the first row of L and the first column of U; thus,

$$3 = [1 \quad 0 \quad 0][u_{11} \quad 0 \quad 0]^T = 1 \cdot u_{11},$$

where 1 is prescribed . Thus, $u_{11} = 3$. Similarly, $a_{12} = 5 = 1 \cdot u_{12} + 0 \cdot u_{22} + 0 \cdot 0 = u_{12}$; thus $u_{12} = 5$. Finally, $a_{13} = 2 = u_{13}$. This takes care of the first row of A. In connnection with the second row of A you have to consider the second row of L, which involves m_{21} and 1. You obtain

$$a_{21} = 0 = m_{21} u_{12} + 0 \quad + 0 \ = m_{21} \cdot 5, \qquad \text{hence } m_{21} = 0$$
$$a_{22} = 8 = m_{21} u_{12} + 1 \cdot u_{22} + 0 \ = u_{22}, \qquad \text{hence } u_{22} = 8$$
$$a_{23} = 2 = m_{21} u_{13} + 1 \cdot u_{23} + 0 \ = u_{23}, \qquad \text{hence } u_{23} = 2.$$

In connection with the third row of A you have to consider the third row of L, consisting of $m_{31}, m_{32}, 1$. You obtain

$$a_{31} = 6 = m_{31} u_{11} + 0 + 0 \qquad = m_{31} \cdot 3, \qquad \text{hence } m_{31} = 2$$
$$a_{32} = 2 = m_{31} u_{12} + m_{32} u_{22} + 0 \quad = 2 \cdot 5 + m_{32} \cdot 8, \qquad \text{hence } m_{32} = -1$$
$$a_{33} = 8 = m_{31} u_{13} + m_{32} u_{23} + 1 \cdot u_{33} = 2 \cdot 2 - 1 \cdot 2 + u_{33}, \qquad \text{hence } u_{33} = 6.$$

In (4) on p. 896 the first line concerns the first row of A and the second line concerns the first column of A; hence in that respect the order of calculation is slightly different from that in Example 1.

Problem Set 18.2. Page 899

7. Cholesky's method. You see that the given matrix A is symmetric. Its Cholesky factorization is

$$\begin{bmatrix} 9 & 6 & 12 \\ 6 & 13 & 11 \\ 12 & 11 & 26 \end{bmatrix} = \begin{bmatrix} l_{11} & 0 & 0 \\ l_{21} & l_{22} & 0 \\ l_{31} & l_{32} & l_{33} \end{bmatrix} \begin{bmatrix} l_{11} & l_{21} & l_{31} \\ 0 & l_{22} & l_{32} \\ 0 & 0 & l_{33} \end{bmatrix}.$$

This matrix A is positive definite. For a larger matrix this may be difficult to check, although in some cases it may be concluded from the kind of physical (or other) application. However, it is not necessary to check for definiteness because all that might happen is that you obtain a complex triangular matrix L and would then probably choose another method. Going through A row by row and applying matrix multiplication (Row times Column) as just before you calculate the following.

$$a_{11} = 9 = l_{11}^2 + 0 + 0 \quad = l_{11}^2, \qquad \text{hence} \quad l_{11} = 3$$

$$a_{12} = 6 = l_{11}l_{21} + 0 + 0 \quad = 3l_{21}, \qquad \text{hence} \quad l_{21} = 2$$

$$a_{13} = 12 = l_{11}l_{31} + 0 + 0 \quad = 3l_{31}, \qquad \text{hence} \quad l_{31} = 4.$$

In the second row of $\mathbf{A}$ you have $a_{21} = a_{12}$ (symmetry!) and need only two calculations,

$$a_{22} = 13 = l_{21}^2 + l_{22}^2 + 0 \quad = 4 + l_{22}^2, \qquad \text{hence} \quad l_{22} = 3$$

$$a_{23} = 11 = l_{21}l_{31} + l_{22}l_{32} + 0 = 2 \cdot 4 + 3l_{22}, \qquad \text{hence} \quad l_{32} = 1.$$

In the third row of $\mathbf{A}$ you have $a_{31} = a_{13}$ and $a_{32} = a_{23}$ and need only one calculation,

$$a_{33} = 26 = l_{31}^2 + l_{32}^2 + l_{33}^2 = 16 + 1 + l_{33}^2, \qquad \text{hence} \quad l_{33} = 3.$$

Now solve $\mathbf{Ax} = \mathbf{b}$, where $\mathbf{b} = [17.4 \quad 23.6 \quad 30.8]^T$. You first use $\mathbf{L}$ and solve $\mathbf{Ly} = \mathbf{b}$, where $\mathbf{y} = [y_1 \ y_2 \ y_3]^T$. Since $\mathbf{L}$ is triangular, you just do back substitution as in the Gauss algorithm. Now since $\mathbf{L}$ is *lower* triangular, whereas the Gauss elimination produces an *upper* tringular matrix, begin with the first equation and obtain y_1. Then obtain y_2 and finally y_3. This simple calculation is written to the right of the corresponding equations.

$$\begin{bmatrix} 3 & 0 & 0 \\ 2 & 3 & 0 \\ 4 & 1 & 3 \end{bmatrix} \begin{bmatrix} y_1 \\ y_2 \\ y_3 \end{bmatrix} = \begin{bmatrix} 17.4 \\ 23.6 \\ 30.8 \end{bmatrix} \qquad \begin{array}{l} y_1 = \frac{1}{3} \cdot 17.4 = 5.8 \\ y_2 = \frac{1}{3}(23.6 - 2y_1) = 4 \\ y_3 = \frac{1}{3}(30.8 - 4y_1 - y_2) = 1.2 \end{array}.$$

In the second part of the procedure you solve $\mathbf{L}^T\mathbf{x} = \mathbf{y}$ for $\mathbf{x}$. This is another back substitution. Since $\mathbf{L}^T$ is *upper* triangular, just as in the Gauss method after the elimination has been completed, the present back substitution is exactly as in the Gauss method, beginning with the last equation, which gives x_3, then using the second equation to get x_2, and finally the first equation to obtain x_1. These calculations are again written to the right of the corresponding equations.

$$\begin{bmatrix} 3 & 2 & 4 \\ 0 & 3 & 1 \\ 0 & 0 & 3 \end{bmatrix} \begin{bmatrix} x_1 \\ x_2 \\ x_3 \end{bmatrix} = \begin{bmatrix} 5.8 \\ 4 \\ 1.2 \end{bmatrix} \qquad \begin{array}{l} x_1 = \frac{1}{3}(5.8 - 2x_2 - 4x_3) = 0.6 \\ x_2 = \frac{1}{3}(4 - x_3) = 1.2 \\ x_3 = \frac{1}{3} \cdot 1.2 = 0.4 \end{array}.$$

Check the solution by substituting it into the given linear system.

17. **Matrix inversion.** The method suggested in this section is illustrated by Example 1 in Sec. 6.7, at which you may perhaps look first. The matrix in the present problem is

$$\begin{bmatrix} -2 & 4 & -1 \\ -2 & 3 & 0 \\ 7 & -12 & 2 \end{bmatrix}.$$

To find its inverse, apply the Gauss-Jordan method to the 3×6 matrix

$$\mathbf{G} = \left[\begin{array}{ccc|ccc} -2 & 4 & -1 & 1 & 0 & 0 \\ -2 & 3 & 0 & 0 & 1 & 0 \\ 7 & -12 & 2 & 0 & 0 & 1 \end{array} \right].$$

The left 3×3 submatrix is the given matrix. The right 3×3 submatrix is the 3×3 unit matrix. At the end of the process the left 3×3 submatrix will be the 3×3 unit matrix, and the right 3×3 submatrix will be the inverse of the given matrix. Leave Row 1 of $\mathbf{G}$ unchanged. Replace Row 2 by Row 2 − Row 1. Replace Row 3 by Row 3 + 3.5 Row 1. This gives the new matrix

$$\mathbf{H} = \begin{bmatrix} -2 & 4 & -1 & | & 1 & 0 & 0 \\ 0 & -1 & 1 & | & -1 & 1 & 0 \\ 0 & 2 & -1.5 & | & 3.5 & 0 & 1 \end{bmatrix}.$$

Now leave Rows 1 and 2 of $\mathbf{H}$ unchanged. Replace Row 3 by Row 3 + 2 Row 2. The new matrix is

$$\mathbf{J} = \begin{bmatrix} -2 & 4 & -1 & | & 1 & 0 & 0 \\ 0 & -1 & 1 & | & -1 & 1 & 0 \\ 0 & 0 & 0.5 & | & 1.5 & 2 & 1 \end{bmatrix}.$$

This was Gauss. The given matrix is triangularized. Now comes Jordan and diagonalizes it. Multiply Row 1 by $-1/2$, Row 2 by -1, and Row 3 by 2. This gives the matrix

$$\mathbf{K} = \begin{bmatrix} 1 & -2 & 0.5 & | & -0.5 & 0 & 0 \\ 0 & 1 & -1 & | & 1 & -1 & 0 \\ 0 & 0 & 1 & | & 3 & 4 & 2 \end{bmatrix}.$$

Now eliminate 0.5 and -1 from the third column of $\mathbf{K}$. Replace Row 1 by Row 1 $-$ 0.5 Row 3. Replace Row 2 by Row 2 + Row 3. Leave Row 3 unchanged. The new matrix is

$$\mathbf{M} = \begin{bmatrix} 1 & -2 & 0 & | & -2 & -2 & -1 \\ 0 & 1 & 0 & | & 4 & 3 & 2 \\ 0 & 0 & 1 & | & 3 & 4 & 2 \end{bmatrix}.$$

Finally eliminate -2 in the second column of $\mathbf{M}$. Replace Row 1 of $\mathbf{M}$ by Row 1 + 2 Row 2. The new matrix is

$$\mathbf{N} = \begin{bmatrix} 1 & 0 & 0 & | & 6 & 4 & 3 \\ 0 & 1 & 0 & | & 4 & 3 & 2 \\ 0 & 0 & 1 & | & 3 & 4 & 2 \end{bmatrix}.$$

The last three columns constitute the inverse of the given matrix. The following discussion may perhaps help you to a better understanding of the method. Follow the discussion with paper and pencil and an example of your own or in terms of the problem just solved. From $\mathbf{Ax} = \mathbf{b}$ you have $\mathbf{x} = \mathbf{A}^{-1}\mathbf{b}$, the existence of the inverse being assumed. The Gauss elimination converts the given system to $\mathbf{Ux} = \mathbf{b}^*$ with upper triangular $\mathbf{U}$ and $\mathbf{b}^*$ obtained from $\mathbf{b}$ in the calculations. $\mathbf{x} = \mathbf{U}^{-1}\mathbf{b}^*$ is then obtained by back substitution. In the Gauss-Jordan method you go on and reduce $\mathbf{U}$ to the identity matrix $\mathbf{I}$, so that the system becomes $\mathbf{Ix} = \mathbf{b}^{**}$ with $\mathbf{b}^{**}$ obtained from $\mathbf{b}^*$ in this process. Since $\mathbf{Ix} = \mathbf{x}$, you have directly $\mathbf{x} = \mathbf{b}^{**}$, that is, $\mathbf{b}^{**}$ is the solution, and back substitution is avoided. Now comes the crucial point of this discussion. If for $\mathbf{b}$ you chose the first column of the unit matrix, call it $\mathbf{b}_1$, you are dealing with $\mathbf{Ax} = \mathbf{b}_1$, hence with $\mathbf{x} = \mathbf{A}^{-1}\mathbf{b}_1$. But by the usual matrix multiplication, $\mathbf{A}^{-1}\mathbf{b}_1$ is simply the first column of the inverse matrix because this multiplication picks from each row of $\mathbf{A}^{-1}$ the first entry. And by Gauss-Jordan, this solution $\mathbf{x}$ appears as the transform of $\mathbf{b}_1$ in this process, that is, you have actually obtained the first column of the inverse (as Column 4 of your above 3×6 matrix $\mathbf{G}$). Similarly, if you choose the second column of the unit matrix, call it $\mathbf{b}_2$, Gauss-Jordan will give you as solution the second column of the inverse matrix as the transform of $\mathbf{b}_2$. And so on.

Sec. 18.3   Linear Systems: Solution by Iteration

Problem Set 18.3. Page 905

5. Gauss-Seidel iteration. This is a case in which you reorder the equations so that the large entries stand on the main diagonal in order to obtain convergence. That is, the third equation becomes the first and is solved for x_1. The first equation becomes the second and is solved for x_2. The second equation becomes the third and is solved for x_3. With this rearrangement you can expect convergence. Indeed, $\mathbf{C}$ in (7) can be shown to have the eigenvalues 0, 0.151, and −0.061, approximately. (In verifying this, don't forget to divide the rows of the coefficient matrix of the rearranged system by 6, 9, 8, respectively.) Hence you can expect rapid convergence (see the discussion between formulas (7) and (8) in the text). In contrast, if you left the given order and solved the first equation for x_1, the second for x_2, and the third for x_3, you do not get convergence because then the eigenvalues of $\mathbf{C}$ are 0, 8.5, and −51, approximately.

7. **Effect of starting values.** The point of the problem is to show that there is surprisingly little difference between corresponding values, as the answer on p. A40 in Appendix 2 shows, although the starting values differ considerably. Hence it is hardly necessary to search extensively for "good" starting values.

13. **Convergence.** The matrix of the system is

$$\begin{bmatrix} 4 & 0 & 5 \\ 1 & 6 & 2 \\ 8 & 2 & 1 \end{bmatrix}.$$

To obtain convergence, reorder the rows as shown.

$$\begin{bmatrix} 8 & 2 & 1 \\ 1 & 6 & 2 \\ 4 & 0 & 5 \end{bmatrix}.$$

Then divide the rows by 8, 6, and 5, respectively, as required in (13) (see $a_{jj} = 1$ at the end of the formula). This gives

$$\begin{bmatrix} 1 & 1/4 & 1/8 \\ 1/6 & 1 & 1/3 \\ 4/5 & 0 & 1 \end{bmatrix}.$$

You now have to consider

$$\mathbf{B} = \mathbf{I} - \mathbf{A} = \begin{bmatrix} 0 & -1/4 & -1/8 \\ -1/6 & 0 & -1/3 \\ -4/5 & 0 & 0 \end{bmatrix}.$$

The eigenvalues are obtained as the solutions of the characteristic equation

$$\det(\mathbf{B} - \lambda\mathbf{I}) = \begin{vmatrix} -\lambda & -1/4 & -1/8 \\ -1/6 & -\lambda & -1/3 \\ -4/5 & 0 & -\lambda \end{vmatrix}$$

$$= -\lambda^3 + \frac{17}{120}\lambda - \frac{1}{15} = 0.$$

A plot shows that there is a real root near −0.5, but there are no further real roots because for large $|\lambda|$ the curve comes closer and closer to the curve of $-\lambda^3$. Hence the other eigenvalues must be complex conjugates. A root-finding method (Sec. 17.2) gives a more accurate value −0.5196. Division of the characteristic equation by $\lambda + 0.5196$ gives the quadratic equation

$$-\lambda^2 + 0.5196\lambda - 0.1283 = 0.$$

The roots are $0.2598 \pm 0.2466i$. Since all the roots are less than 1 in absolute value, the spectral radius is less than 1, by definition. This is necessary and sufficient for convergence (see at the end of the section).

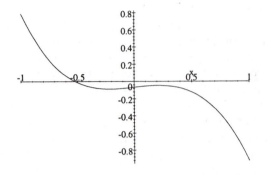

Section 18.3. Problem 13. Curve of the characteristic polynomial

15. **Matrix norm.** This simple problem illustrates that the three norms usually tend to give similar values. The same is true with Prob. 19. Hence one often chooses the norm that is most convenient from a computational point of view. See, however, in the next section that a matrix norm often results from a choice of a vector norm, so that in that respect, one is not completely free to choose.

Sec. 18.4   Linear Systems: Ill-conditioning, Norms

Problem Set 18.4. Page 912

7. **Matrix norms and condition numbers.** You have to consider the given matrix

$$\mathbf{A} = \begin{bmatrix} 4 & 1 \\ 0 & 2 \end{bmatrix} \text{ and its inverse } \mathbf{A}^{-1} = \begin{bmatrix} 1/4 & -1/8 \\ 0 & 1/2 \end{bmatrix}.$$

Begin with the l_1-norm. You have to remember that the l_1-vector norm gives for matrices the column "sum" norm (the "..." indicating that we take sums of absolute values). This gives 4 for $\mathbf{A}$ (the first column) and 5/8 for $\mathbf{A}^{-1}$ (the second column). Hence $\kappa(\mathbf{A}) = 4(5/8) = 2.5$. Now turn to the l_∞ norm. You have to remember that this vector norm gives for matrices the row "sum" norm. This gives 5 for $\mathbf{A}$ (the first row) and 1/2 for $\mathbf{A}^{-1}$ (the second row). Hence $\kappa(\mathbf{A}) = 5(1/2) = 2.5$. This is the same value as before. (Is this the case for all triangular real 2×2 matrices? For all real 2×2 matrices? How would you start to experiment on these questions?)

15. **Ill-conditioning.** The given system is

$$4.50x_1 + 3.55x_2 = 5.20$$
$$3.55x_1 + 2.80x_2 = 4.10.$$

Its coefficient matrix

$$\mathbf{A} = \begin{bmatrix} 4.50 & 3.55 \\ 3.55 & 2.80 \end{bmatrix}$$

consists of two almost proportional rows. Indeed, the system is very ill-conditioned. Its column sum norm is $4.50 + 3.55 = 8.05$. Its row sum norm is the same because $\mathbf{A}$ is symmetric. The inverse of $\mathbf{A}$ is

$$\mathbf{A}^{-1} = \begin{bmatrix} -1120 & 1420 \\ 1420 & -1800 \end{bmatrix}$$

and has the column sum norm $1420 + 1800 = 3220$, which equals the row sum norm, again for reasons of symmetry. The product of the norms is the condition number

$$\kappa(\mathbf{A}) = 8.05 \cdot 3220 = 25921.$$

This is very large and makes it plausible that the small change from $\mathbf{b}_1$ to $\mathbf{b}_2$ by 0.1 in the second component causes the solution to change from $-2, 4$ to $-144, 184$, a change of about one thousand times that of that component. Also, if you try to solve the system by the Gauss elimination with a small number of decimals, you obtain nonsensical results, whereas for calculations with 8 or 9 decimals the results are satisfactory.

17. Small residuals for poor solutions. In the present case, formula (1) with the suggested $\tilde{x} = [a \quad y]^T$ (see p. A40 in Appendix 2 of the book) is

$$\mathbf{r} = \begin{bmatrix} 5.2 \\ 4.1 \end{bmatrix} - \begin{bmatrix} 4.50 & 3.55 \\ 3.55 & 2.80 \end{bmatrix} \begin{bmatrix} a \\ y \end{bmatrix}.$$

In components this can be written

$$r_1 = 5.2 - 4.50\,a - 3.55\,y \tag{A}$$

$$r_2 = 4.1 - 3.55\,a - 2.80\,y.$$

If you set $r_1 = 0, r_2 = 0$, you would get the exact solution because this would be the given system in Prob. 15. Following the suggestion on p. A40, choose an a, say, a large a such as $a = 100$, and solve each equation $r_1 = 0$ and $r_2 = 0$ separately, with a_1 and a_2 as given in (A). You obtain

$$5.2 - 450 - 3.55\,y = 0, \quad \text{solution } y = \frac{1}{3.55}(5.2 - 450) = -125.296$$

$$4.1 - 355 - 2.80\,y = 0, \quad \text{solution } y = \frac{1}{2.80}(4.1 - 355) = -125.321.$$

From this you see that you can expect a small residual if you set

$$x_1 = a = 100, \quad x_2 = y = -125.3.$$

Indeed, you obtain

$$r_1 = 5.2 - 450 - 3.55(-125.3) = 0.015$$

$$r_2 = 4.1 - 355 - 2.8(-125.3) = -0.060.$$

Sec. 18.5   Method of Least Squares

Problem Set 18.5. Page 916

1. Fitting by a straight line. As in Example 1 in the text, the straight line for fitting the points by the method of least squares is obtained by solving the normal equations (4). In working with paper and pencil, it is best to first make up an orderly table of the given data and auxiliary quantities needed in (4). This may look as follows.

x_j	y_j	x_j^2	$x_j y_j$
0	3	0	0
2	1	4	2
3	−1	9	−3
5	−2	25	−10
Sum 10	1	38	−11

Since you have $n = 4$ pairs of values, with the sums in your table the augmented matrix of (4) has the form

$$\begin{bmatrix} 4 & 10 & 1 \\ 10 & 38 & -11 \end{bmatrix}.$$

The solution is $a = 37/13 = 2.84615, b = -27/26 = -1.03846$. Hence the desired straight line is
$y = 2.84615 - 1.03846x$.

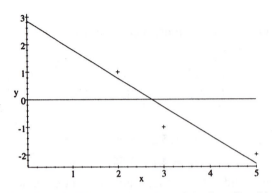

Section 18.6. Problem 1. Given data and straight line fitted by least squares

11. **Fitting by a quadratic parabola.** A quadratic parabola is uniquely determined by three given points. In
this problem, five points are given. You can fit a quadratic parabola by solving the normal equations (8).
Arrange the data and auxiliary quantities in (8) again in a table.

x	y	x^2	x^3	x^4	xy	x^2y
2	0	4	8	16	0	0
3	3	9	27	81	9	27
5	4	25	125	625	20	100
6	3	36	216	1296	18	108
7	1	49	343	2401	7	49
Sum 23	11	123	719	4419	54	284

Hence the augmented matrix of the system of normal equations is

$$\begin{bmatrix} 5 & 23 & 123 & 11 \\ 23 & 123 & 719 & 54 \\ 123 & 719 & 4419 & 284 \end{bmatrix}.$$

The solution obtained, for instance, by Gauss elimination is

$$b_0 = -8.357, \quad b_1 = 5.446, \quad b_2 = -0.589.$$

Hence the desired quadratic parabola that fits the data by the least squares principle is
$y = -8.357 + 5.446x - 0.589x^2$.

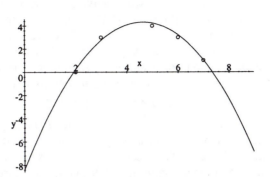

Section 18.6. Problem 11. Given points and quadratic parabola fitted by least squares

13. Comparison of linear and quadratic fit. The figure shows that a straight line obviously is not sufficient. The quadratic parabola gives a much better fit. It depends on the physical or other law underlying the data whether the fit by a quadratic polynomial is satisfactory and whether the remaining discrepancies can be attributed to chance variations, such as inaccuracy of measurement. Calculation shows that the augmented matrix of the normal equations for the straight line is

$$\begin{bmatrix} 5 & 10 & 8.3 \\ 10 & 30 & 17.5 \end{bmatrix}$$

and gives $y = 1.48 + 0.09x$. The augmented matrix for the quadratic polynomial is

$$\begin{bmatrix} 5 & 10 & 30 & 8.3 \\ 10 & 30 & 100 & 17.5 \\ 30 & 100 & 354 & 56.31 \end{bmatrix}$$

and gives $y = 1.896 - 0.741x + 0.208x^2$.

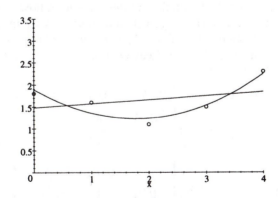

Section 18.6. Problem 13. Fit by a straight line and by a quadratic parabola

Sec. 18.7   Inclusion of Matrix Eigenvalues

Problem Set 18.7. Page 924

1. Gerschgorin circles. Gerschgorin's theorem is one of the earliest theorems on the numerical determination of matrix eigenvalues. The application of the theorem to a real or complex square matrix is very simple. In the present problem the Gerschgorin disks have the centers 5.1, 4.9, and −6.8 and the radii 0.5, 0.7, and 0.4, respectively. These disks consist of two disjoint parts (see the figure). The right part is formed by the union of two disks; hence it contains two eigenvalues. The left part, consisting of a single disk, must contain a single eigenvalue. This follows from Theorem 2. Note that the matrix is *not* skew-symmetric because its main diagonal entries are not zero. Hence you cannot apply Theorem 5 in Sec. 18.6 and conclude that its eigenvalues are pure imaginary or zero. 3D-values of the eigenvalues are -6.791 and $4.996 \pm 0.387i$. These can be determined, for instance, by sketching or plotting the curve of the characteristic polynomial, which by developing the characteristic determinant is found to be

$$- \lambda^3 + 3.2\lambda^2 + 42.75\lambda - 170.512.$$

The curve intersects the x-axis only once, near $\lambda = -7$, a value that can be improved to -6.791 by Newton's method. Then division by $\lambda + 6.791$ gives a quadratic equation whose roots are complex conjugates, as given before. You see that the left Gerschgorin disk does contain just one eigenvalue, whereas the other two eigenvalues lie in the union of the other two Gerschgorin disks with centers at 5.1 and 4.9.

Section 18.7. Problem 1. Gerschgorin circles

7. **Similarity transformation.** The matrix in Prob. 3 shows a typical situation. It may have resulted from a numerical method of diagonalization which left off-diagonal entries of various sizes but not exceeding 10^{-2} in absolute value. Gerschgorin's theorem then gives circles of radius 2×10^{-2}. These furnish upper bounds for the deviation of the eigenvalues from the main diagonal entries. This describes the starting situation for the present problem. Now in various applications, one is often interested in the eigenvalue of largest or smallest absolute value. In your matrix, the smallest eigenvalue is about 5, with a maximum possible deviation of 2×10^{-2}, as given by Gerschgorin's theorem. You now wish to decrease the size of this Gerschgorin disk as much as possible. Example 2 in the text shows how you should proceed. The entry 5 stands in the first row and column. Hence you should apply to **A** a similarity transformation involving a diagonal matrix **T** with main diagonal $a, 1, 1$, where a is as large as possible. The inverse of **T** is the diagonal matrix with main diagonal $1/a, 1, 1$. Leave a arbitrary and first determine the result of the similarity transformation (as in Example 2)

$$\mathbf{B} = \mathbf{T}^{-1}\mathbf{A}\mathbf{T} = \begin{bmatrix} 1/a & 0 & 0 \\ 0 & 1 & 0 \\ 0 & 0 & 1 \end{bmatrix} \begin{bmatrix} 5 & 0.01 & 0.01 \\ 0.01 & 8 & 0.01 \\ 0.01 & 0.01 & 9 \end{bmatrix} \begin{bmatrix} a & 0 & 0 \\ 0 & 1 & 0 \\ 0 & 0 & 1 \end{bmatrix}$$

$$= \begin{bmatrix} 5 & 0.01/a & 0.01/a \\ 0.01a & 8 & 0.01 \\ 0.01a & 0.01 & 9 \end{bmatrix}.$$

You see that the Gerschgorin disks of the transformed matrix **B** are

Center	Radius
5	$0.02/a$
8	$0.01(a + 1)$
9	$0.01(a + 1)$

The last two disks must be small enough so that they do not touch or even overlap the first disk. Since $8 - 5 = 3$, the radius of the second disk after the transformation must be less than $3 - 0.02/a$, that is,

$$0.01(a + 1) < 3 - 0.02/a.$$

Multiplication by $100a \, (> 0)$ gives $a^2 + a < 300\,a - 2$. If you replace the inequality sign by an equality sign, you obtain the quadratic equation $a^2 - 299a + 2 = 0$. Hence a must be less than the larger root 298.9933 of this equation, say, for convenience, $a = 298$. Then the radius of the second disk is $0.01(a + 1) = 2.99$, so that the disk will not touch the first one, and neither will the third, which is farther away from the first. The first disk is substantially reduced in size, by a factor of almost 300, the radius of the reduced disk being

$$0.02/298 = 0.000067114.$$

The choice of $a = 100$ would give a reduction by a factor 100, as requested in the problem. Your systematic approach shows that you can do better.

11. **Spectral radius.** By definition, the spectral radius of a square matrix $\mathbf{A}$ is the absolute value of an eigenvalue of $\mathbf{A}$ that is largest in absolute value. Since every eigenvalue of $\mathbf{A}$ lies in a Gerschgorin disk, for every eigenvalue of $\mathbf{A}$ you must have (make a sketch)

$$|a_{jj}| + \sum |a_{jk}| \geqq |\lambda_j|$$

where you sum over all off-diagonal entries in Row j (and the eigenvalues of $\mathbf{A}$ are numbered suitably). By taking this for a largest $|\lambda_j|$, you accomplish two things. First, on the right you obtain the spectral radius. Second, on the left you obtain the row "sum" norm. This proves the statement.

19. **Collatz's theorem.** The matrix $\mathbf{A}$ has equal row sums 7; hence 7 must be an eigenvalue of $\mathbf{A}$. The other eigenvalue of $\mathbf{A}$ is 4; it has algebraic multiplicity 2, that is, it is a double root of the characteristic equation. In the present case the characteristic equation can be solved by subtracting Row 2 of the characteristic matrix from Row 1, then Row 3 from Row 2, obtaining

$$\begin{bmatrix} 4 - \lambda & -4 + \lambda & 0 \\ 0 & 4 - \lambda & -4 + \lambda \\ 1 & 1 & 5 - \lambda \end{bmatrix}.$$

Then add Column 1 to Column 2 and develop the determinant of this matrix by the first row (which now contains two zeros); this gives

$$\begin{vmatrix} 4 - \lambda & 0 & 0 \\ 0 & 4 - \lambda & -4 + \lambda \\ 1 & 2 & 5 - \lambda \end{vmatrix} = (4 - \lambda)[(4 - \lambda)(5 - \lambda) + 2(4 - \lambda)] = (4 - \lambda)^2[5 - \lambda + 2].$$

This shows that the eigenvalues are 7 and 4, as claimed, and 4 has the algebraic multiplicity 2.

Sec. 18.8   Eigenvalues by Iteration (Power Method)

Example 1. Six vectors are listed. The first was scaled. The others were obtained by multiplication by $\mathbf{A}$ and subsequent scaling. You can use any of these vectors for obtaining a corresponding Rayleigh quotient q as an approximate value of an (unknown) eigenvalue of $\mathbf{A}$ and a corresponding error bound δ for q. Hence you have six possibilities using one of the given vectors (and many more if you want to compute further vectors). You must not use two of the given vectors because of the scaling, but just one vector, for instance, $\mathbf{x}_1$, and then its product $\mathbf{A}\mathbf{x}_1$. That is,

$$\mathbf{A} = \begin{bmatrix} 0.49 & 0.02 & 0.22 \\ 0.02 & 0.28 & 0.20 \\ 0.22 & 0.20 & 0.40 \end{bmatrix}, \quad \mathbf{x}_1 = \begin{bmatrix} 0.890244 \\ 0.609756 \\ 1 \end{bmatrix}, \quad \mathbf{A}\mathbf{x}_1 = \begin{bmatrix} 0.668415 \\ 0.388537 \\ 0.717805 \end{bmatrix}.$$

From these data you calculate the inner products

$$m_0 = \mathbf{x}_1^T\mathbf{x}_1 \qquad = 2.164337$$
$$m_1 = \mathbf{x}_1^T\mathbf{A}\mathbf{x}_1 \qquad = 1.549770$$
$$m_2 = (\mathbf{A}\mathbf{x}_1)^T\mathbf{A}\mathbf{x}_1 \quad = 1.112983.$$

These now give the Rayleigh quotient q and error bound δ of q

$$q = m_1/m_0 \qquad = 0.716048$$
$$\delta = \sqrt{m_2/m_0 - q^2} \quad = 0.038887 .$$

q approximates the eigenvalue 0.72 of $\mathbf{A}$, so that the error of q is

$$\epsilon = 0.72 - q \qquad = 0.003952.$$

These values agree with those in the table on p. 928 of the book.

Problem Set 18.8. Page 928

1. **Power method without scaling.** Without scaling the components of the vectors successively obtained will generally keep growing (or decreasing). Since the given matrix is symmetric, you can apply Theorem 1, which yields error bounds for the approximations of λ (usually the largest eigenvalue in absolute value, but no general statements can be made). Our simple matrix has the eigenvalues 11 and 1, as can readily be computed, and the problem serves only to explain the method in a very simple case in which you can see what is going on in each step. The computation of the vectors needed gives the following results.

$\mathbf{A}$	$\mathbf{x}_0$	$\mathbf{x}_1$	$\mathbf{x}_2$	$\mathbf{x}_3$
$\begin{bmatrix} 9 & 4 \\ 4 & 3 \end{bmatrix}$	$\begin{bmatrix} 1 \\ 1 \end{bmatrix}$	$\begin{bmatrix} 13 \\ 7 \end{bmatrix}$	$\begin{bmatrix} 145 \\ 73 \end{bmatrix}$	$\begin{bmatrix} 1597 \\ 799 \end{bmatrix}$

In each step of the further calculations take two adjacent vectors $\mathbf{x}$ and $\mathbf{y} = \mathbf{A}\mathbf{x}$, that is, $\mathbf{x}_0$ and $\mathbf{x}_1$, then $\mathbf{x}_1$ and $\mathbf{x}_2$, then $\mathbf{x}_2$ and $\mathbf{x}_3$. For the first of these three pairs $\mathbf{x}$, $\mathbf{y}$, namely, for $\mathbf{x}_0$ and $\mathbf{x}_1$, now compute (see Theorem 1)

$$m_0 = \mathbf{x}_0^T\mathbf{x}_0 = 1^2 + 1^2 = 2,$$
$$m_1 = \mathbf{x}_0^T\mathbf{x}_1 = 1\cdot13 + 1\cdot7 = 20$$
$$m_2 = \mathbf{x}_1^T\mathbf{x}_1 = 13^2 + 7^2 = 218$$
$$q = m_1/m_0 = 20/2 = 10 \qquad \text{(approximation of an eigenvalue)}$$
$$\delta^2 = m_2/m_0 - q^2 = 218/2 - 10^2 = 9$$
$$\delta = 3 \qquad \text{(error bound; see Theorem 1).}$$

Since $\mathbf{A}$ is symmetric, its eigenvalues are real. Hence conclude from Theorem 1 that an eigenvalue of $\mathbf{A}$ must lie in the closed interval

$$q - \delta = 7 \leq \lambda \leq q + \delta = 13.$$

It is typical that the error bound is much larger than the actual error, but the interval $q - \delta \leq \lambda \leq q + \delta$ is best possible, that is, for values q and δ calculated from a given symmetric matrix $\mathbf{A}$ (of any size) there is a symmetric matrix $\mathbf{B}$ for which the endpoints of that interval are eigenvalues of $\mathbf{B}$. (Another question would be how to actually find such a matrix $\mathbf{B}$ in a concrete case. Problem 10 contributes to this in a very special case.) For the next pair $\mathbf{x}_1$ and $\mathbf{x}_2$ you obtain

$$m_0 = \mathbf{x}_1^T \mathbf{x}_1 = 13^2 + 7^2 = 218 \qquad \text{(this is } m_2 \text{ of the previous step)}$$

$$m_1 = \mathbf{x}_1^T \mathbf{x}_2 = 13 \cdot 145 + 7 \cdot 73 = 2396$$

$$m_2 = \mathbf{x}_2^T \mathbf{x}_2 = 145^2 + 73^2 = 26354$$

$$q = m_1/m_0 = 10.99083 \qquad \text{(approximation of an eigenvalue)}$$

$$\delta^2 = m_2/m_0 - q^2 = 0.091659$$

$$\delta = 0.302752 \qquad \text{(error bound for } q\text{)}.$$

From this and Theorem 1 conclude that an eigenvalue of **A** must lie in the interval

$$q - \delta = 10.68807 \le \lambda \le q + \delta = 11.29358.$$

You see that the approximation has substantially increased in quality. The same is true for the error bound, which is smaller than in the first step by a factor 10. But, again, the error is much smaller than the bound, so that the latter does not give an indication of the size of the error. In the third step compute

$$m_0 = \mathbf{x}_2^T \mathbf{x}_2 = 26354 \qquad \text{(this is } m_2 \text{ of the previous step)}$$

$$m_1 = \mathbf{x}_2^T \mathbf{x}_3 = 145 \cdot 1597 + 73 \cdot 799 = 289892$$

$$m_2 = \mathbf{x}_3^T \mathbf{x}_3 = 1597^2 + 799^2 = 3188810$$

$$q = m_1/m_0 = 10.99992 \qquad \text{(approximation of an eigenvalue)}$$

$$\delta^2 = m_2/m_0 - q^2 = 0.0007589$$

$$\delta = 0.027548 \qquad \text{(error bound for } q\text{)}.$$

From this and Theorem 1 conclude that the interval

$$q - \delta = 10.97237 \le \lambda \le q + \delta = 11.02747$$

must contain an eigenvalue of **A**. Again, the error 0.00008 of q is much smaller than the error bound 0.027548.

13. Power method with scaling. The given matrix is

$$\mathbf{A} = \begin{bmatrix} 3.6 & -1.8 & 1.8 \\ -1.8 & 2.8 & -2.6 \\ 1.8 & -2.6 & 2.8 \end{bmatrix}.$$

Use the same notation as in Example 1 in the text. From $\mathbf{x}_0 = [1 \quad 1 \quad 1]^T$ calculate $\mathbf{Ax}_0$ and then scale it as indicated in the problem, calling the resulting vector $\mathbf{x}_1$. This is the first step. In the second step calculate $\mathbf{Ax}_1$ and then scale it, calling the resulting vector $\mathbf{x}_2$. And so on. The numerical results are as follows.

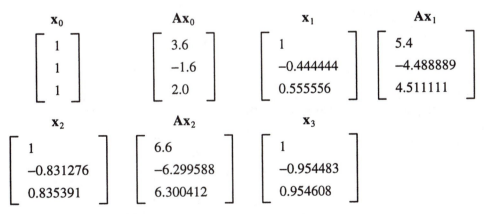

The calculation of approximations q (Rayleigh quotients) and error bounds δ proceeds similarly to the method without scaling (see the previous problem in this Manual). However, in the first step use

$$\mathbf{x} = \mathbf{x}_0 \quad \text{and} \quad \mathbf{y} = \mathbf{A}\mathbf{x} = \mathbf{A}\mathbf{x}_0 \quad (\text{not } \mathbf{x}_1).$$

In the second step use

$$\mathbf{x} = \mathbf{x}_1 \quad \text{and} \quad \mathbf{y} = \mathbf{A}\mathbf{x} = \mathbf{A}\mathbf{x}_1 \quad (\text{not } \mathbf{x}_2).$$

In the third step use

$$\mathbf{x} = \mathbf{x}_2 \quad \text{and} \quad \mathbf{y} = \mathbf{A}\mathbf{x} = \mathbf{A}\mathbf{x}_2 \quad (\text{not } \mathbf{x}_3).$$

The calculations that give approximations q (Rayleigh quotients) and error bounds are as follows.

m_0	$\mathbf{x}_0^T \mathbf{x}_0 = 3$	$\mathbf{x}_1^T \mathbf{x}_1 = 1.506173$	$\mathbf{x}_2^T \mathbf{x}_2 = 2.388897$
m_1	$\mathbf{x}_0^T \mathbf{A}\mathbf{x}_0 = 4$	$\mathbf{x}_1^T \mathbf{A}\mathbf{x}_1 = 9.901235$	$\mathbf{x}_2^T \mathbf{A}\mathbf{x}_2 = 17.100002$
m_2	$(\mathbf{A}\mathbf{x}_0)^T \mathbf{A}\mathbf{x}_0 = 19.52$	$(\mathbf{A}\mathbf{x}_1)^T \mathbf{A}\mathbf{x}_1 = 69.660247$	$(\mathbf{A}\mathbf{x}_2)^T \mathbf{A}\mathbf{x}_2 = 122.94000$
$q = m_1/m_0$	1.333333	6.573770	7.158115
$\delta^2 = m_2/m_0 - q^2$	4.728889	3.035378	0.224465
δ	2.174601	1.742234	0.47377
$q - \delta$	-0.841267	4.831536	6.684337
$q + \delta$	3.507935	8.316004	7.631893

Solving the characteristic equation shows that the matrix has the eigenvalues 0.2, 1.8, and 7.2. Corresponding eigenvectors are

$$\mathbf{z}_1 = [0 \quad 1 \quad 1]^T, \quad \mathbf{z}_2 = [2 \quad 1 \quad -1]^T, \quad \mathbf{z}_3 = [1 \quad -1 \quad 1]^T,$$

respectively. You see that the interval obtained in the first step includes the eigenvalues 0.2 and 1.8. Only in the second step and third step of the iteration did you obtain intervals that include the largest eigenvalue, as is usually the case from the beginning on. The reason for this interesting observation is the fact that $\mathbf{x}_0$ is a linear combination of all three eigenvectors,

$$\mathbf{x}_0 = \mathbf{z}_1 + \frac{1}{3}(\mathbf{z}_2 + \mathbf{z}_3),$$

as can be easily verified, and its needs several iterations until the powers of the largest eigenvalue make the iterate $\mathbf{x}_j$ come close to $\mathbf{z}_3$, the eigenvector corresponding to $\lambda = 7.2$. This siutation occurs quite frequently, and and one needs the more steps for obtaining satisfactory results the closer in absolute value the other eigenvalues are to the absolutely largest one. See also Prob. 11 for some further explanation.

Sec. 18.9   Tridiagonalization and QR-Factorization

Example 2. The tridiagonalized matrix is (p.936)

$$\mathbf{B} = \begin{bmatrix} 6 & -\sqrt{18} & 0 \\ -\sqrt{18} & 7 & \sqrt{2} \\ 0 & \sqrt{2} & 6 \end{bmatrix}.$$

We use the abbreviations c_2, s_2, and t_2 for $\cos \theta_2$, $\sin \theta_2$, and $\tan \theta_2$, respectively. We multiply $\mathbf{B}$ from the left by

$$\mathbf{C}_2 = \begin{bmatrix} c_2 & -s_2 & 0 \\ -s_2 & c_2 & 0 \\ 0 & 0 & 0 \end{bmatrix}.$$

The purpose of this multiplication is to obtain a matrix $\mathbf{C}_2 \mathbf{B} = [b_{jk}^{(2)}]$ for which the off-diagonal entry $b_{21}^{(2)}$ is zero. Now this entry is the inner product of Row 2 of $\mathbf{C}_2$ times Column 1 of $\mathbf{B}$, that is,

$$-s_2 \cdot 6 + c_2(-\sqrt{18}) = 0, \quad \text{thus} \quad t_2 = -\sqrt{18}/6 = -\sqrt{1/2}.$$

From this and the formulas that express cos and sin in terms of tan we obtain

$$c_2 = 1/\sqrt{1 + t_2^2} = \sqrt{2/3} = 0.816496581,$$

$$s_2 = t_2/\sqrt{1 + t_2^2} = -\sqrt{1/3} = -0.577350269.$$

θ_3 is determined similarly, with the purpose of obtaining $b_{32}^{(3)} = 0$ in $\mathbf{C}_3 \mathbf{C}_2 \mathbf{B} = [b_{jk}^{(3)}]$.

Problem Set 18.9. Page 937

1. Tridiagonalization. The given matrix

$$\mathbf{A} = \begin{bmatrix} 0.49 & 0.02 & 0.22 \\ 0.02 & 0.28 & 0.20 \\ 0.22 & 0.20 & 0.40 \end{bmatrix}$$

is symmetric. Hence you can apply Householder's method for obtaining a tridiagonal matrix (which will have two zeros instead of the entries 0.22). Proceed as in Example 1 of the text. Since $\mathbf{A}$ is of size $n = 3$, you have to perform $n - 2 = 1$ step. (In Example 1 we had $n = 4$ and needed $n - 2 = 2$ steps.) Calculate the vector $\mathbf{v}_1$ from (4). Denote it simply by $\mathbf{v}$ and its components by $v_1(= 0)$, v_2, v_3 because you do only one step. Similarly, denote S_1 in (4c) by S. Compute

$$S = \sqrt{a_{21}^2 + a_{31}^2} = \sqrt{0.02^2 + 0.22^2} = 0.2209072203.$$

If you compute, using, say, 6 digits, you may expect that instead of those two zeros in the tridiagonalized matrix you obtain entries of the order 10^{-6} or even larger in absolute value. You always have $v_1 = 0$. From (4a) you obtain the second component

$$v_2 = \sqrt{\frac{1 + a_{21}/S}{2}} = \sqrt{\frac{1 + 0.02/0.2209072203}{2}} = 0.7384225572.$$

From (4b) with $j = 3$ and sgn $a_{21} = +1$ (because a_{21} is positive) you obtain the third component

$$v_3 = a_{31}/(2v_2 S) = 0.22/(2v_2 S) = 0.6743382884.$$

With these values you now compute $\mathbf{P}_r$ from (2), where $r = 1, .., n - 2$, so that you have only $r = 1$ and can denote $\mathbf{P}_1$ simply by $\mathbf{P}$. Note well that $\mathbf{v}^T\mathbf{v}$ would be the dot product of the vector by itself (thus the square of its length), whereas $\mathbf{v}\mathbf{v}^T$ is a 3×3 matrix because of the usual matrix multiplication. You thus obtain from (2)

$$\mathbf{P} = \mathbf{I} - 2\mathbf{v}\mathbf{v}^T$$

$$= \mathbf{I} - 2\begin{bmatrix} v_1^2 & v_1 v_2 & v_1 v_3 \\ v_2 v_1 & v_2^2 & v_2 v_3 \\ v_3 v_1 & v_3 v_2 & v_3^2 \end{bmatrix}$$

$$= \begin{bmatrix} 1 - 2v_1^2 & -2v_1 v_2 & -2v_1 v_3 \\ -2v_2 v_1 & 1 - 2v_2^2 & -2v_2 v_3 \\ -2v_3 v_1 & -2v_3 v_2 & 1 - 2v_3^2 \end{bmatrix}$$

$$= \begin{bmatrix} 1 & 0 & 0 \\ 0 & -0.090\,535\,7460 & -0.995\,893\,2066 \\ 0 & -0.995\,893\,2064 & 0.090\,535\,7456 \end{bmatrix}.$$

Finally use $\mathbf{P}$ and its inverse $\mathbf{P}^{-1} = \mathbf{P}$ for the similarity transformation that will produce the tridiagonal matrix

$$\mathbf{B} = \mathbf{PAP} = \mathbf{P}\begin{bmatrix} 0.49 & -0.220\,907\,2204 & -10^{-10} \\ 0.02 & -0.224\,528\,6502 & -0.260\,742\,9487 \\ 0.22 & -0.416\,464\,4318 & -0.162\,964\,3431 \end{bmatrix}$$

$$= \begin{bmatrix} 0.49 & -0.220\,907\,2204 & -10^{-10} \\ -0.220\,907\,2204 & 0.435\,081\,9672 & 0.185\,901\,6396 \\ -10^{-10} & 0.185\,901\,6396 & 0.244\,918\,0330 \end{bmatrix}.$$

The point of the use of similarity transformations is that they preserve the spectrum of $\mathbf{A}$, consisting of the eigenvalues

$$0.09, \quad 0.36, \quad 0.72,$$

which can be found, for instance, by plotting the characteristic polynomial of $\mathbf{A}$ and applying Newton's method for improving the values obtained.

9. **QR-factorization.** The purpose of this factorization is the determination of approximate values of all the eigenvalues of a given matrix. To save work, one usually begins by tridiagonalizing a given matrix, which must be symmetric. The given matrix

$$\mathbf{B}_0 = [b_{jk}] = \begin{bmatrix} 7.0 & 0.5 & 0 \\ 0.5 & 3.5 & 0.1 \\ 0 & 0.1 & -1.5 \end{bmatrix}$$

is tridiagonal. Hence QR can begin. Proceed as in Example 2 on p. 935 of the book. (See also in this Manual above.) Write c_2, s_2, t_2 for $\cos\theta_2, \sin\theta_2, \tan\theta_2$, respectively. Consider the matrix

$$\mathbf{C}_2 = \begin{bmatrix} c_2 & s_2 & 0 \\ -s_2 & c_2 & 0 \\ 0 & 0 & 1 \end{bmatrix}$$

with the angle of rotation θ_2 determined so that in the product $\mathbf{W}_0 = \mathbf{C}_2\mathbf{B}_0 = [w_{jk}]$ the entry w_{21} is zero. By the usual matrix multiplication (row times column) w_{21} is the inner product of Row 2 of $\mathbf{C}_2$ times Column 1 of $\mathbf{B}_0$, that is,

$$-s_2 b_{11} + c_2 b_{21} = 0, \quad \text{hence} \quad t_2 = s_2/c_2 = b_{21}/b_{11}.$$

From this and the formulas for cos and sin in terms of tan (usually discussed in calculus) you obtain

$$c_2 = 1/\sqrt{1 + (b_{21}/b_{11})^2} = 0.9974586997 \tag{I/1}$$

$$s_2 = \frac{b_{21}}{b_{11}}/\sqrt{1 + (b_{21}/b_{11})^2} = 0.0712470500.$$

Now calculate $\mathbf{C}_2\mathbf{B}_0$ (as in the middle of p. 936). Denote this matrix (which has no notation on p. 936) by $\mathbf{W} = [w_{jk}]$. Thus

$$\mathbf{W} = [w_{jk}] = \mathbf{C}_2\mathbf{B}_0 = \begin{bmatrix} 7.017834423 & 0.7480940249 & 0.0071247050 \\ 0 & 3.455481924 & 0.09974587000 \\ 0 & 0.1 & -1.5 \end{bmatrix}.$$

C_2 has served its purpose: instead of $b_{21} = 0.5$ you now have $w_{21} = 0$. (Instead of $w_{21} = 0$ on the computer you may get -10^{-10} or another very small entry.) Now use the abbreviations c_3, s_3, t_3 for $\cos\theta_3$, $\sin\theta_3$, $\tan\theta_3$. Consider the matrix

$$\mathbf{C}_3 = \begin{bmatrix} 1 & 0 & 0 \\ 0 & c_3 & s_3 \\ 0 & -s_3 & c_3 \end{bmatrix}$$

with the angle of rotation θ_3 such that the product matrix $\mathbf{R}_0 = [r_{jk}] = \mathbf{C}_3\mathbf{W} = \mathbf{C}_3\mathbf{C}_2\mathbf{B}_0$ the entry r_{32} is zero. This entry is the inner product of Row 3 of $\mathbf{C}_3$ times Column 2 of $\mathbf{W}$. Hence

$$-s_3 w_{22} + c_3 w_{32} = 0, \quad \text{so that} \quad t_3 = s_3/c_3 = w_{32}/w_{22} = 0.02893952340.$$

This gives for c_3 and s_3

$$c_3 = 1/\sqrt{1 + t_3^2} = 0.9995815152, \qquad s_3 = t_3/\sqrt{1 + t_3^2} = 0.0289274127. \tag{II/1}$$

Using this, you obtain

$$\mathbf{R}_0 = \mathbf{C}_3\mathbf{W} = \mathbf{C}_3\mathbf{C}_2\mathbf{B}_0 = \begin{bmatrix} 7.017834423 & 0.7480940249 & 0.0071247050 \\ 0 & 3.456928598 & 0.0563130089 \\ 0 & 0 & -1.502257663 \end{bmatrix}.$$

(Instead of 0 you may obtain 10^{-10} or another very small term. Similarly in the further calculations.)
Finally, multiply $\mathbf{R}_0$ from the right by $\mathbf{C}_2^T\mathbf{C}_3^T$. This gives

$$\mathbf{B}_1 = \mathbf{R}_0\mathbf{C}_2^T\mathbf{C}_3^T = \mathbf{C}_3\mathbf{C}_2\mathbf{B}_0\mathbf{C}_2^T\mathbf{C}_3^T$$

$$= \begin{bmatrix} 7.053299490 & 0.2462959646 & 0 \\ 0.2462959646 & 3.448329498 & -0.0434564273 \\ 0 & -0.0434564273 & -1.501628991 \end{bmatrix}.$$

The given matrix $\mathbf{B}_0$ (and thus, also the matrix $\mathbf{B}_1$) has the eigenvalues

$$7.070049927, \quad 3.431961091, \quad -1.502011018.$$

You see that the main diagonal entries of $\mathbf{B}_1$ are approximations that are not very accurate, a fact that you could have concluded from the relatively large size of the off-diagonal entries of $\mathbf{B}_1$. In practice, one would perform further steps of the iteration until all off-diagonal elements have decreased in absolute value to less than a given bound. The answer on p. A41 in Appendix 2 gives the results of two more steps, which are obtained by the following calculations.

Step 2. The calculations are the same as before, with $\mathbf{B}_0 = [b_{jk}]$ replaced by $\mathbf{B}_1 = [b_{jk}^{(1)}]$. Hence, instead of (I/1) you now have

$$c_2 = 1/\sqrt{1 + (b_{21}^{(1)}/b_{11}^{(1)})^2} = 0.9993908803 \tag{I/2}$$

$$s_2 = (b_{21}^{(1)}/b_{11}^{(1)})/\sqrt{1 + (b_{21}^{(1)}/b_{11}^{(1)})^2} = 0.0348979852.$$

You can now write the matrix $\mathbf{C}_2$, which has the same general form as before, and calculate the product

$$\mathbf{W}_1 = [w_{jk}^{(1)}] = \mathbf{C}_2\mathbf{B}_1$$

$$= \begin{bmatrix} 7.057598419 & 0.3664856925 & -0.0015165418 \\ 0 & 3.437633820 & -0.0434299571 \\ 0 & -0.0434564273 & -1.501628991 \end{bmatrix}.$$

Now calculate the entries of $\mathbf{C}_3$ from (II/1) with $t_3 = w_{32}/w_{22}$ replaced by $t_3^{(1)} = w_{32}^{(1)}/w_{22}^{(1)}$, that is,

$$c_3 = 1/\sqrt{1 + (t_3^{(1)})^2} = 0.9999201074 \tag{II/2}$$

$$s_3 = t_3^{(1)}/\sqrt{1 + (t_3^{(1)})^2} = -0.0126403677.$$

You can now write $\mathbf{C}_3$, which has the same general form as in Step 1, and calculate

$$\mathbf{R}_1 = \mathbf{C}_3\mathbf{W}_1 = \mathbf{C}_3\mathbf{C}_2\mathbf{B}_1$$

$$= \begin{bmatrix} 7.057598419 & 0.3664856925 & -0.0015165418 \\ 0 & 3.437908483 & -0.0244453448 \\ 0 & 0 & -1.502057993 \end{bmatrix}.$$

This gives the next result

$$\mathbf{B}_2 = [b_{jk}^{(2)}] = \mathbf{R}_1\mathbf{C}_2^T\mathbf{C}_3^T = \mathbf{C}_3\mathbf{C}_2\mathbf{B}_1\mathbf{C}_2^T\mathbf{C}_3^T$$

$$= \begin{bmatrix} 7.066089109 & 0.1199760792 & 0 \\ 0.1199760792 & 3.4358484889 & 0.0189865653 \\ 0 & 0.0189865653 & -1.501937990 \end{bmatrix}.$$

The approximations of the eigenvalues have improved. The off-diagonal entries are smaller than in $\mathbf{B}_1$. Nevertheless, in practice the accuracy would still not be sufficient, so that one would do several more steps. Do one more step, whose result is also given on p. A41 in Appendix 2 of the book.

Step 3. The calculations are the same as in Step 2, with $\mathbf{B}_1 = [b_{jk}^{(1)}]$ replaced by $\mathbf{B}_2 = [b_{jk}^{(2)}]$. Hence calculate the entries of $\mathbf{C}_2$ from

$$c_2 = 1/\sqrt{1 + (b_{21}^{(2)}/b_{11}^{(2)})^2} = 0.9998558858 \tag{I/3}$$

$$s_2 = (b_{21}^{(2)}/b_{11}^{(2)})/\sqrt{1 + (b_{21}^{(2)}/b_{11}^{(2)})^2} = 0.0169766878.$$

You can now write the matrix $\mathbf{C}_2$ and calculate the product

$$\mathbf{W}_2 = [w_{jk}^{(2)}] = \mathbf{C}_2\mathbf{B}_2$$

$$= \begin{bmatrix} 7.067107581 & 0.1782881229 & 0.0003223290 \\ 0 & 3.433316936 & 0.0189838291 \\ 0 & 0.0189865653 & -1.501937990 \end{bmatrix}.$$

Now calculate the entries of $\mathbf{C}_3$ from (II/2) with $t_2^{(1)}$ replaced by $t_3^{(2)} = w_{32}^{(2)}/w_{32}^{(2)}$, that is,

$$c_3 = 1/\sqrt{1 + (t_3^{(2)})^2} = 0.9999847092 \tag{II/3}$$

$$s_3 = t_3^{(2)}/\sqrt{1 + (t_3^{(2)})^2} = 0.0055300094.$$

Write $\mathbf{C}_3$ and calculate

$$\mathbf{R}_2 = \mathbf{C}_3\mathbf{W}_2 = \mathbf{C}_3\mathbf{C}_2\mathbf{B}_2$$

$$= \begin{bmatrix} 7.067107581 & 0.1782881229 & 0.0003223290 \\ 0 & 3.433369434 & 0.0106778076 \\ 0 & 0 & -1.502020005 \end{bmatrix}$$

and, finally,

$$\mathbf{B}_3 = \mathbf{R}_2\mathbf{C}_2^T\mathbf{C}_3^T = \mathbf{C}_3\mathbf{C}_2\mathbf{B}_2\mathbf{C}_2^T\mathbf{C}_3^T$$

$$= \begin{bmatrix} 7.069115852 & 0.0582872409 & 0 \\ 0.0582872409 & 3.432881194 & -0.0083061848 \\ 0 & -0.0083061848 & -1.501997039 \end{bmatrix}.$$

This is again an improvement over the result of Step 2.

CHAPTER 19. Numerical Methods for Differential Equations

Sec. 19.1 Methods for First-Order Differential Equations

Problem Set 19.1. Page 951

3. **Euler method.** This method is hardly used in practice because it is not accurate enough for most purposes, and there are other methods (Runge-Kutta methods, in particular) that give much more accurate values without too much more work. However, the Euler method explains the principle underlying this class of methods in the simplest possible form, and this is the purpose of the present problem. The latter has the advantage that it concerns a differential equation that can easily be solved exactly, so that you can observe the behavior of the error as the computation is progressing from step to step. The given inital value problem is

$$y' + 5x^4 y^2 = 0, \qquad y(0) = 1.$$

For the Euler method you have to write the differential equation in the form

$$y' = f(x, y) = -5x^4 y^2. \tag{A}$$

The required step size is $h = 0.2$, so that 10 steps will give approximate solution values from 0 to 2.0. Because of (A) the formula (3) for the Euler method takes the form

$$y_{n+1} = y_n + 0.2 \, (-5 x_n^4 y_n^2) = y_n - x_n^4 y_n^2. \tag{B}$$

Because of the initial condition $y(0) = 1$ your starting values are

$$x = x_0 = 0 \quad \text{and} \quad y = y_0 = 1.$$

The exact solution is obtained by separating variables. Dividing (A) by y^2 on both sides and integrating, you obtain

$$y'/y^2 = -5x^4, \quad -\frac{1}{y} = -x^5 + c.$$

Taking the reciprocal and multiplying by -1 gives

$$y = \frac{1}{x^5 + c^*} \qquad (c^* = -c).$$

From this and the initial condition $y(0) = 1$ you obtain $c^* = 1$. Hence the solution of the problem is

$$y = \frac{1}{x^5 + 1}. \tag{C}$$

Use (C) in computing the error of the approximations obtained from (B). The computations with 10S rounded to 6S give the values shown in the following table.

Table for Problem 3. Computations with Euler's Method

n	x_n	y_n	y_n^2	$x_n^4 y_n^2$	Exact	Error
0	0	1	1	0	1	0
1	0.2	1	1	0.001600	0.999680	−0.000320
2	0.4	0.998400	0.996803	0.025518	0.989864	−0.008536
3	0.6	0.972882	0.946499	0.122666	0.927850	−0.045031
4	0.8	0.850216	0.722867	0.296086	0.753194	−0.097022
5	1.0	0.554129	0.307059	0.307059	0.500000	−0.054129
6	1.2	0.247070	0.061044	0.126580	0.286671	+0.039601
7	1.4	0.120490	0.014518	0.055772	0.156783	0.036293
8	1.6	0.064718	0.004188	0.027449	0.087064	0.022346
9	1.8	0.037269	0.001389	0.014581	0.050262	0.012993
10	2.0	0.022688	0.000515	0.008236	0.030303	0.007615

It is interesting that the error is neither monotone increasing nor of constant sign, as you might have expected. Of course, this has to do with the particular form of the equation and its solution, which approaches zero as x approaches infinity. The figure shows the behavior of the solution and the approximate values marked as points.

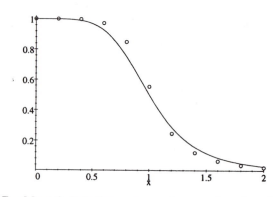

Section 19.1. Problem 3. Solution curve and approximations by Euler's method

13. Classical Runge-Kutta method. This is perhaps the most popular method. The given initial value problem is (see Prob. 11)

$$y' = \frac{2}{x}\sqrt{y - \ln x} + \frac{1}{x}, \qquad y(1) = 0. \tag{A}$$

In Prob. 11 this was solved by Euler's method with $h = 0.1$ for 8 steps from 1.0 to 1.8. The error was determined from the exact solution and was found to increase from 0 to 0.05, approximately. The exact solution can be obtained as follows. The form of the differential equation suggests introducing the new unknown function

$$z = y - \ln x. \quad \text{Then} \quad z' = y' - \frac{1}{x} = \frac{2}{x}\sqrt{z}, \tag{B}$$

where the last equality sign follows by using (A). You can now separate the variables, obtaining

$$\frac{z'}{\sqrt{z}} = \frac{2}{x}.$$

By integration, $2\sqrt{z} = 2\ln x + c^*$, hence $\sqrt{z} = \ln x + c$. Squaring and then using (B), you have

$$z = (\ln x + c)^2, \quad y = z + \ln x = (\ln x + c)^2 + \ln x.$$

Since $\ln 1 = 0$, you obtain from this and the initial condition $y(1) = c^2 = 0$. Hence the solution is $y = (\ln x)^2 + \ln x$, as shown in Prob. 11. The point of Prob. 13 is a comparison of the accuracy of Euler's method with that of the Runge-Kutta method. Now in the latter you have to compute four auxiliary quantities k_1, k_2, k_3, k_4 per step; hence in the required two steps this amounts to eight such computations, compared to eight steps in the Euler method in Prob. 11; in this sense, the comparison seems fair. The error will turn out to be about half of that of Euler's method. The results of the Runge-Kutta calculations (10S, rounded to 5S) are shown in the table.

Table for Problem 13. Computations with the Runge-Kutta method

x_n	y_n	k_1	k_2	k_3	k_4	Exact	Error
1.0	0	0.4	0.42197	0.44621	0.47501	0	0
1.4	0.43522	0.46529	0.47241	0.47440	0.47436	0.44969	0.01446
1.8	0.90744					0.93328	0.02584

Sec. 19.2   Multistep Methods

Problem Set 19.2. Page 955

1. Adams-Moulton method. The initial value problem to be solved is
$$y' = f(x,y) = x + y, \quad y(0) = 0. \tag{A}$$
The differential equation is linear. Hence you can solve it exactly, so that no numerical method would be needed. Indeed, write the equation in (A) in the standard form (1), Sec. 1.6,
$$y' - y = x,$$
and solve it by (4), Sec. 1.6, with $p = -1$, hence $h = -x$, obtaining
$$y(x) = e^x \left(\int e^{-x} x \, dx + c \right) = ce^x - x - 1.$$
The initial condition gives $y(0) = c - 0 - 1 = 0, c = 1$. Hence the solution of the initial value problem (A) is
$$y(x) = e^x - x - 1. \tag{B}$$
You can later use (B) for determining the errors of the approximate values obtained by the Adams-Moulton method. Now begin with the computation. From (A) you have
$$f_n = f(x_n, y_n) = x_n + y_n, \quad f_{n-1} = f(x_{n-1}, y_{n-1}) = x_{n-1} + y_{n-1}$$
and similarly for the other terms in (7a). Hence (7a) takes the form
$$y_{n+1}^* = y_n + \frac{0.1}{24}[55(x_n + y_n) - 59(x_{n-1} + y_{n-1}) + 37(x_{n-2} + y_{n-2}) - 9(x_{n-3} + y_{n-3})].$$
This gives the predictor. Similarly, the corrector (7b) takes the form
$$y_{n+1} = y_n + \frac{0.1}{24}[9(x_{n+1} + y_{n+1}^*) + 19(x_n + y_n) - 5(x_{n-1} + y_{n-1}) + (x_{n-2} + y_{n-2})].$$
Arrange the numerical values obtained as in Table 19.9 on p. 955.

x_n	Starting y_n	Predicted y_n^*	Corrected y_n	Exact Values	Error $\times 10^{-8}$
0	0				
0.1	0.005 170 83				
0.2	0.021 402 6				
0.3	0.049 858 5				
0.4		0.091 820 10	0.091 824 54	0.091 824 70	16
0.5		0.148 716 45	0.148 721 31	0.148 721 27	−4
0.6		0.222 113 67	0.222 119 08	0.222 118 80	−28
0.7		0.313 747 30	0.313 753 27	0.313 752 71	−56
0.8		0.425 535 24	0.425 541 83	0.425 540 93	−91
0.9		0.559 597 13	0.559 604 42	0.559 603 11	−131
1.0		0.718 275 57	0.718 283 62	0.718 281 83	−180

You see that the differences between predictor and corrector are of the order 10^{-6} to 10^{-5}. These differences are monotone increasing, namely, in terms of the last three digits shown,

$$444, \ 486, \ 541, \ 597, \ 659, \ 729, \ 805.$$

The errors of the corrected values are much less, of the order 10^{-7} to 10^{-6}. This shows that the process of correcting the predicted values is definitely worthwhile. From $x = 0.5$ on, the error is negative and is monotone increasing in absolute value. Monotonicity is typical of many cases, but other behavior also appears, for instance, if the solution happens to be periodic. The solution of the present problem is monotone increasing (see the figure) because its derivative is nonnegative. In the figure the approximate values obtained lie practically on the curve; the errors are much too small for exhibiting them graphically.

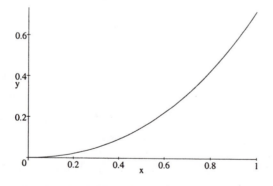

Section 19.2. Problem 1. Solution curve

Sec. 19.3   Methods for Systems and Higher Order Equations

Problem Set 19.3. Page 961

1. Euler's method for systems. The given system is
$$y_1' = f_1(x, y) = 2y_1 - 4y_2, \qquad y_2' = f_2(x, y) = y_1 - 3y_2. \tag{A}$$
Hence the recursion relation (5) with $h = 0.1$ takes the form (in components)

$$y_{1,n+1} = y_{1,n} + 0.1(2y_{1,n} - 4y_{2,n}), \qquad y_{2,n+1} = y_{2,n} + 0.1(y_{1,n} - 3y_{2,n}). \tag{B}$$

You see that this is not more complicated than Euler's method for a single equation, except for the fact that in each step you now have to work on two equations and each of them involves the result of both $y_{1,n}$ and $y_{2,n}$ of the previous step. Solve the system exactly, so that you can calculate the errors and judge the accuracy of the approximate values obtained by Euler's method. Proceed as in Sec. 3.3. The matrix of the system is

$$\mathbf{A} = \begin{bmatrix} 2 & -4 \\ 1 & -3 \end{bmatrix}.$$

Its characteristic equation is

$$\det(\mathbf{A} - \lambda\mathbf{I}) = (2 - \lambda)(-3 - \lambda) - (-4)1 = \lambda^2 + \lambda - 2 = (\lambda + 2)(\lambda - 1) = 0.$$

Hence the eigenvalues of $\mathbf{A}$ are -2 and 1. Corresponding eigenvectors are found by inserting the eigenvalues into the system of equations. Thus, for $\lambda = -2$ you have

$$(2 + 2)z_1 - 4z_2 = 0, \qquad \text{hence} \qquad z_2 = z_1,$$

and you can take $[1 \quad 1]^T$ as the first eigenvector. Note here that you have used z instead of x in Sec. 3.3 because x is supposed to be used as the independent variable. Similarly for $\lambda = 1$ you obtain

$$(2 - 1)z_1 - 4z_2 = 0, \quad \text{hence} \quad z_1 = 4z_2$$

and you can take $[4 \quad 1]^T$ as the second eigenvector. This gives the general solution

$$\mathbf{y}(x) = c_1 \begin{bmatrix} 1 \\ 1 \end{bmatrix} e^{-2x} + c_2 \begin{bmatrix} 4 \\ 1 \end{bmatrix} e^x. \tag{C}$$

The given initial values are $y_1(0) = 3$, $y_2(0) = 0$, in vectorial form $\mathbf{y}(0) = [3 \quad 0]^T$. From this and (C) you obtain

$$\mathbf{y}(0) = c_1 \begin{bmatrix} 1 \\ 1 \end{bmatrix} + c_2 \begin{bmatrix} 4 \\ 1 \end{bmatrix} = \begin{bmatrix} 3 \\ 0 \end{bmatrix}, \quad \text{in components,} \quad \begin{array}{l} c_1 + 4c_2 = 3 \\ c_1 + c_2 = 0 \end{array}.$$

The solution obtained by inspection, by elimination, or by Cramer's rule is $c_1 = -1$, $c_2 = 1$. Hence the given initial value problem has the solution

$$\mathbf{y}(x) = -\begin{bmatrix} 1 \\ 1 \end{bmatrix} e^{-2x} + \begin{bmatrix} 4 \\ 1 \end{bmatrix} e^x,$$

in components

$$y_1(x) = -e^{-2x} + 4e^x$$
$$y_2(x) = -e^{-2x} + e^x.$$

The recursion formula (B) gives the approximate values for $x = 0, 0.1, 0.2, ..., 1.0$ shown in the table. The errors ϵ_1 and ϵ_2 of y_1 and y_2, respectively, are obtained by using (C). The first figure shows $y_1(x)$ as a curve in the xy_1-plane and the values obtained by Euler's method marked by crosses. The second figure shows $y_2(x)$ in the xy_2-plane and the approximate values as crosses. The third figure shows y_2 as a function of y_1. Hence this is the trajectory of the initial value problem plotted in the phase plane (the y_1y_2-plane). The values obtained by Euler's method are again marked by crosses. From the table and from the figures you see that the values are too inaccurate for practical purposes. This agrees with your experience in the application of Euler's method to a single differential equation.

x	y_1	y_2	$\varepsilon(y_1)$	$\varepsilon(y_2)$
0	3	0	0	0
0.1	3.6	0.3	0.0	0.0
0.2	4.20	0.57	0.02	−0.02
0.3	4.812	0.819	0.039	−0.018
0.4	5.4468	1.0545	0.0712	−0.0120
0.5	6.11436	1.28283	0.11265	−0.00199
0.6	6.824100	1.509417	0.163181	0.011508
0.7	7.585153	1.739002	0.223261	0.028154
0.8	8.406583	1.975817	0.293684	0.047828
0.9	9.297573	2.223730	0.375541	0.070574
1.0	10.267596	2.486368	0.470964	0.096578

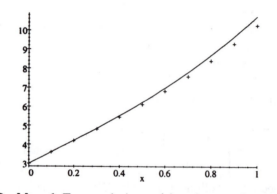

Section 19.3. Problem 1. Exact solution $y_1(x)$ and approximate values (crosses)

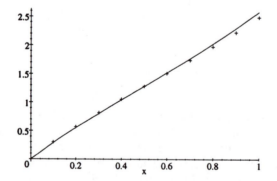

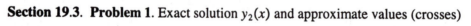

Section 19.3. Problem 1. Exact solution $y_2(x)$ and approximate values (crosses)

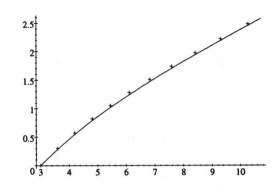

Section 19.3. Problem 1. Exact solution as a curve (trajectory) in the y_1y_2-plane and approximate values

5. **Classical Runge-Kutta method.** The initial value problem is the same as in Prob. 1. Two steps with the required (very large) $h = 0.5$ will give approximations for $x = 0.5$ and 1.0, which you can compare with the values in Prob. 1 obtained by Euler's method. The system is (in component form)

$$y_1' = f_1(x,y) = 2y_1 - 4y_2 \qquad \text{(A)}$$
$$y_2' = f_2(x,y) = y_1 - 3y_2.$$

The initial values are $y_1(0) = 3$, $y_2(0) = 0$. Now use the formula (6) for the classical **Runge-Kutta method** in Sec. 19.3. Formula (6b) consists of four vector formulas. Since (A) consists of two equations, each of the vector functions in (6) has two components. This will give you $4 \cdot 2 = 8$ formulas for the components. You could write

$$\mathbf{k}_1 = [k_{11} \ \ k_{12}]^T \qquad \text{(B)}$$
$$\mathbf{k}_2 = [k_{21} \ \ k_{22}]^T$$
$$\mathbf{k}_3 = [k_{31} \ \ k_{32}]^T$$
$$\mathbf{k}_4 = [k_{41} \ \ k_{42}]^T.$$

However, a simpler notation is that in Example 2 of Sec. 19.3 of the text, namely, instead of $\mathbf{k}_1, ..., \mathbf{k}_4$ write

$$\mathbf{a} = [a_1 \ \ a_2]^T \qquad \text{(C)}$$
$$\mathbf{b} = [b_1 \ \ b_2]^T$$
$$\mathbf{c} = [c_1 \ \ c_2]^T$$
$$\mathbf{d} = [d_1 \ \ d_2]^T.$$

(The text of the book uses the notation (B), which more distinctly shows the transition from the "scalar case" of a single differential equation to the "vector case" of a system of differential equations. (If you prefer (B), which forces you to carry along *two indices*, use it.) A further notational simplification giving the computation a simpler look is

$$y_1 = y, \quad y_2 = z \qquad \text{(D)}$$

because then instead of $y_{1,n}$ and $y_{2,n}$ (as in the text) you simply have $y_{1,n} = y_n$ and $y_{2,n} = z_n$. This will help in work by hand as well as in writing up a computer program. (Again, if you don't like (D), use y_1 and y_2.) From (6c), written in the notations (C) and (D), you obtain the following equations for calculating the auxiliary quantities occurring in the classical Runge-Kutta method

$$a_1 = hf_1(x_n, y_n, z_n)$$
$$a_2 = hf_2(x_n, y_n, z_n)$$

and so on. Inserting

$$f_1 = 2y_1 - 4y_2 = 2y - 4z$$
$$f_2 = y_1 - 3y_2 = y - 3z$$

from (A) and $h = 0.5$, as required, gives

$$a_1 = hf_1(x_n, y_n, z_n) = 0.5(2y_n - 4z_n)$$

$$a_2 = hf_2(x_n, y_n, z_n) = 0.5(y_n - 3z_n).$$

Note that in the system the independent variable x does not appear explicitly. This is an advantage because you need not pay attention to the place at which x is taken in each step (x_n or $x_n + h/2$ or $x_n + h$). Similarly,

$$b_1 = hf_1(x_n + h/2, y_n + a_1/2, z_n + a_2/2) = 0.5[2(y_n + a_1/2) - 4(z_n + a_2/2)]$$

$$b_2 = hf_2(x_n + h/2, y_n + a_1/2, z_n + a_2/2) = 0.5[(y_n + a_1/2) - 3(z_n + a_2/2)].$$

Note that in the general formula, b_1 and b_2 differ only by f_1 and f_2, so that from b_1 you can immediately see the form of b_2 by looking at the given system. Furthermore,

$$c_1 = hf_1(x_n + h/2, y_n + b_1/2, z_n + b_2/2) = 0.5[2(y_n + b_1/2) - 4(z_n + b_2/2)]$$

$$c_2 = hf_2(x_n + h/2, y_n + b_1/2, z_n + b_2/2) = 0.5[y_n + b_1/2 - 3(z_n + b_2/2)]$$

and finally

$$d_1 = hf_1(x_n + h, y_n + c_1, z_n + c_2) = 0.5[2(y_n + c_1) - 4(z_n + c_2)]$$

$$d_2 = hf_2(x_n + h, y_n + c_1, z_n + c_2) = 0.5[y_n + c_1 - 3(z_n + c_2)].$$

The recursion for x is

$$x_{n+1} = x_n + h.$$

The next values for $y_1 = y$ and $y_2 = z$ are given by (6c), that is,

$$y_{n+1} = y_n + \frac{1}{6}(a_1 + 2b_1 + 2c_1 + d_1)$$

$$z_{n+1} = z_n + \frac{1}{6}(a_2 + 2b_2 + 2c_2 + d_2).$$

The computation gives

x	y_1	y_2	Error of y_1	Error of y_2
0.5	6.218750	1.273438	0.008256	0.007404
1.0	10.728760	2.576721	0.009032	0.006225

The errors are much smaller than the corresponding errors in Prob. 1. Of course, further accuracy can be gained if we reduce $h = 0.5$ to, say, $h = 0.1$ (the value used in Prob. 1). The computation (10S rounded to 6D) gives the following values, exhibiting a very substantial reduction of the errors.

x	y_1	y_2	Error of y_1 $\times 10^6$	Error of y_2 $\times 10^6$
0.0	3	0	0	0
0.1	3.601 50	0.286 438	3	3
0.2	4.215 286	0.551 078	5	4
0.3	4.850 617	0.801 042	6	5
0.4	5.517 962	1.042 490	7	6
0.5	6.226 997	1.280 835	8	6
0.6	6.987 272	1.520 918	9	7
0.7	7.808 404	1.767 149	10	7
0.8	8.700 257	2.023 638	11	6
0.9	9.673 102	2.294 298	11	6
1.0	10.737 779	2.582 940	13	6

Sec. 19.4　Methods for Elliptic Partial Differential Equations

Problem Set 19.4. Page 969

1. **Derivation of (6c).** For this derivation you have to know Taylor's formula

$$u(x+h, y+k) = u + hu_x + ku_y + \frac{1}{2}(h^2 u_{xx} + 2hk u_{xy} + k^2 u_{yy}) + ... \tag{A}$$

where the function u and all the derivatives on the right are evaluated at (x, y). The further derivation is now automatic. If you replace h by $-h$ on the left, you get corresponding minus signs on the right, that is, a minus sign in the second and fifth term,

$$u(x-h, y+k) = u - hu_x + ku_y + \frac{1}{2}(h^2 u_{xx} - 2hk u_{xy} + k^2 u_{yy}) + ... \tag{B}$$

The right side of (6c) tells you what further expressions you should consider, namely,

$$u(x+h, y-k) = u + hu_x - ku_y + \frac{1}{2}(h^2 u_{xx} - 2hk u_{xy} + k^2 u_{yy}) + ... \tag{C}$$

and

$$u(x-h, y-k) = u - hu_x - ku_y + \frac{1}{2}(h^2 u_{xx} + 2hk u_{xy} + k^2 u_{yy}) + ... \tag{D}$$

The idea of the proof of (6c) now is to combine (A) ... (D) so that the derivative u_{xy} will remain, whereas all the other derivatives as well as the function u itself will drop out. In (A) minus (B) the function u and the derivatives u_y, u_{xx}, and u_{yy} drop out and you are left with

$$2hu_x + 2hk u_{xy}. \tag{E}$$

In (D) minus (C) the function u and the derivatives u_y, u_{xx}, and u_{yy} drop out and you are left with

$$-2hu_x + 2hk u_{xy}. \tag{F}$$

Addition (E) plus (F) gives $4hk u_{xy}$. Division by $4hk$ gives u_{xy}, the left side of (6c). Now

$$(E) + (F) = (A) - (B) - (C) + (D)$$

and this is precisely the right side of (6c). This completes the derivation.

3. **Potential. Liebmann's method (Gauss-Seidel iteration).** Proceed as in Example 1. Sketch the square and the grid and indicate the boundary potential as well the notation for the four interior points at which

you have to find the potential. The linear system to be solved is (we indicate at each equation the point from which it results)

$$(P_{11}) \quad -4u_{11} \; + \; u_{21} \; + \; u_{12} \qquad\qquad = \; -330$$
$$(P_{21}) \quad\;\; u_{11} \; -4u_{21} \qquad\quad + \; u_{22} \; = \; -210$$
$$(P_{12}) \quad\;\; u_{11} \qquad\qquad -4u_{12} \; + \; u_{22} \; = \; -330$$
$$(P_{22}) \qquad\qquad\quad u_{21} \; + \; u_{12} \; -4u_{22} \; = \; -210.$$

The augmented matrix of this system is

$$\begin{bmatrix} -4 & 1 & 1 & 0 & -330 \\ 1 & -4 & 0 & 1 & -210 \\ 1 & 0 & -4 & 1 & -330 \\ 0 & 1 & 1 & -4 & -210 \end{bmatrix}.$$

Apply Gauss elimination. Row 1 is the pivot row. The next matrix is

$$\begin{bmatrix} -4 & 1 & 1 & 0 & -330 \\ 0 & -3.75 & 0.25 & 1 & -292.5 \\ 0 & 0.25 & -3.75 & 1 & -412.5 \\ 0 & 1 & 1 & -4 & -210 \end{bmatrix} \begin{array}{l} \\ \text{Row } 2 + 0.25\,\text{Row } 1 \\ \text{Row } 3 + 0.25\,\text{Row } 1 \\ \text{Row } 4. \end{array}$$

Row 2 is the pivot row. The next matrix is

$$\begin{bmatrix} -4 & 1 & 1 & 0 & -330 \\ 0 & -3.75 & 0.25 & 1 & -292.5 \\ 0 & 0 & -3.733\,333 & 1.066\,667 & -432 \\ 0 & 0 & 1.066\,667 & -3.733\,333 & -288 \end{bmatrix} \begin{array}{l} \\ \\ \text{Row } 3 + (1/15)\,\text{Row } 2 \\ \text{Row } 4 + (4/15)\,\text{Row } 2. \end{array}$$

Row 3 is the next (and last) pivot row. Row 4 of the next matrix is Row 4 plus 2/7 times Row 3 of the previous matrix, namely,

$$[0 \quad 0 \quad 0 \quad -3.428571 \quad -411.428572].$$

Back substitution gives, in this order,

$$u_{22} = 120, \quad u_{12} = 150, \quad u_{21} = 120, \quad u_{11} = 150.$$

The result reflects a symmetry (can you see it now?), which you could have used to reduce your 4×4 system to a 2×2 system and solve the latter. We emphasize again that in practice, such systems are very large, due to much finer grids, and our problem serves to explain and illustrate the principle. For fine grids the matrix is sparse, so that the Liebmann method (Gauss-Seidel iteration) becomes advantageous. In our problem the iteration uses the system in the form (as in Example 1 of the text)

$$u_{11} = \qquad\quad 0.25u_{21} + 0.25u_{12} \qquad\qquad + 82.5$$
$$u_{21} = 0.25u_{11} \qquad\qquad\qquad + 0.25u_{22} \; + 52.5$$
$$u_{12} = 0.25u_{11} \qquad\qquad\qquad + 0.25u_{22} \; + 82.5$$
$$u_{22} = \qquad\quad 0.25u_{21} + 0.25u_{12} \qquad\qquad + 52.5.$$

The iteration, starting with the suggested values 100, 100, 100, 100, gives the following values.

	Step 1	Step 2	Step 3	Step 4	Step 5	Step 10
100	132.5	145.3125	148.828125	149.707031	149.926758	149.999928
100	110.625	117.65625	119.414062	119.853516	119.963379	119.999964
100	140.625	147.65625	149.414062	149.853516	149.963379	149.999964
100	115.3125	118.828125	119.707031	119.926758	119.981690	119.999982

Fifteen steps give accurate 10S-values.

Sec 19.5   Neumann and Mixed Problems. Irregular Boundary

Problem Set 19.5. Page 975

3. **Mixed boundary conditions for the Laplace equation.** Proceed as in Example 1. The situation is simpler because you are dealing with the Laplace equation, whereas Example 1 in the text concerns a Poisson equation. Equation (1) in Example 1 is a list of boundary values, in which the values 0 are not included. There are 10 grid points on the boundary. For Prob. 3 the 10 boundary values, beginning at 0 and going around counterclockwise, are

$$1, \quad 1, \quad 1, \quad 1, \quad 1, \quad 1, \quad u_n(P_{22}) = 1, \quad u_n(P_{12}) = 1, \quad 1, \quad 1. \qquad (A)$$

Here $u_n(P_{22}) = u_y(P_{22})$ and $u_n(P_{12}) = u_y(P_{12})$, just as in Example 1; that is, the outer normal direction at the upper edge of the rectangle is the positive y-direction. For the two inner points you obtain two equations, corresponding to (2a) in Example 1, which you label (P_{11}) and (P_{21}); these are the inner points from which (and from whose neighbors) you get the two equations. The left sides of the equations are the same as in Example 1. The right sides differ; they are -2 (-1 from point P_{10} and -1 from point P_{01}) and -2 (-1 from point P_{20} and -1 from P_{31}), respectively. Hence these equations are

$$\begin{array}{ll} (P_{11}) & -4u_{11} + u_{21} + u_{12} = -2 \\ (P_{21}) & u_{11} - 4u_{21} + u_{22} = -2. \end{array} \qquad (B)$$

(Labeling the equations would not be absolutely necessary because the term with the coefficient -4 indicates whose stencil you are considering.) u_{12} and u_{22} are unknown because at P_{12} and P_{22} the normal derivative is given, not the function value u. As in Example 1 you extend the rectangle and the grid in the positive y-direction , introducing the points P_{13} and P_{23} as in Fig. 426b and assuming that the Laplace equation continues to hold in the extended rectangle. Then you can write down two more equations (the analog of (2b) in the example), namely,

$$\begin{array}{ll} (P_{12}) & u_{11} - 4u_{12} + u_{22} + u_{13} = -1 \\ (P_{22}) & u_{21} + u_{12} - 4u_{22} + u_{23} = -1. \end{array} \qquad (C)$$

In (P_{12}) the -1 on the right comes from P_{02}. In (P_{22}) the -1 comes from P_{32}. You now have these two additional equations, at the price of two new unknowns u_{13} and u_{23}, so it looks as if you have gained nothing. However, you have not yet used the condition on the upper edge of the original rectangle (the normal derivative being 1 there), and this is what you do next, just as in Example 1. This gives (since $h = 0.5$)

$$1 = \frac{\partial}{\partial y} u_{12} \approx \frac{u_{13} - u_{11}}{2h} = u_{13} - u_{11},$$

hence

$$u_{13} = u_{11} + 1$$

and

$$1 = \frac{\partial}{\partial y} u_{22} \approx \frac{u_{23} - u_{21}}{2h} = u_{23} - u_{21}.$$

hence

$$u_{23} = u_{21} + 1.$$

Now substitute the expressions for u_{13} and u_{23} into (C) and simplify. In the first of these equations you have $u_{11} + u_{13} = 2u_{11} + 1$. In the second equation, $u_{21} + u_{21} = 2u_{21} + 1$. Taking the terms 1 to the right, the equations in (C) thus become

$$2u_{11} \qquad - 4u_{12} + u_{22} = -2 \tag{D}$$

$$2u_{21} \quad + u_{12} - 4u_{22} = -2.$$

Your system to be solved consists of the four equations in (B) and (D). Its augmented matrix looks as follows. (For better orientation write the unknowns in a row above the matrix, where rs denotes the right side.)

$$
\begin{array}{ccccc}
u_{11} & u_{21} & u_{12} & u_{22} & rs \\
\end{array}
$$

$$
\begin{bmatrix}
-4 & 1 & 1 & 0 & -2 \\
1 & -4 & 0 & 1 & -2 \\
2 & 0 & -4 & 1 & -2 \\
0 & 2 & 1 & -4 & -2
\end{bmatrix}
$$

You can solve this by Gauss elimination. To eliminate u_{11}, use Row 1 as pivot row. Then compute a new matrix

$$
\begin{bmatrix}
-4 & 1 & 1 & 0 & -2 \\
0 & -3.75 & 0.25 & 1 & -2.5 \\
0 & 0.5 & -3.5 & 1 & -3 \\
0 & 2 & 1 & -4 & -2
\end{bmatrix}
\quad
\begin{array}{l}
\\
\text{Row } 2 + 0.25\,\text{Row } 1 \\
\text{Row } 3 + 0.5\,\text{Row } 1 \\
\text{Row } 4
\end{array}
$$

Row 2 is the pivot row and is left unchanged. The next matrix is

$$
\begin{bmatrix}
-4 & 1 & 1 & 0 & -2 \\
0 & -3.75 & 0.25 & 1 & -2.5 \\
0 & 0 & -3.466667 & 1.133333 & -3.333333 \\
0 & 0 & 1.133333 & -3.466667 & -3.333333
\end{bmatrix}
\quad
\begin{array}{l}
\\
\\
\text{Row } 3 + \{0.5/3.75\}\text{Row } 2 \\
\text{Row } 4 + \{2/3.75\}\text{Row } 2
\end{array}
$$

Finally, Row 3 is the pivot row and Row 4 + (1.133333/3.466667) Row 3 is the new Row 4, which is of the form

$$[0 \quad 0 \quad 0 \quad -3.096154 \quad -4.423077].$$

Back substitution now gives, in this order,

$$u_{22} = 1.428571, \quad u_{12} = 1.428571, \quad u_{21} = 1.142857, \quad u_{11} = 1.142857.$$

The result indicates that you could have saved much of this work by using symmetry. In practice, this will not happen too often because it requires that the region shows symmetry and, in addition, the given boundary values exhibit the same symmetry. Our present problem satisfies these conditions; the rectangle and the boundary values are both symmetric with respect to the vertical line $x = 0.75$.

13. Irregular boundary. First make a sketch of your own, showing the region, the grid, and the numerical values of the given boundary data. (Use a red pencil for the latter, in order not to confuse them with notations for the boundary points.) On the x-axis the boundary values at the grid points are $u_{00} = 0$ (which will not be needed), $u_{10} = 3$, $u_{20} = 6$, $u_{30} = 9$ (not needed). On the y-axis you have $u_{01} = u_{02} = u_{03} = 0$. Furthermore, $u = 9 - 3y$ on the right vertical boundary gives $u_{31} = 9 - 3 \cdot 1 = 6$. On the upper horizontal portion of the boundary the potential is 0, hence $u_{13} = 0$. The sloping portion of the boundary is given by

$$y = 4.5 - x. \quad \text{Hence} \quad x = 4.5 - y.$$

For the lowest point of it you have $x = 3$; hence $y = 4.5 - 3 = 1.5$. For the next grid point on it you see that $y = 2$, hence $4.5 - 2 = 2.5$. The next grid point corresponds to $x = 2$, as you see from the figure; hence $y = 4.5 - 2 = 2.5$. For the highest point, $y = 3$, hence $x = 4.5 - 3 = 1.5$. Hence the four points just considered on the sloping portion of the boundary have the coordinates and potential $u(x,y) = x^2 - 1.5x = x(x - 1.5)$ as follows.

$$(3, 1.5) \quad \text{hence} \quad u = 3(3 - 1.5) = 4.5, \tag{A}$$

$$(2.5, 2) \quad " \quad u = 2.5(2.5 - 1.5) = 2.5$$

$$(2, 2.5) \quad " \quad u = 2(2 - 1.5) = 1$$

$$(1.5, 3) \quad " \quad u = 1.5(1.5 - 1.5) = 0.$$

You will need only the second and third of these points and potentials. You are now ready to set up the linear system of equations. You have 4 inner points $P_{11}, P_{21}, P_{12}, P_{22}$. For the first three of these, you obtain equations of the usual form, namely (see the figure and the boundary values given or derived from the given formula referring to the portion of the boundary on the x-axis)

$$(P_{11} :) \quad -4u_{11} + u_{21} + u_{12} \quad\quad = -u_{10} - u_{01} = -3 - 0 = -3$$

$$(P_{21} :) \quad u_{11} - 4u_{21} \quad\quad + u_{22} = -u_{20} - u_{31} = -6 - 6 = -12$$

$$(P_{12} :) \quad u_{11} \quad\quad\quad - 4u_{12} + u_{22} = -u_{02} - u_{13} = -0 - 0 = 0.$$

For P_{22} the situation is as in Fig. 427 in the text with $a = 1/2$ and $b = 1/2$. This case is given as a particular case of (5) on p. 973 at the bottom. From the stencil you see that the two points closer to P_{22} get a weight greater than 1, namely, 4/3, whereas the other two points (the ones that are at the usual distance h from the center P_{22} of the stencil) now each have the reduced weight 2/3 instead of 1; this is physically understandable. Accordingly, your fourth equation changes its usual form

$$u_{21} + u_{12} - 4u_{22} = \ldots$$

to the form

$$(P_{22}:) \quad \frac{2}{3}u_{21} + \frac{2}{3}u_{12} - 4u_{22} = \frac{4}{3}(-2.5) + \frac{4}{3}(-1) = -4.666667.$$

If you had forgotten 4/3 on the right, you could have discovered it by checking whether the sum of the coefficients of all terms when taken to the left equals 0, that is,

$$\frac{2}{3} + \frac{2}{3} - 4 + \frac{4}{3} + \frac{4}{3} = 0.$$

Multiplication of the equation (P_{22}) by 3 gives the simpler form

$$(P_{22}:) \quad 2u_{21} + 2u_{12} - 12u_{22} = -14.$$

The augmented matrix of the linear system thus obtained is

$$\begin{bmatrix} -4 & 1 & 1 & 0 & -3 \\ 1 & -4 & 0 & 1 & -12 \\ 1 & 0 & -4 & 1 & 0 \\ 0 & 2 & 2 & -12 & -14 \end{bmatrix}.$$

By Gauss elimination you obtain the solution

$$u_{11} = 2, \quad u_{21} = 4, \quad u_{12} = 1, \quad u_{22} = 2.$$

Take a look at the solution. Insert these four values in your sketch. Although you do not have too many values inside the region (just four), you can still obtain a qualitative picture of the equipotential lines (curves) in the region. The highest potential (9) is at the right lower corner. Now find $u = 8$ on the x-axis and on the vertical portion and draw the line $u = 8$ in the region; this curve looks like a quarter-circle. Do the same for $u = 7, 6, 5, 4$. For $u = 4$ you have the help that it must pass through P_{21}, that is, $(x, y) = (2, 1)$. Also, you can locate the endpoint of the curve on the sloping portion of the boundary. Next find $u = 3, 2, 1$ on the x-axis as well as on the sloping portion of the boundary. Draw the curves $u = 3, 2$ (passing through the points $(1, 1)$ and $(2, 2)$), and $u = 1$ (passing through the point $(1, 2)$). This gives you a good qualitative picture of the potential as well as the impression that the values obtained by our calculations are reasonable approximate values of the potential at the four inner points.

Sec. 19.6   Methods for Parabolic Equations

Problem Set 19.6. Page 981

1. **Nondimensional form of the heat equation.** $\tilde{x}$ ranges from 0 to L. Hence $x = \tilde{x}/L$ ranges from 0 to 1. Now apply the chain rule, obtaining

$$\tilde{u}_{\tilde{t}} = u_t \frac{dt}{d\tilde{t}} = u_t \frac{c^2}{L^2}$$

and

$$\tilde{u}_{\tilde{x}\tilde{x}} = u_{xx}\left(\frac{dx}{d\tilde{x}}\right)^2 = u_{xx} \cdot \frac{1}{L^2}.$$

Now multiply the heat equation by L^2 and divide it by c^2.

3. **Explicit method for the heat equation.** $h = 1$ is the given step in x-direction. $k = 0.5$ is the given step in t-direction. Hence, to reach $t = 2$, you have to do 4 time steps. The initial temperature is $f(x) = x - 0.1x^2 = x(1 - 0.1x)$. It satisfies the conditions $f(0) = 0$ and $f(10) = 0$ at the ends of the bar. At the grid points $x = 0, 1, 2, ..., 10$ the initial temperature u is

$$x = 0 \quad 1 \quad\; 2 \quad\; 3 \quad\; 4 \quad\; 5 \quad\; 6 \quad\; 7 \quad\; 8 \quad\; 9 \quad\; 10$$
$$u = 0 \quad 0.9 \quad 1.6 \quad 2.1 \quad 2.4 \quad 2.5 \quad 2.4 \quad 2.1 \quad 1.6 \quad 0.9 \quad 0 \quad .$$

In (5) you have $r = k/h^2 = 0.5$. Hence the first term of (5) on the right is 0, and (5) takes the form

$$u_{i,j+1} = 0.5(u_{i+1,j} + u_{i-1,j}). \tag{A}$$

i runs in the x-direction and j in the t-direction (the time direction). In each time row the first value and the last value are 0; this is the temperature at which the ends of the bar are kept at all times. From (A) you see that for obtaining the new value you have to take the arithmetic mean of two values in the preceding time row; one is one x-step to the left and the other one x-step to the right of the value which you want to calculate. Hence the simple computation looks as follows.

$x = 0$	0	1	2	3	4	5	6	7	8	9	10
$t = 0$	0	0.9	1.6	2.1	2.4	2.5	2.4	2.1	1.6	0.9	0
$t = 0.5$	0	0.8	1.5	2.0	2.3	2.4	2.3	2.0	1.5	0.8	0
$t = 1.0$	0	0.75	1.4	1.9	2.2	2.3	2.2	1.9	1.4	0.75	0
$t = 1.5$	0	0.7	1.325	1.8	2.1	2.2	2.1	1.8	1.325	0.7	0
$t = 2.0$	0	0.6625	1.25	1.7125	2.0	2.1	2.0	1.7125	1.25	0.6625	0

9. **Crank-Nicolson method.** Formula (9) was obtained by taking $r = k/h^2 = 1$. Since $h = 0.2$ is required, you must take

$$k = h^2 = 0.04.$$

This is the same value as in Example 1 in the text. Hence you must do 5 time steps to reach $t = 0.2$. The initial temperature is "triangular", $f(x) = x$ if $0 \leq x \leq 0.5$ and $f(x) = 1 - x$ if $0.5 \leq x \leq 1$. Since $h = 0.2$, you need the initial temperature at $x = 0, 0.2, 0.4, 0.6, 0.8, 1.0$. The values are

$$
\begin{array}{cccccc}
i = & 0 & 1 & 2 & 3 & 4 & 5 \\
x = & 0 & 0.2 & 0.4 & 0.6 & 0.8 & 1.0 \\
u = & 0 & 0.2 & 0.4 & 0.4 & 0.2 & 0.
\end{array}
\tag{B}
$$

Step 1 ($t = 0.04$). The use of (9) is required. For $j = 0$ this formula is

$$
4u_{i,1} - u_{i+1,1} - u_{i-1,1} = u_{i+1,0} + u_{i-1,0}.
\tag{C}
$$

In each time row you have 6 values. Now 2 of them are 0 for all t. You have to determine the temperature at the remaining 4 inner points $x = 0.2, 0.4, 0.6, 0.8$, corresponding to $i = 1, 2, 3, 4$. For these values of i you obtain from (C) the system

$$
\begin{aligned}
(i = 1) \quad & 4u_{1,1} - u_{2,1} - u_{0,1} && = u_{2,0} + u_{0,0} \\
(i = 2) \quad & 4u_{2,1} - u_{3,1} - u_{1,1} && = u_{3,0} + u_{1,0} \\
(i = 3) \quad & 4u_{3,1} - u_{4,1} - u_{2,1} && = u_{4,0} + u_{2,0} \\
(i = 4) \quad & 4u_{4,1} - u_{5,1} - u_{3,1} && = u_{5,0} + u_{3,0}.
\end{aligned}
$$

(Perhaps you will find it helpful that we have retained the commas in the indices, although this would not have been absolutely necessary, as Example 1 in the book illustrates.) Since the initial temperature is symmetric with respect to $x = 0.5$ (and the temperature is 0 at both ends!), so is the temperature at the 4 inner points for all t. In formulas,

$$
u_{3,j} = u_{2,j} \quad \text{and} \quad u_{4,j} = u_{1,j}.
\tag{D}
$$

If you insert (D) into your system of four equations and use that $u_{5,j} = u_{0,j} = 0$, you see that the third equation becomes identical with the second one, and the fourth one with the first one. Hence you can restrict yourself to the first and second equations. In these equations, $u_{0,1} = 0$, $u_{0,0} = 0$, and $u_{3,0} = u_{2,0}$ (see (B) or (C) with $j = 0$), so that these equations take the form

$$
\begin{aligned}
(i = 1) \quad & 4u_{1,1} - u_{2,1} = u_{2,0} = 0.4 \tag{E} \\
(i = 2) \quad & -u_{1,1} + 3u_{2,1} = u_{2,0} + u_{1,0} = 0.4 + 0.2 = 0.6,
\end{aligned}
$$

where $3u_{2,1}$ results from $4u_{2,1} - u_{3,1} = 4u_{2,1} - u_{2,1}$. The augmented matrix of this system is

$$
\begin{bmatrix} 4 & -1 & 0.4 \\ -1 & 3 & 0.6 \end{bmatrix}.
$$

By Gauss elimination you obtain the solution

$$
u_{1,1} = 0.163636 = u_{4,1}, \qquad u_{2,1} = 0.254545 = u_{3,1}.
\tag{F}
$$

Step 2 ($t = 0.08$). The matrix of the system remains the same; only the right sides change. 0.4 was $u_{2,0}$. Hence you now have to take $u_{2,1}$ since you are now dealing with $j = 1$. The term 0.6 was the sum of $u_{2,0}$ and $u_{1,0}$. Hence you now have to take $u_{2,1} + u_{1,1}$ as the right side of the second equation. This gives the augmented matrix

$$
\begin{bmatrix} 4 & -1 & 0.254545 \\ -1 & 3 & 0.418182 \end{bmatrix}.
$$

The solution is

$$
u_{1,2} = 0.107438 = u_{4,2}, \qquad u_{2,2} = 0.175207 = u_{3,2}.
$$

Step 3 ($t = 0.12$). The augmented matrix is

$$
\begin{bmatrix} 4 & -1 & 0.175207 \\ -1 & 3 & 0.282645 \end{bmatrix}.
$$

The solution is

$$u_{1,3} = 0.0734786 = u_{4,3}, \qquad u_{2,3} = 0.118708 = u_{3,3}.$$

Step 4 ($t = 0.16$). The augmented matrix is

$$\begin{bmatrix} 4 & -1 & 0.118\,708 \\ -1 & 3 & 0.192\,186 \end{bmatrix}.$$

The solution is

$$u_{1,4} = 0.049846 = u_{4,4}, \qquad u_{2,4} = 0.080678 = u_{3,4}.$$

Step 5 ($t = 0.20$). The augmented matrix is

$$\begin{bmatrix} 4 & -1 & 0.080\,678 \\ -1 & 3 & 0.130\,524 \end{bmatrix}.$$

The solution is

$$u_{1,5} = 0.033869 = u_{4,5}, \qquad u_{2,5} = 0.054798 = u_{3,5}. \qquad \text{(G)}$$

The series in Example 3 of Sec. 11.5 with $L = 1$ and $c = 1$ [the equation is assumed to be $u_t = u_{xx}$, hence $c = 1$; see (1)] is

$$u(x,t) = \frac{4}{\pi^2}\left(\sin \pi x\, e^{-\pi^2 t} - \frac{1}{9} \sin 3\pi x\, e^{-9\pi^2 t} + - \ldots \right).$$

Hence for $t = 0.2$ this becomes

$$u(x,0.2) = \frac{4}{\pi^2}\left(\sin \pi x\, e^{-0.2\pi^2} - \frac{1}{9} \sin 3\pi x\, e^{-1.8\pi^2} + - \ldots \right). \qquad \text{(H)}$$

The second term is already very small because π^2 is about 10 and e^{-18} is about 10^{-8}, and the exponential function in the next term would be e^{-49}, which is about 10^{-21}. Hence the values obtained from the sum of the first two terms of (G) are very accurate. The computation gives

	$x = 0.2$	$x = 0.4$
u in (G)	0.033869	0.054798
u in (H)	0.033091	0.053543
Error of u in G	−0.000778	−0.001255

Your calculation illustrates that, although our h is relatively large, the values obtained by the Crank-Nicolson method are rather accurate. Furthermore, you see that the series in Sec. 11.5 is very useful numerically.

Sec. 19.7 Methods for Hyperbolic Equations

Problem Set 19.7. Page 984

1. **Vibrating string problem (1)-(4).** From (4) you see that the ends of the string are at $x = 0$ and $x = 1$. The initial displacement is given by the parabola

$$f(x) = 0.1x(1 - x).$$

(The factor 0.1 has been included as a reminder that the wave equation (1) was derived under the assumptions that the string has a small displacement and makes small angles with the horizontal x-axis at all times.) The initial velocity of the string is assumed to be zero. Since $h = 0.2$, you need $f(x)$ at 0, 0.2, 0.4, 0.6, 0.8, 1.0, that is,

$$x = \quad 0 \quad 0.2 \quad 0.4 \quad 0.6 \quad 0.8 \quad 1.0$$

$$u(x,0) = f(x) = \quad 0 \quad 0.016 \quad 0.024 \quad 0.024 \quad 0.016 \quad 0 \ . \tag{A}$$

From this and (8) with $g_i = 0$ (the initial velocity is zero!) and $k = 0.2$, as required in the problem, you obtain [also using (4)]

$$u(x,0.2) \quad 0 \quad 0.012 \quad 0.020 \quad 0.020 \quad 0.012 \quad 0 \ . \tag{B}$$

For the remaining calculations you have to use (6), whose right side in each step includes values from *two* preceding time rows. The numerical values obtained [including those in (A) and (B)] are as follows.

$x =$	0	0.2	0.4	0.6	0.8	1.0
$t = 0$	0	0.016	0.024	0.024	0.016	0
$t = 0.2$	0	0.012	0.020	0.020	0.012	0
$t = 0.4$	0	0.004	0.008	0.008	0.004	0
$t = 0.6$	0	−0.004	−0.008	−0.008	−0.004	0
$t = 0.8$	0	−0.012	−0.020	−0.020	−0.012	0
$t = 1.0$	0	−0.016	−0.024	−0.024	−0.016	0
$t = 1.2$	0	−0.012	−0.020	−0.020	−0.012	0
$t = 1.4$	0	−0.004	−0.008	−0.008	−0.004	0
$t = 1.6$	0	0.004	0.008	0.008	0.004	0
$t = 1.8$	0	0.012	0.020	0.020	0.012	0
$t = 2.0$	0	0.016	0.024	0.024	0.016	0

You see that these values correspond to one full cycle because the last line equals the first, so that for continuing t the string starts its next cycle. The reason can be seen from (11*) and (11) in Sec. 11.3 because for $c = 1$ and $L = 1$ you have $\lambda_n t = (cn\pi/L)t = n\pi t$ and for $t = 2$ this equals $2n\pi$, which is a period of the cosine and sine in (11).

7. **Nonzero initial displacement and velocity**. The initial displacement is
$$f(x) = 1 - \cos 2\pi x.$$
Since $h = 0.1$, you need its values at $x = 0, 0.1, ..., 1.0$. Now the curve of $f(x)$ is symmetric with respect to $x = 1/2$, as is clear by inspection; formally it is obtained from the addition formula for the cosine by calculating
$$\cos(2\pi(1-x)) = \cos(2\pi - 2\pi x) = \cos 2\pi \cos 2\pi x + \sin 2\pi \sin 2\pi x$$
$$= 1 \cdot \cos 2\pi x + 0.$$
Hence you may calculate $f(0.1), ..., f(0.5)$ and then use $f(0.6) = f(0.4)$, etc. The values are (6D)

$x =$	0	0.1	0.2	0.3	0.4	0.5
$f(x)$	0	0.190983	0.690983	1.309017	1.809017	2.000000

The initial velocity is
$$g(x) = x - x^2 = x(1-x).$$
Its values for the same x will be needed in (8) to get started. $g(x)$ is also symmetric with respect to $x = 1/2$, so that it suffices to calculate $kg(x) = 0.1g(x)$ for $x = 0, 0.1, ..., 0.5$. We include $f(x)$ for convenience. $u(x,0.1)$ is then calculated from (8) and $u(x,0.2)$ from (6), with 10S and then rounded to 6D. Calculations of $u(x, 0.1)$ and $u(x, 0.2)$ are given after the table.

x	0	0.1	0.2	0.3	0.4	0.5
$f(x)$	0	0.190983	0.690983	1.309017	1.809017	2.000000
$0.1g(x)$	0	0.009	0.016	0.021	0.024	0.025
$u(x, 0.1)$	0	0.354492	0.766000	1.271000	1.678508	1.834017
$u(x, 0.2)$	0	0.575017	0.934509	1.135491	1.296000	1.357017

For $u(x, 0.1)$, formula (8) with $i = 1, 2, \ldots$ gives (we set again commas between the two indices)

$$u_{1,1} = \tfrac{1}{2}(u_{0,0} + u_{2,0}) + 0.1\, g_1 = \tfrac{1}{2}(0 + 0.690983) + 0.009 = 0.354492$$

$$u_{2,1} = \tfrac{1}{2}(u_{1,0} + u_{3,0}) + 0.1\, g_2 = \tfrac{1}{2}(0.190983 + 1.309017) + 0.016 = 0.766000$$

$$u_{3,1} = \tfrac{1}{2}(u_{2,0} + u_{4,0}) + 0.1\, g_3 = \tfrac{1}{2}(0.690983 + 1.809017) + 0.021 = 1.271000$$

$$u_{4,1} = \tfrac{1}{2}(u_{3,0} + u_{5,0}) + 0.1\, g_4 = \tfrac{1}{2}(1.309017 + 2.000000) + 0.024 = 1.678508$$

$$u_{5,1} = \tfrac{1}{2}(u_{4,0} + u_{6,0}) + 0.1\, g_5 = \tfrac{1}{2}(1.809017 + 1.809017) + 0.025 = 1.834017.$$

For the next time row ($t = 0.2$) you have to use (6), obtaining

$$u_{1,2} = u_{0,1} + u_{2,1} - u_{1,0} = 0 + 0.766000 - 0.190983 = 0.575017$$

$$u_{2,2} = u_{1,1} + u_{3,1} - u_{2,0} = 0.354492 + 1.271000 - 0.690983 = 0.934509$$

$$u_{3,2} = u_{2,1} + u_{4,1} - u_{3,0} = 0.766000 + 1.678508 - 1.309017 = 1.135491$$

$$u_{4,2} = u_{3,1} + u_{5,1} - u_{4,0} = 1.271000 + 1.834017 - 1.809017 = 1.296000$$

$$u_{5,2} = u_{4,1} + u_{6,1} - u_{5,0} = 1.678508 + 1.678508 - 2.000000 = 1.357016.$$

PART F.   OPTIMIZATION, GRAPHS

CHAPTER 20.   Unconstrained Optimization. Linear Programming

Sec. 20.1   Basic Concepts. Unconstrained Optimization

Problem Set 20.1. Page 993

3. Cauchy's method of steepest descent. The given function is

$$f(\mathbf{x}) = 2\,(x_1^2 + x_2^2) + x_1 x_2 - 5\,(x_1 + x_2). \tag{I}$$

The given starting value is $\mathbf{x}_0 = [1 \quad -2]^T$. Proceed as in Example 1, beginning with the general formulas and using the starting value later. To simplify notations, denote the components of the gradient of f by f_1 and f_2. The gradient of f is

$$\nabla f(x) = [f_1 \quad f_2]^T = [4x_1 + x_2 - 5 \quad 4x_2 + x_1 - 5]^T.$$

In terms of components,

$$f_1 = 4x_1 + x_2 - 5, \qquad f_2 = 4x_2 + x_1 - 5. \tag{II}$$

Furthermore,

$$\mathbf{z}(t) = [z_1 \quad z_2]^T = \mathbf{x} - t\nabla f(\mathbf{x}) = [x_1 - tf_1 \quad x_2 - tf_2]^T.$$

In terms of components,

$$z_1(t) = x_1 - tf_1, \qquad z_2(t) = x_2 - tf_2. \tag{III}$$

Now obtain $g(t) = f(z(t))$ from $f(\mathbf{x})$ in (I) by replacing x_1 with z_1 and x_2 with z_2. This gives

$$g(t) = 2\,(z_1^2 + z_2^2) + z_1 z_2 - 5\,(z_1 + z_2).$$

Calculate the derivative of $g(t)$ with respect to t, obtaining

$$g'(t) = 4(z_1 z_1' + z_2 z_2') + z_1' z_2 + z_1 z_2' - 5(z_1' + z_2').$$

From (III) you see that $z_1' = -f_1$ and $z_2' = -f_2$. Substitute this and z_1 and z_2 from (III) into $g'(t)$, obtaining

$$g'(t) = 4[(x_1 - tf_1)\,(-f_1) + (x_2 - tf_2)\,(-f_2)] + (-f_1)\,(x_2 - tf_2) + (x_1 - tf_1)\,(-f_2) - 5\,(-f_1 - f_2).$$

Order the terms as follows. Collect the terms containing t and denote their sum by D (suggesting "denominator" in what follows). This gives

$$tD = t[4f_1^2 + 4f_2^2 + f_1 f_2 + f_1 f_2]. \tag{IV}$$

Denote the sum of the other terms by N (suggesting "numerator"), obtaining

$$N = -4x_1 f_1 - 4x_2 f_2 - f_1 x_2 - x_1 f_2 + 5f_1 + 5f_2. \tag{V}$$

With these notations you have $g'(t) = tD + N$. Solving $g'(t) = 0$ for t gives

$$t = -\frac{N}{D}.$$

Step 1. For the given $\mathbf{x} = \mathbf{x}_0 = [1 \quad -2]^T$ you have $x_1 = 1, x_2 = -2$ and from (II)

$$f_1 = -3, \qquad f_2 = -12, \qquad tD = 684\,t, \qquad N = -153,$$

so that

$$t = t_0 = -N/D = 0.223\,684\,211.$$

From this and (II) and (III) you obtain the next approximation $\mathbf{x}_1$ of the desired solution in the form

$$\mathbf{x}_1 = \mathbf{z}(t_0) = [1 - t_0\,(-3) \quad -2 - t_0\,(-12)]^T = [1 + 3t_0 \quad -2 + 12t_0]^T = [1.671\,052\,632 \quad 0.684\,210\,52$$

This completes the first step.

Step 2. Instead of $\mathbf{x}_0$ now use $\mathbf{x}_1$, in terms of components,

$$x_1 = 1.671\,052\,632, \qquad x_2 = 0.684\,210\,526.$$

From this and (II) you obtain $f_1 = 2.368\,421\,053$, $f_2 = -0.592\,105\,263$. From this and (IV) it follows that $tD = 21.035\,318\,56t$, and (V) gives $N = -5.960\,006\,925$. The corresponding solution of $g'(t) = 0$ is

$$t = t_1 = 0.283\,333\,3333.$$

From this and (III) calculate

$$\mathbf{x}_2 = \mathbf{z}(t_1) = [x_1 - t_1 f_1 \quad x_2 - t_2 f_2]^T = [1.000\,000\,000 \quad 0.851\,973\,684]^T.$$

Step 3. Using (II), (IV), (V), and (III), in this order, with $\mathbf{x} = \mathbf{x}_2$, calculate

$$f_1 = -0.148\,016\,316, \quad f_2 = -0.592\,105\,263, \quad tD = 1.665\,296\,316t, \quad N = -0.372\,500\,433,$$

hence $t = t_2 = -N/D = 0.223\,684\,211$ and

$$\mathbf{x}_3 = \mathbf{z}(t_2) = [1.033\,111\,15 \quad 0.984\,418\,283]^T.$$

The further steps give

$$\mathbf{x}_4 = [1.000\,000\,000 \quad 0.992\,696\,070]^T$$
$$\mathbf{x}_5 = [1.001\,633\,774 \quad 0.999\,231\,165]^T$$
$$\mathbf{x}_6 = [0.999\,999\,999 \quad 0.999\,639\,6088]^T$$

and so on. The exact solution is $\mathbf{x} = [1 \quad 1]^T$. This can be seen by substituting $x_1 = y_1 + 1$, $x_2 = y_2 + 1$ into $f(\mathbf{x})$, which transforms it into

$$2\,(y_1^2 + y_2^2) + y_1 y_2 - 5. \qquad \text{(VI)}$$

Except for the -5 this is a quadratic form with the symmetric coefficient matrix

$$\begin{bmatrix} 2 & 1/2 \\ 1/2 & 2 \end{bmatrix}.$$

The eigenvalues of this matrix are 2.5 and 1.5, with eigenvectors $[1 \quad 1]^T$ and $[1 \quad -1]^T$, respectively (derive!). Geometrically, this means that the principal axes of the ellipses in the figure make 45 degree angles with the coordinate axes. Hence if you apply a 45-degree rotation (see p. 320 of the book) to (VI), given by

$$y_1 = (X_1 - X_2)/\sqrt{2}$$
$$y_2 = (X_1 + X_2)/\sqrt{2}$$

(note that cos 45 degrees = sin 45 degrees = $1/\sqrt{2}$), you obtain the function

$$h(\mathbf{X}) = 2(X_1^2 + X_2^2) + \frac{1}{2}(X_1^2 - X_2^2) - 5 = \frac{5}{2}X_1^2 + \frac{3}{2}X_2^2 - 5$$

and the curves $h(\mathbf{X}) = c = const$ (with $c > 5$) are ellipses whose principal axes lie in the $X_1 X_2$-coordinate axes.

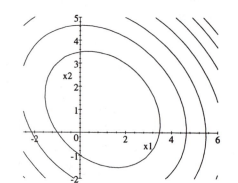

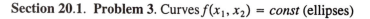

Section 20.1. Problem 3. Curves $f(x_1, x_2) = const$ (ellipses)

Sec. 20.2   Linear Programming

Problem Set 20.2. Page 997

1. **Position of maximum.** Consider what happens as you move the straight line $z = c = const$, beginning with its position when $c = 0$ (which is shown in Fig. 442) and increasing c continuously.

5. **Region, constraints.** The given inequalities are

$$- 0.5x_1 + x_2 \leq 2 \tag{A}$$

$$x_1 + x_2 \geq 2 \tag{B}$$

$$- x_1 + 5x_2 \geq 5. \tag{C}$$

Consider (A). The equation $-0.5x_1 + x_2 = 2$ gives a straight line. Putting $x_2 = 0$ gives $x_1 = -4$ as the intersection point with the x_1-axis. Putting $x_1 = 0$ gives $x_2 = 2$ as the intersection point with the x_2-axis. Putting $x_1 = x_2 = 0$ in the inequality gives $0 + 0 \leq 2$, which is true. Hence the region to be determined extends from the line downward. Similarly for the line $x_1 + x_2 = 2$ in (B), which intersects the axes at $x_1 = 2$ and $x_2 = 2$. Since for $x_1 + x_2 = 0$ the inequality $0 + 0 \geq 2$ is false, the region to be obtained extends from the line upward (away from the origin, leaving the origin outside). Similarly for (C), which gives the line $-x_1 + 5x_2 = 5$, intersecting the axes at $x_1 = -5$ (put $x_2 = 0$) and $x_2 = 1$ (put $x_1 = 0$) and the region extending upward. Hence the region is bounded by a portion of (A) (above), (B) (on the left), and (C) (below), and is unbounded (extends to infinity) on the right. Note that it lies entirely in the first quadrant of the $x_1 x_2$-plane, so that the conditions $x_1 \geq 0$, $x_2 \geq 0$ (often imposed by the kind of application, for instance, number of items produced, time or quantity of raw material needed, etc.) are automatically satisfied.

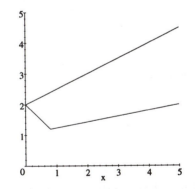

Section 20.2. Problem 5. Region determined by the three inequalities

17. **Maximum profit.** The profit per lamp L_1 is 15 and that per lamp L_2 is 10. Hence if you produce x_1 lamps L_1 and x_2 lamps L_2, the total profit is

$$f(x_1, x_2) = 15x_1 + 10x_2.$$

You wish to determine x_1 and x_2 such that the profit $f(x_1, x_2)$ is as large as possible. Limitations arise due to the available work force. For the sake of simplicity the problem is talking about two workers W_1 and W_2, but it is rather obvious how the corresponding constraints could be modified if teams of workers were involved or if additional constraints arose from raw material. The assumption is that for this kind of high-quality work, W_1 is available 100 hours per month and that he or she assembles 3 lamps L_1 per hour or 2 lamps L_2 per hour. Hence W_1 needs 1/3 hour for assembling a lamp L_2 and 1/2 hour for assembling a lamp L_2. For a production of x_1 lamps L_1 and x_2 lamps L_2 this gives the restriction (constraint)

$$\frac{1}{3}x_1 + \frac{1}{2}x_2 \leq 100. \tag{A}$$

(As in other applications, it is essential to measure time or other physical quantities by the same units throughout a calculation.) (A) with equality sign gives a straight line which intersects the x_1-axis at 300 (put $x_2 = 0$) and the x_2-axis at 200 (put $x_1 = 0$); see the figure. If you put both $x_1 = 0$ and $x_2 = 0$, the inequality becomes $0 + 0 \leq 100$, which is true. This means that the region to be determined extends from that line downward. Worker W_2 paints the lamps, namely, 3 lamps L_1 per hour and 6 lamps L_2 per hour. Hence painting a lamp L_1 takes 1/3 hour, and painting a lamp L_2 takes 1/6 hour. W_2 is available 80 hours per month. Hence if x_1 lamps L_1 and x_2 lamps L_2 are produced per month, his or her availability gives the constraint

$$\frac{1}{3}x_1 + \frac{1}{6}x_2 \leq 80. \tag{B}$$

(B) with the equality sign gives a straight line which intersects the x_1-axis at 240 (put $x_2 = 0$) and the x_2-axis at 480 (put $x_1 = 0$); see the figure. If you put $x_1 = 0$ and $x_2 = 0$ the inequality (B) becomes $0 + 0 \leq 80$, which is true. Hence the region to be determined extends from that line downward. And the region must lie in the first quadrant because you must have $x_1 \geq 0$ and $x_2 \geq 0$. The intersection of those two lines is at $(210, 60)$. This gives the maximum profit $f = 210 \cdot 15 + 60 \cdot 10 = 3750$. The straight line $f = 3750$ (the middlemost of the three lines in the figure) is given by $x_2 = 375 - 1.5x_1$. And by varying c in the line $f = const$, that is, in $x_2 = c - 1.5x_1$, which corresponds to moving the line up and down, it becomes obvious that $(210, 60)$ does give the maximum profit.

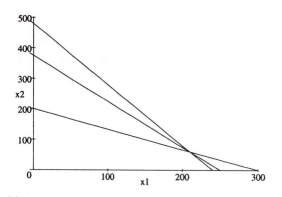

Section 20.1. Problem 17. Constraints (A) (lower line) and (B)

Sec. 20.3   Simplex Method

Problem Set 20.3. Page 1001

1. Maximization by the simplex method. The objective function to be maximized is

$$z = f(x_1, x_2) = 30x_1 + 20x_2. \tag{A}$$

The constraints are

$$-x_1 + x_2 \leq 5, \tag{B}$$
$$2x_1 + x_2 \leq 10.$$

Begin by writing this in normal form (see (1) and (2) in Sec. 20.3). The inequalities are converted to equations by introducing slack variables, one slack variable per inequality. In (A) and (B) you have the variables x_1 and x_2. Hence denote the slack variables by x_3 (for the first inequality in (B)) and x_4. This gives the normal form (with the objective function written as an equation)

$$
\begin{aligned}
z - 30x_1 - 20x_2 &= 0 \\
-x_1 + x_2 + x_3 &= 5 \\
2x_1 + x_2 + x_4 &= 10.
\end{aligned}
\tag{C}
$$

This is a linear system of equations. The corresponding augmented matrix (a concept you should know!–see Sec. 6.3) is called the *initial simplex table* and is denoted by $\mathbf{T}_0$. Obviously it is

$$\mathbf{T}_0 = \begin{bmatrix} 1 & -30 & -20 & 0 & 0 & 0 \\ 0 & -1 & 1 & 1 & 0 & 5 \\ 0 & 2 & 1 & 0 & 1 & 10 \end{bmatrix}. \tag{D}$$

Take a look at (4) on p. 999, which has an extra line on top showing z, the variables, and b (denoting the terms on the right side in (C)). Perhaps you may add such a line in (D) and also draw the dashed lines, which separate the first row of $\mathbf{T}_0$ from the others as well as the columns corresponding to z, to the given variables, to the slack variables, and to the right sides. Perform Operation O_1. The first column with a negative entry in Row 1 is Column 2, the entry being -30. This is the column of the first pivot. Perform Operation O_2. Divide the right sides by the corresponding entries of the column just selected. This gives $5/(-1) = -5$ and $10/2 = 5$. The smallest positive of these two quotients is 5. It corresponds to Row 3. Hence select Row 3 as the row of the pivot. (You cannot choose Row 2 because to eliminate -30, you would have to take Row 1 -30 Row 2 as the new Row 1, which would give $z = 0 - 30 \cdot 5 = -150$ as the value of z, which is impossible since $x_1 \geq 0$ and $x_2 \geq 0$ by assumption.) Perform Operation O_3, that is, create zeros in Column 2 by the row operations

$$\text{Row } 1 + 15 \, \text{Row } 3$$

$$\text{Row } 2 + 0.5 \, \text{Row } 3.$$

This gives the new simplex table (with Row 3 as before)

$$\mathbf{T}_1 = \begin{bmatrix} 1 & 0 & -5 & 0 & 15 & 150 \\ 0 & 0 & 3/2 & 1 & 1/2 & 10 \\ 0 & 2 & 1 & 0 & 1 & 10 \end{bmatrix}.$$

This was the first step. Now comes the second step, which is necessary because of the negative entry -5 in Row 1 of $\mathbf{T}_1$. Hence the column of the pivot is Column 3 of $\mathbf{T}_1$. Calculate $10/(3/2)$ and $10/1$. The first of these is the smallest. Hence the pivot row is Row 2. To create zeros in Column 3 you have to do the row operations Row 1 $+5/(3/2)$ Row 2 Row 3 $-(2/3)$Row 2 , leaving Row 2 unchanged. This gives the simplex table

$$\mathbf{T}_2 = \begin{bmatrix} 1 & 0 & 0 & 10/3 & 50/3 & 550/3 \\ 0 & 0 & 3/2 & 1 & 1/2 & 10 \\ 0 & 2 & 0 & -2/3 & 2/3 & 10/3 \end{bmatrix}.$$

Since no more negative entries appear in Row 1, you are finished. From Row 1 you see that $f_{max} = 550/3$. Row 3 gives the corresponding x_1-value $(10/3)/2 = 5/3$. Row 2 gives the corresponding x_2-value $10/(3/2) = 20/3$. Hence the maximum value of $z = f(x_1, x_2)$ is reached at the point $P : (5/3, 20/3)$ in the $x_1 x_2$-plane.

Draw a sketch of the region determined by the constraints and convince yourself that the maximum value is taken at one of the vertices of the quadrangle determined by the constraints, with vertices at $(0, 0)$, $(5, 0)$, $(0, 5)$, and P.

7. Minimization by the simplex method. The given problem in normal form (with $z = f(x_1, x_2)$ written as an equation) is

$$z - 5x_1 + 20x_2 \qquad\qquad = 0$$

$$- 2x_1 + 10x_2 + x_3 \qquad = 5$$

$$2x_1 + 5x_2 + \qquad x_4 = 10.$$

From this you see that the initial simplex table is

$$\mathbf{T}_0 = \begin{bmatrix} 1 & -5 & 20 & 0 & 0 & 0 \\ 0 & -2 & 10 & 1 & 0 & 5 \\ 0 & 2 & 5 & 0 & 1 & 10 \end{bmatrix}.$$

Since you minimize (instead of maximizing), consider the columns whose first entry is positive (instead of negative). There is only one such column, namely, Column 3. The quotients are $5/10 = 1/2$ (from Row 2) and $10/5 = 2$ (from Row 3). The smallest of these is $1/2$. Hence you have to choose Row 2 as pivot row and 10 as the pivot. Create zeros by the row operations Row $1 - 2$ Row 2 (this gives the new Row 1) and Row $3 - (1/2)$ Row 2 (this gives the new Row 3), leaving Row 2 unchanged. The result is

$$\mathbf{T}_1 = \begin{bmatrix} 1 & -1 & 0 & -2 & 0 & -10 \\ 0 & -2 & 10 & 1 & 0 & 5 \\ 0 & 3 & 0 & -1/2 & 1 & 15/2 \end{bmatrix}.$$

Since in the first row there are no further positive entries, you are done. From Row 1 of $\mathbf{T}_1$ you see that $f_{\min} = -10$. From Row 2 (and columns 3 and 6) you see that $x_2 = 5/10 = 1/2$. From Row 3 (and columns 5 and 6) you see that $x_4 = (15/2)/1 = 15/2$. Now x_4 appears in the second constraint, written as equation, that is,

$$2x_1 + 5x_2 + x_4 = 10.$$

Inserting $x_2 = 1/2$ and $x_4 = 15/2$ gives $2x_1 + 10 = 10$, hence $x_1 = 0$. Hence the minimum -10 of $z = f(x_1, x_2)$ occurs at the point $(0, 1/2)$. Since this problem involves only two variables (not counting the slack variables), as a control and to better understand the problem, you can plot the constraints, which determine a quadrangle, and calculate the values of f at the four vertices, obtaining 0 at $(0, 0)$, 25 (at $(5, 0)$, -7.5 at $(2.5, 1)$, and -10 at $(0, 0.5)$. This confirms your result.

Sec. 20.4   Simplex Method: Degeneracy, Difficulties in Starting

Problem Set 20.4. Page 1007

3. Degeneracy. The given problem is

$$z = x_1 + x_2$$
$$2x_1 + 3x_2 \leqq 130$$
$$3x_1 + 8x_2 \leqq 300$$
$$4x_1 + 3x_2 \leqq 140.$$

Its normal form (with $z = f(x_1, x_2)$ written as an equation) is

$$\begin{aligned} z - x_1 - x_2 &= 0 \\ 2x_1 + 3x_2 + x_3 &= 130 \\ 3x_1 + 8x_2 \qquad + x_4 &= 300 \\ 4x_1 + 2x_2 \qquad\qquad + x_5 &= 140. \end{aligned}$$

From this you see that the initial simplex table is

$$\mathbf{T}_0 = \begin{bmatrix} 1 & -1 & -1 & 0 & 0 & 0 & 0 \\ 0 & 2 & 3 & 1 & 0 & 0 & 130 \\ 0 & 3 & 8 & 0 & 1 & 0 & 300 \\ 0 & 4 & 2 & 0 & 0 & 1 & 140 \end{bmatrix}.$$

The first pivot must be in Column 2 because of the entry -1 in this column. Determine the row of the first pivot by calculating

$$130/2 = 65 \quad \text{(from Row 2)}$$
$$300/3 = 100 \quad \text{(from Row 3)}$$
$$140/4 = 35 \quad \text{(from Row 4).}$$

Since 35 is smallest, Row 4 is the pivot row and 4 the pivot. With this the next simplex table becomes

$$\mathbf{T_1} = \begin{bmatrix} 1 & 0 & -0.5 & 0 & 0 & 0..25 & 35 \\ 0 & 0 & 2 & 1 & 0 & -0.5 & 60 \\ 0 & 0 & 6.5 & 0 & 1 & -0.75 & 195 \\ 0 & 4 & 2 & 0 & 0 & 1 & 140 \end{bmatrix} \begin{array}{l} \text{Row 1 + 0.25 Row 4} \\ \text{Row 2 - 0.5 Row 4} \\ \text{Row 3 - 0.75 Row 4} \\ \text{Row 4} \end{array} .$$

You have reached a point at which $z = 35$. To find the point, calculate

$$x_1 = 140/4 = 35 \quad \text{(from Row 4 and Column 2)}$$
$$x_3 = 60/1 = 60 \quad \text{(from Row 2 and Column 4).}$$

From this and the first constraint you obtain

$$2x_1 + 3x_2 + x_3 = 70 + 3x_2 + 60 = 130, \quad \text{hence} \quad x_2 = 0.$$

(More simply: x_1, x_3, x_4 are basic. x_2, x_5 are nonbasic. Equating the latter to zero gives $x_2 = 0, x_5 = 0$.) Thus $z = 35$ at the point $(35, 0)$ on the x_1-axis.

Column 3 of $\mathbf{T_1}$ contains the negative entry -0.5. Hence this column is the column of the next pivot. To obtain the row of the pivot, calculate

$$60/2 = 30 \quad \text{(from Row 2 and Column 3)}$$
$$195/6.5 = 30 \quad \text{(from Row 3 and Column 3)}$$
$$140/2 = 70 \quad \text{(from Row 4 and Column 3).}$$

Hence you could take 2 or 6.5. Take the first of the two, so that Row 2 is the pivot row. With this calculate the next simplex table

$$\mathbf{T_2} = \begin{bmatrix} 1 & 0 & 0 & 0.25 & 0 & 0.125 & 50 \\ 0 & 0 & 2 & 1 & 0 & -0.5 & 60 \\ 0 & 0 & 0 & -3.25 & 1 & 0.875 & 0 \\ 0 & 4 & 0 & -1 & 0 & 1.5 & 80 \end{bmatrix} \begin{array}{l} \text{Row 1 + 0.25 Row 2} \\ \text{Row 2} \\ \text{Row 3 - 3.25 Row 2} \\ \text{Row 4 - Row 2} \end{array} .$$

There are no more negative entries in Row 1. Hence you have reached the maximum $z_{max} = 50$. You see that x_1, x_2, x_4 are basic, and x_3, x_5 are nonbasic. z_{max} occurs at $(20, 30)$ because $x_1 = 80/4 = 20$ (from Row 4 and Column 2) and $x_2 = 60/2 = 30$ (from Row 2 and Column 3). The point $(20, 30)$ corresponds to a degenerate solution because $x_4 = 0/1 = 0$ from Row 3 and Column 5, in addition to $x_3 = 0$ and $x_5 = 0$. Geometrically, this means that the straight line $3x_1 + 8x_2 = 300$ resulting from the second constraint, also passes through $(20, 30)$ because $3 \cdot 20 + 8 \cdot 30 = 300$. Now in Example 1 of Sec. 20.4 we reached a degenerate solution before we reached the maximum (the optimal solution), and for this reason we had to do an additional step (Step 2). In contrast, in the present problem you reached the maximum when you reached a degenerate solution. Hence no additional work is necessary.

CHAPTER 21.   Graphs and Combinatorial Optimization

Sec. 21.1   Graphs and Digraphs

Problem Set 21.1. Page 1014

7. **Adjacency matrix of a digraph**. The given figure shows 4 vertices, denoted by 1, 2, 3, 4, and 4 edges e_1, e_2, e_3, e_4. This is a digraph (directed graph), not just a graph, because each vertex has a direction. indicated by an arrow head. Thus, edge e_1 goes from vertex 1 to vertex 2, and so on. And there are two edges connecting vertices 1 and 3. These have opposite directions (e_2 from vertex 1 to vertex 3, and e_3 from vertex 3 to vertex 1). In a graph there cannot be two edges connecting the same pair of vertices because we have excluded this (as well as edges going from a vertex back to the same vertex, as well as isolated vertices; see Fig. 446). An adjacency matrix $\mathbf{A}$ of a graph or digraph is always square, $n \times n$, where n is the number of vertices. This is the case, regardless of the number of edges. Hence in the present problem, $\mathbf{A}$ is 4×4. The definition on p. 1013 shows that in our case, $a_{12} = 1$ because the digraph has an edge (namely, e_1), which goes from vertex 1 to vertex 2. Now comes a point about which you should think a little. a_{12} is the entry in Row 1 and Column 2. Since e_{12} goes *from* 1 *to* 2, by definition, the row number is the number of the vertex at which an edge *begins*, and the column number is the number of the vertex at which the edge *ends*. Think this over and look at the matrix in Example 2 on p. 1013. Since there are three edges that begin at 1 and end at 2, 3, 4, and since there is no edge that begins at 1 and ends at 1 (no loop), the first row of $\mathbf{A}$ is

$$0 \quad 1 \quad 1 \quad 1.$$

Since the digraph has 4 edges, the matrix $\mathbf{A}$ must have 4 ones, the three we have just listed and a fourth resulting from the edge that goes from 3 to 1. Obviously, this gives the entry $a_{31} = 1$. In this way you obtain the matrix shown in the answer on p. A44 of the book.

11. **Graph for a given adjacency matrix**. The matrix is 3×3. Hence the graph has 3 vertices. The diagonal entries are always zero since there are no loops. Since all the other entries are 1, every vertex is joined to every other vertex; such a graph is called **complete** (see also Prob. 16). Hence the present graph looks like a triangle. It has 3 edges, in agreement with the fact that each edge in an adjacency matrix creates 2 ones, and our matrix has $3 \cdot 2 = 6$ ones.

19. **Incidence matrix $\widetilde{\mathbf{B}}$ of a digraph**. The incidence matrix of a graph or digraph is an $n \times m$ matrix, where n is the number of vertices and m is the number of edges. Each row corresponds to one of the vertices, and each column to one of the edges. Hence each column contains two ones (in the case of a graph), or a one and a minus one (in the case of a digraph). In Prob. 19 the matrix is square, but just by chance because the number of edges equals the number of vertices. The first column corresponds to the edge e_1, which goes from vertex 1 to vertex 2. Hence, by definition, $\widetilde{b}_{11} = -1$ and $\widetilde{b}_{21} = 1$. Similarly for the other columns shown on p. A45 of the book.

Sec. 21.2   Shortest Path Problems. Complexity

Problem Set 21.2. Page 1019

1. **Shortest path**. s is a vertex that belongs to a hexagon. s has two adjacent vertices, which get the label 1. Each of the latter has one adjacent vertex, which gets the label 2. These two vertices now labeled 2 are adjacent to the last still unlabeled vertex of the hexagon, which thus gets the label 3. This leaves five vertices still unlabeled. Two of these five are adjacent to the vertex labeled 3 and thus get the label 4. The left one labeled 4 is adjcent to the vertex t, which thus gets labeled 5, provided that there is no shorter way for reaching t. There is no shorter way. You could reach t from the right. But the adjacent vertex to the

right of t gets the label 4 because the vertex below it is labeled 3 since it is adjacent to a vertex of the hexagon labeled 2.

9. **Hamiltonian cycle.** For instance, start at s downward, take 3 more vertices on the hexagon H, then the vertex outside H labeled 4, then the vertex inside H, then t, then the vertex to the right of t, then the vertex below it. Then return to H, taking the remaining two vertices of H and return to s.

17. **Postman problem.** The idea is that the postman goes through all the streets (the edges) and returns to the vertex (the post office) from where he came. In the present case the situation is rather obvious. Either he serves vertices 1, 2, 3 first, and then 4, 5, 6 (this corresponds to the solution given on p. A45 in Appendix 2), or conversely. In either case he has to traverse 3 - 4 twice in order to return to the point he started from. This is vertex 1 in the given solution, but any other vertex will do it and will give a shortest walk of the same length (see Prob. 16).

19. **The symbol O.** By definition, an algorithm that is $O(m)$ involves $am + b$ operations. The symbol O is used for conveniently characterizing the behavior of algorithms for large m. Hence you can drop b as being small compared to am. Let k be the number of operations per hour done by the old computer. Then $am = k$ or $m = m_1 = k/a$. The new computer does $100k$ operations per hour. Hence for it, $am = 100k$ or $m = m_2 = 100k/a = 100m_1$, as claimed. For an algorithm that is $O(m^2)$ you can consider cm^2, dropping the term in m and the constant term possibly present. Then you have $cm^2 = k$ for the old computer, hence $m = m_1 = \sqrt{k/c}$. For the new computer you have $cm^2 = 100k$, hence

$$m = m_2 = \sqrt{100\,k/c} = 10\sqrt{k/c} = 10m_1.$$

Similarly in the other cases.

Sec. 21.3   Bellman's Optimality Principle. Dijkstra's Algorithm

Problem Set 21.3. Page 1023

1. **Shortest path.** By inspection:

> Drop 20 because $6 + 14$ does the same.
> Drop 18 because $6 + 8$ is shorter.
> Drop 14 because $8 + 4$ is shorter.

Dijkstra's algorithm runs as follows. (Sketch the figure yourself and keep it handy while you are working.)

Step 1

1. $L_1 = 0$, $\tilde{L}_2 = 6, \tilde{L}_3 = 20, \tilde{L}_4 = 18$. Hence $\mathcal{PL} = \{1\}$, $\mathcal{TL} = \{2,3,4\}$. No ∞ appears because each of the vertices 2, 3, 4 is adjacent to 1, that is, is connected to vertex 1 by a single edge.

2. $L_2 = \min (\tilde{L}_2, \tilde{L}_3, \tilde{L}_4) = \min (6, 20, 18) = 6$. Hence $k = 2$, $\mathcal{PL} = \{1,2\}$, $\mathcal{TL} = \{3,4\}$. Thus you started from vertex 1, as always, and added to the set $\mathcal{PL}$ the vertex which is closest to vertex 1, namely vertex 2. This leaves 3 and 4 with temporary labels. These must now be updated. This is Operation 3 of the algorithm (see the table on p. 1022).

3. Update the temporary label $\tilde{L}_3$ of vertex 3,
   $$\tilde{L}_3 = \min (20, 6 + l_{23}) = \min (20, 6 + 14) = 20.$$
   20 is the old temporary label of vertex 3, and 14 is the distance from vertex 2 to vertex 3, to which you have to add the distance 6 from vertex 1 to vertex 2, which is the permanent label of vertex 2. Update the temporary label $\tilde{L}_4$ of vertex 4,
   $$\tilde{L}_4 = \min (18, 6 + l_{24}) = \min (18, 6 + 8) = 14.$$
   18 is the old temporary label of vertex 4, and 8 is the distance from vertex 2 to vertex 4. Vertex 2 belongs to the set of permanently labeled vertices, and 14 shows that vertex 4 is now closer to this set $\mathcal{PL}$ than it had been before. This is the end of Step 1.

Step 2

1. Extend the set $\mathcal{PL}$ by including that vertex of $\mathcal{TL}$ that is cloest to a vertex in $\mathcal{PL}$, that is, add to $\mathcal{PL}$ the vertex with the smallest temporary label. Now vertex 3 has the temporary label 20, and vertex 4 has the temporary label 14. Accordingly, include vertex 4 in $\mathcal{PL}$. Its permanent label is

$$L_4 = \min \, (\widetilde{L}_3, \widetilde{L}_4) = \min \, (20, 14) = 14.$$

Hence you now have $k = 4$, so that $\mathcal{PL} = \{1, 2, 4\}$ and $\mathcal{TL} = \{3\}$.

2. Update the temporary label $\widetilde{L}_3$ of vertex 3,

$$\widetilde{L}_3 = \min \, (20, 14 + l_{43}) = \min \, (20, 14 + 4) = 18.$$

20 is the old temporary label of vertex 3, and 4 is the distance from vertex 4 (which alredy belongs to $\mathcal{PL}$) to vertex 3.

Step 3

Since only a single vertex, 3, is left in $\mathcal{TL}$, you finally assign the temporary label 18 as the permanent label to vertex 3.

Hence the remaining roads are

                         from vertex 1 to vertex 2   Length 6
                         from vertex 2 to vertex 4   Length 8
                         from vertex 4 to vertex 3   Length 4.

The total length of the remaining roads is 18 and these roads satisfy the condition that they connect all four communities.

Since Dijkstra's algorithm gives a shortest path from vertex 1 to each other vertex, it follows that these shortest paths also provide paths from any of these vertices to every other vertex, as required in the present problem. The solution agrees with the above solution by inspection.

7. **Dijkstra's algorithm.** The procedure is the same as in Example 1 in Sec. 21.3 and as in Prob. 1 just considered. Make a sketch of the graph and use it during your work.

Step 1

1. Vertex 1 gets the permanent label 0. The other vertices get the temporary labels 2 (vertex 2), ∞ (vertex 3), 5 (vertex 4) and ∞ (vertex 5).

The further work is an application of Operation 2 (assigning a permanent label to the (or a) vertex closest to $\mathcal{PL}$) and Operation 3 (updating the temporary labels of the vertices that are still in the set $\mathcal{TL}$ of the temporarily labeled vertices), in alternating order.

2. $L_2 = 2$ (the minimum of 2, 5, and ∞) .

3. $\widetilde{L}_3 = \min \, (\infty, 2 + 3) = 5$.
   $\widetilde{L}_4 = \min \, (5, 2 + 1) = 3$.
   $\widetilde{L}_5 = \min \, (\infty, \infty) = \infty$.

Step 2

1. $L_4 = \min \, (5, 3, \infty) = 3$. Thus $\mathcal{PL} = \{1, 2, 4\}$, $\mathcal{TL} = \{3, 5\}$. Two vertices are left in $\mathcal{TL}$; hence you have to make two updatings.

2. $\widetilde{L}_3 = \min \, (5, 3 + 1) = 4.$
   $\widetilde{L}_5 = \min \, (\infty, 3 + 4) = 7$.

Step 3

1. $L_3 = \min \, (4, 7) = 4$.

2. $\widetilde{L}_5 = \min \, (7, 4 + 2) = 6$.

Step 4

1. $L_5 = \widetilde{L}_5 = 6.$

Your result is as follows.

Step	Vertex added to $\mathcal{PL}$	Permanent label	Edge added to the path	Length of edge
1	1, 2	0, 2	(1, 2)	2
2	4	3	(2, 4)	1
3	3	4	(4, 3)	1
4	5	6	(3, 5)	2

The permanent label of a vertex is the length of the shortest path from vertex 1 to that vertex. Mark the shortest path from vertex 1 to vertex 5 in your sketch and convince yourself that you have omitted three edges of lengths 3, 4, and 5, the edges retained being shorter.

Sec. 21.4   Shortest Spanning Trees. Kruskal's Greedy Algorithm

Example 1. Application of Kruskal's algorithm (p. 1025) We reproduce the list of double labels (Table 21.5, obtained from the rather simple Table 21.4) and give some further explanations to it.

Vertex	Choice 1 (3, 6)	Choice 2 (1, 2)	Choice 3 (1, 3)	Choice 4 (4, 5)	Choice 5 (3, 4)
1		(1, 0)			
2		(1, 1)			
3	(3, 0)		(1, 2)		
4				(4, 0)	(1, 3)
5				(4, 4)	(1, 4)
6	(3, 3)		(1, 3)		

You can now see what the shortest spanning tree looks like, by going through the table line by line. Do this, simultaneously sketching your findings.

Line 1. (1, 0) shows that 1 is a root.

Line 2. (1, 1) shows that 2 is in a subtree with root 1 and is preceded by 1. (This tree consists of the single edge (1, 2).)

Line 3. (3, 0) means that 3 first is a root, and (1, 1) shows that later it is in a subtree with root 1, and then is preceded by 1, that is, joined to the root by a single edge (1, 3).

Line 4. (4, 0) shows that 4 first is a root, and (1, 3) shows that later it is in a subtree with root 1 and is preceded by 3.

Line 5. (4, 4) shows that 5 first belongs to a subtree with root 4 and is preceded by 4, and (1, 4) shows that later 5 is in a (larger) subtree with root 1 and is still preceded by 4. This subtree actually is the whole tree to be found because we are now dealing with Choice 5.)

Line 6. (3, 3) shows that 6 is first in a subtree with root 3 and is preceded by 3, and then later is in a subtree with root 1 and is still preceded by 3.

Problem Set 21.4. Page 1027

1. **Kruskal's algorithm.** Trees constitute a very important type of graph. Kruskal's algorithm is very straightforward. It begins by ordering the edges of a given graph G in ascending order of length. The length of an edge (i, j) is denoted by l_{ij}. Arrange the result in a table similar to Table 21.4 on p. 1026. The

given graph G has $n = 5$ vertices. Hence a spanning tree in G has $n - 1 = 4$ edges, so that you can terminate your table when 4 edges have been chosen.

Edge	Length	Choice
(2, 3)	1	1st
(2, 5)	2	2nd
(1, 3)	3	3rd
(2, 4)	4	4th

The list of double labels looks as follows.

Vertex	Choice 1 (2, 3)	Choice 2 (2, 5)	Choice 3 (1, 3)	Choice 4 (2, 4)
1			(2, 3)	
2	(2, 0)			
3	(2, 2)			
4				(2, 2)
5		(2, 2)		

Explanation:

Choice 1. (2, 0) because 2 is a root of the subtree consisting of edge (2, 3). Furthermore, (2, 2) because 2 is the root of the subtree to which 3 belongs, and 2 is the predecessor of 3 in this subtree.

Choice 2. (2, 2) because 2 is the root of the subtree to which 5 belongs, and 2 is the predecessor of 5 in this subtree. (Sketch it.)

Choice 3. (2, 3) because 2 is the root of the subtree to which 1 belongs, and 3 is the predecessor of 1 in this subtree. (Sketch it.)

Choice 4. (2, 2) because 2 is the root of the subtree (actually, the tree to be determined) to which 4 now belongs, and 2 is the predecessor of 4 in this subtree. Sketch the tree obtained and compute its length (the sum of the lengths of its edges).

15. **Trees that are paths**. Let T be a tree with exactly two vertices of degree 1. Suppose that T is not a path. Then it must have at least one vertex v of degree $d \geq 3$. Each of the d edges incident with v will eventually lead to a vertex of degree 1 (at least one such vertex) because T is a tree, so it cannot have cycles (definition on p. 1015). This contradicts the assumption that T has but *two* vertices of degree 1.

Sec. 21.5   Prim's Algorithm for Shortest Spanning Trees

Problem Set 21.5. Page 1030

7. **Shortest spanning tree obtained by Prim's algorithm**. In each step, U is the set of vertices of the tree T to be grown, and S is the set of edges of T. The beginning is at vertex 1, as always. The table is similar to that in Example 1 on p. 1030. It contains the initial labels and then in each column the effect of relabeling. Explanations follow after the table.

Vertex	Initial	Relabeling		
		(I)	(II)	(III)
2	$l_{12} = 16$	$l_{24} = 4$	$l_{24} = 4$	-
3	$l_{13} = 8$	$l_{34} = 2$	-	-
4	$l_{14} = 4$	-	-	-
5	$l_{15} = \infty$	$l_{45} = 14$	$l_{35} = 10$	$l_{35} = 10$

1. $i(k) = 1$, $U = \{1\}$, $S = \emptyset$. Vertices 2, 3, 4 are adjacent to vertex 1. This gives their inital labels equal to the length of the edges connecting them with vertex 1 (see the table). Vertex 5 gets the initial label ∞ because the graph has no edge $(1, 5)$; that is, vertex 5 is not adjacent to vertex 1.

2. $\lambda_4 = l_{14} = 4$ is the smallest of the initial labels. Hence include vertex 4 in U and edge $(1, 4)$ as the first edge of the growing tree T. Thus, $U = \{1,4\}$, $S = \{(1,4)\}$.

3. Each time you include a vertex in U (and the corresponding edge in S) you have to update labels. This gives the three numbers in column (I) because vertex 2 is adjacent to vertex 4, with $l_{24} = 4$ (the length of edge $(2, 4)$), and so is vertex 3, with $l_{34} = 2$ (the length of edge $(3, 4)$). Vertex 5 is also adjacent to vertex 4, so that ∞ is now gone and replaced by $l_{45} = 14$ (the length of edge $(4, 5)$).

2. $\lambda_3 = l_{34} = 2$ is the smallest of the labels in (I). Hence include vertex 3 in U and edge $(3, 4)$ in S. You now have $U = \{1,3,4\}$ and $S = \{(1,4),(3,4)\}$.

3. Column (II) shows the next updating. $l_{24} = 4$ remains because vertex 2 is not closer to the new vertex 3 than to vertex 4. Vertex 5 is closer to vertex 3 than to vertex 4, hence the update is $l_{35} = 10$, replacing 14.

2. The end of the procedure is now quite simple. l_{24} is smaller than l_{35} in column (II), so that you set $\lambda_2 = l_{24} = 4$ and include vertex 2 in U and edge $(2, 4)$ in S. You thus have $U = \{1,2,3,4\}$ and $S = \{(1,4),(3,4),(2,4)\}$.

3. Updating gives no change because vertex 5 is closer to vertex 3, whereas it is not even adjacent to vertex 2.

2. $\lambda_5 = l_{35} = 10$. $U = \{1,2,3,4,5\}$, so that your spanning tree T consists of the edges $S = \{(1,4),(3,4),(2,4),(3,5)\}$.

13. Prim's algorithm gives the same tree as Kruskal's algorithm did in Sec. 21.4, but the edges appear in a different order, namely,

$$\text{Prim} \quad (1,3) \quad (2,3) \quad (2,5) \quad (2,4)$$

$$\text{Kruskal} \quad (2,3) \quad (2,5) \quad (1,3) \quad (2,4) \ .$$

The table for Prim's algorithm is as follows.

Vertex	Initial label	Relabeling		
		(I)	(II)	(III)
2	$l_{12} = 5$	$l_{12} = 5$	-	-
3	$l_{13} = 3$	-	-	-
4	$l_{14} = 6$	$l_{14} = 6$	$l_{24} = 4$	$l_{24} = 4$
5	$l_{15} = \infty$	$l_{15} = \infty$	$l_{25} = 2$	-

This table is obtained by the algorithm in the following steps (whose explanation is similar to that in Prob. 1).

1. $i(k) = 1$, $U = \{1\}$, $S = \emptyset$
2. $\lambda = l_{13} = 3$, $U = \{1,3\}$, $S = \{(1,3)\}$
3. No change, continue using the initial labels for vertices 2, 4, 5; see column (I)

2. $\lambda_2 = l_{23} = 1, U = \{1,2,3\}, S = \{(1,3),(2,3)\}$
3. $l_{24} = 4, l_{25} = 2$; see column (II)
2. $\lambda_5 = l_{25} = 2, U = \{1,2,3,5\}, S = \{(1,3),(2,3),(2,5)\}$
3. $l_{24} = 4$ remains; see column (III)
2. $\lambda_4 = l_{24} = 4$. This gives the vertex set U and the edge set S of the spanning tree obtained, namely,
   $U = \{1,2,3,4,5\},\ \ S = \{(1,3),(2,3),(2,5),(2,4)\}$.

Sec. 21.6   Networks. Flow Augmenting Paths

Problem Set 21.6. Page 1037

1. **Cut set, capacity.** $S = \{1,2,3\}$ is given and T consists of the other vertices, that is, $T = \{4,5,6\}$. Now look at the figure. Draw a curve that separates S from T. Then you will see that the curve cuts the edge $(1,4)$, whose capacity is 10, the edge $(5,2)$, which is a backward edge, the edge $(3,5)$, whose capacity is 5, and the edge $(3,6)$, whose capacity is 13. By definition, the capacity $\text{cap}\,(S, T)$ is the sum of the capacities of the three forward edges,

$$\text{cap}\,(S,T) = 10 + 5 + 13 = 28.$$

The edge $(5, 2)$ is indeed a backward edge because it goes from vertex 5, which belongs to T, to vertex 3, which is an element of S. And backward edges are not included in the capacity of a cut set, by definition, for reasons given in the text.

11. **Flow augmenting paths** for a given flow in a given network exist only if the given flow is not maximum. 1 - 2 - 3 - 6 is not augmenting because the edge $(2, 3)$ is used to capacity. Similarly, 1 - 4 - 5 - 6 is not augmenting since the edge $(5, 6)$ is used to capacity. For such a small network you can find flow augmenting paths by trial and error (if they exist). For large networks you need an algorithm, such as that of Ford and Fulkerson in the next section.

13. **Flow augmenting path**. The given answer is

$$\Delta_{13} = 8 - 5 = 3, \qquad \Delta_{35} = 11 - 7 = 4, \qquad \Delta_{56} = 13 - 9 = 4.$$

From this you see that the path 1 - 3 - 5 - 6 is flow augmenting and admits an additional flow

$$\Delta = \min\,(3,4,4) = 3.$$

Similarly, the path 1 - 4 - 6 is flow augmenting with

$$\Delta = \min\,(6 - 3, 4 - 1) = \min\,(3,3) = 3.$$

Another flow augmenting path is 1 - 2 - 4 - 6, with

$$\Delta = \min\,(5 - 2, 4 - 2, 4 - 1) = \min\,(3,2,3) = 2.$$

Another flow augmenting path is 1 - 2 - 4 - 5 - 6, with

$$\Delta = \min\,(5 - 2, 4 - 2, 5 - 2, 13 - 9) = \min\,(3,2,3,4) = 2.$$

Another flow augmenting path is 1 - 4 - 5 - 6, with

$$\Delta = \min\,(6 - 3, 5 - 2, 13 - 9) = \min\,(3,3,4) = 3.$$

Any path containing $(4, 3)$ as a forward edge is not flow augmenting because this edge is used to capacity.

17. **Maximum flow**. The given flow in the network in Prob. 13 is 10, as you can see by looking at the two edges $(4, 6)$ and $(5, 6)$ that go into the target t (the sink 6), or by looking at the three edges leaving vertex 1 (the source s), whose flow is $2 + 5 + 3 = 10$. To find the maximum flow in Prob. 13 by inspection, note the following. Each of the three edges outgoing from vertex 1 could carry an additional flow of 3. Hence you may augment the given flow by 3 by using the path 1 - 4 - 5 - 6. Then the edges $(1, 4)$ and $(4, 5)$ are used to capacity . This increases the given flow from 10 to 13. Next you can use the path 1 - 2 - 4 - 6, whose capacity is

$$\Delta = \min (5 - 2, 4 - 2, 4 - 1) = 2.$$

This increases the flow from 13 to 15. For this new increased flow the capacity of the path 1 - 3 - 5 - 6 is

$$\Delta = \min (3, 4, 13 - 12) = 1$$

because the first increase increased the flow in the edge $(5, 6)$ from 9 to 12. Hence you can increase our flow from 15 to 16. Finally, consider the path 1 - 3 - 4 - 6. The edge $(4, 3)$ is a backward edge in this path. By decreasing the existing flow in this edge from 2 to 1 you can push a flow 1 through this path. Then the edge $(4, 6)$ is used to capacity, whereas the edge $(1, 3)$ is still not fully used. But since both edges going to vertex 6, namely, $(4, 6)$ and $(5, 6)$, are now used to capacity, you cannot augment the flow further, so that you have reached the maximum flow 17. The flows in the edges are

$$f_{12} = 4, \quad f_{13} = 7, \quad f_{14} = 6$$
$$f_{24} = 4$$
$$f_{35} = 8$$
$$f_{43} = 1, \quad f_{45} = 5, \quad f_{46} = 4$$
$$f_{56} = 13.$$

Sketch the network with the new flow and check that Kirchhoff's law is satisfied at every vertex.

Sec. 21.7   Ford-Fulkerson Algorithm for Maximum Flow

Problem Set 21.7. Page 1040

1. **Maximum flow.** Example 1 in the text shows how you can proceed in applying the Ford-Fulkerson algorithm for obtaining flow augmenting paths until the maximum flow is reached. No algorithms would be needed for the modest problems in our problem sets. Hence the point of this and similar problems is to obtain familiarity with the most important algorithms for basic tasks in this chapter, as they will be needed for solving large-scale real-life problems. Keep this in mind,. to avoid misunderstandings. From time to time look at Example 1 in the text, which is similar and may help you to see what to do next.

   1. The given initial flow is $f = 6$. This can be seen by looking at the flows 2 in edge $(1, 2)$, 1 in edge $(1, 3)$, and 3 in edge $(1, 4)$, whose sum is 6, or, more simply, by looking at the flows 5 and 1 in the two edges $(2, 5)$ and $(3, 5)$, respectively, that end at vertex 5 (the target t).

   2. Label s (= 1) by Ø. Mark the other edges 2, 3, 4, 5 "unlabeled."

   3. Scan 1. This means labeling the vertices 2, 3, and 4 adjacent to vertex 1 as explained in Step 3 of Table 21. 8 (the table of the Ford-Fulkerson algorithm), which in the present case amounts to the following. $j = 2$ is the first unlabeled vertex in this process, which corresponds to the first part of Step 3 in Table 21.8. You have $c_{12} > f_{12}$ and compute

      $$\Delta_{12} = c_{12} - f_{12} = 4 - 2 = 2 \quad \text{and} \quad \Delta_2 = \Delta_{12} = 2.$$

      Label 2 with the forward label $(1^+, \Delta_2) = (1^+, 2)$.

      $j = 3$ is the second unlabeled vertex adjacent to 1, and you compute

      $$\Delta_{13} = c_{13} - f_{13} = 3 - 1 = 2 \quad \text{and} \quad \Delta_3 = \Delta_{13} = 2.$$

      Label 3 with the forward label $(1^+, \Delta_3) = (1^+, 2)$.

      $j = 4$ is the third unlabeled vertex adjacent to 1, and you compute

      $$\Delta_{14} = c_{14} - f_{14} = 10 - 3 = 7 \quad \text{and} \quad \Delta_4 = \Delta_{14} = 7.$$

      Label 4 with the forward label $(1^+, \Delta_4) = (1^+, 7)$.

   4. Scan 2. This is necessary since you have not yet reached t (vertex 5), that is, you have not yet obtained a flow augmenting path. Adjacent to vertex 2 are the vertices 1, 4, and 5. Vertices 1 and 4 are labeled. Hence the only vertex to be considered is 5. Compute

      $$\Delta_{25} = c_{25} - f_{25} = 8 - 5 = 3.$$

The calculation of Δ_5 differs from the corresponding previous ones. From the table you see that

$$\Delta_5 = \min(\Delta_2, \Delta_{25}) = \min(2, 3) = 2.$$

The idea here is that $\Delta_{25} = 3$ is of no help because in the previous edge $(1, 2)$ you can increase the flow only by 2. Label 5 with the forward label $(2^+, \Delta_5) = (2^+, 2)$.

5. You have obtained a first flow augmenting path P: 1 - 2 - 5.
6. You augment the flow by $\Delta_5 = 2$ and set $f = 6 + 2 = 8$.
7. Remove the labels from 2, 3, 4, 5 and go to Step 3. Sketch the given network, with the new flows $f_{12} = 4$ and $f_{25} = 7$. The other flows remain the same as before. You will now obtain a second augmenting path.
3. Scan 1. Adjacent are 2, 3, 4. You have $c_{12} = f_{12}$; edge $(1, 2)$ is used to capacity and is no longer to be considered. For vertex 3 you compute

$$\Delta_{13} = c_{13} - f_{13} = 3 - 1 = 2 \quad \text{and} \quad \Delta_3 = \Delta_{13} = 2.$$

Label 3 with the forward label $(1^+, 2)$. For vertex 4 compute

$$\Delta_{14} = c_{14} - f_{14} = 10 - 3 = 7 \quad \text{and} \quad \Delta_4 = \Delta_{14} = 7.$$

Label 4 with the forward label $(1^+, 7)$.
3. You need not scan 2 because you now have $f_{12} = 4$ so that $c_{12} - f_{12} = 0$; $(1, 2)$ is used to capacity; the condition $c_{12} > f_{12}$ in the algorithm is not satisfied. Scan 3. Adjacent to 3 are the vertices 4 and 5. For vertex 4 you have $c_{43} = 6$ but $f_{43} = 0$, so that the condition $f_{43} > 0$ is violated. Similarly, for vertex 5 you have $c_{35} = f_{35} = 1$, so that the condition $c_{35} > f_{35}$ is violated and you must go on to vertex 4. Scan 4. The only unlabeled vertex adjacent to 4 is 2, for which you compute

$$\Delta_{42} = c_{42} - f_{42} = 5 - 3 = 2$$

and

$$\Delta_2 = \min(\Delta_4, \Delta_{42}) = \min(7, 2) = 2.$$

Label 2 with the forward label $(4^+, 2)$.
4. Scan 2. Unlabeled adjacent to 2 is vertex 5. Compute

$$\Delta_{25} = c_{25} - f_{25} = 8 - 7 = 1$$

and

$$\Delta_5 = \min(\Delta_2, \Delta_{25}) = \min(2, 1) = 1.$$

Label 5 with the forward label $(2^+, 1)$.
5. You have obtained a second flow augmenting path P: 1 −4 −2 −5.
6. Augment the existing flow 8 by $\Delta_5 = 1$ and set $f = 8 + 1 = 9$.
7. Remove the labels from 2, 3, 4, 5 and go to Step 3. Sketch the given network with the new flows, write the capacities and flows in each edge, obtaining edge $(1, 2)$: $(4, 4)$, edge $(1, 3)$: $(3, 1)$, edge $(1, 4)$: $(10, 4)$, edge $(2, 5)$: $(8, 8)$, edge $(3, 5)$: $(1, 1)$, edge $(4, 2)$: $(5, 4)$, and edge $(4, 3)$: $(6, 0)$. You see that the two edges going into vertex 5 are used to capacity, hence the flow is maximum. Indeed, the algorithm shows that vertex 5 can no longer be reached.

Sec. 21.8   Assignment Problems, Bipartite Matching

Problem Set 21.8. Page 1045

3. **A graph that is not bipartite.** Proceed in the order of the numbers of the vertices. Put vertex 1 into S and its adjacent vertices 2, 3, 4 into T. Then consider 2, which is now in T. Hence for the graph to be bipartite, its adjacent vertices 1 and 4 should be in S. But vertex 4 has just been put into T. This contradicts and shows that the graph is not bipartite.

5. **Bipartite graph.** Since graphs can be graphed in different ways, one cannot see immediately what is going on. Hence in the present problem you have to proceed systematically.

1. Put vertex 1 into S and all its adjacent vertices 2, 4, 6 into T.
2. Vertex 2 is in T. Put its adjacent vertices 1, 3, 5 into S.
3. Vertex 3 is in S. For the graph to be bipartite, its adjacent vertices 2, 4, 6 should be in T, as is the case.
4. Vertex 4 is in T. Its adjacent vertices 1, 3, 5 are in S, as it should be for a bipartite graph.
5. Vertex 5 is in S. Hence for the graph to be bipartite, its adjacent vertices 2, 4, 6 should be in T, as is the case.
6. Vertex 6 is in T, and its adjacent vertices 1, 3, 5 are in S. Since none of these six steps gave any contradiction, conclude that the given graph is bipartite. Sketch this graph in a form that one can see immediately that it is bipartite. Try to write this method in the form of an algorithm and apply it to Prob. 3.

11. **Algorithm for maximum cardinality matching.** Let $S = \{1,2,3\}$. Then $T = \{4,5,6\}$. (You could equally well let $S = \{4,5,6\}$ and $T = \{1,2,3\}$.) Use the algorithm in Table 21.9 on p. 1044. The given matching is $M = \{(1,4),(3,6)\}$.
    1. An exposed vertex is one that is not endpoint of an edge in M. Hence the vertices 2 and 5 are exposed. Vertex 2 is in S and gets the label $\emptyset$.
    2. Vertex 1 is in S and (1, 5) is not in M and 5 is unlabeled. Hence label 5 with 1. Vertex 2 is in S and (2, 5), (2, 6) are not in M and 6 is unlabeled (whereas 5 is labeled). Hence label 6 with 2. Vertex 3 is in S and (3, 4), (3, 5) are not in M and 4 is unlabeled. Hence label 4 with 3.
    3. Vertices 4 and 6 are nonexposed in T. Hence label 1 with 4 (recall that (1, 4) is an edge in M), and label 3 with 6 (recall that (3, 6) is an edge in M).
    4. Now comes backtracking. The only exposed vertex in T is 5. Its label is 1. Hence add (1, 5) as the first edge of the augmenting path P to be obtained. Now vertex 1 has the label 4, and you add (1, 4) to P as the next edge. (Recall that (1, 4) is in M). Vertex 4 has the label 3. Hence add (3, 4) to P. Vertex 3 has the label 6 and you add (3, 6) (which is in M) to P. Vertex 6 has the label 2 and you add (2, 6) to P. The augmenting path obtained is
    $$P : (2,6),(3,6),(3,4),(1,4),(1,5).$$
    5. You now augment the given matching by removing its edges (3,6) and (1,4) from P and taking the other three edges of P, namely,
    $$(2,6),(3,4),(1,5),$$
    as your new matching. You now have a matching of cardinality 3. You need not look for further augmenting paths because this is the maximum possible cardinality if S or T consists of 3 vertices each. Since no vertex is left exposed in the new matching, this is a complete matching, a term mentioned on p. 1042.

19. **K_4 *is planar*** because you can graph it as a square A, B, C, D, then add one diagonal, say, A, C, inside, and then join B, D not by a diagonal inside (which would cross) but by a curve outside the square.

PART G.   PROBABILITY AND STATISTICS

CHAPTER 22.   Data Analysis. Probability Theory

Sec. 22.1   Data: Representation, Average, Spread

Problem Set 22.1. Page 1054

1. **Representation of data.** For the present purpose first order the given data

$$12 \quad 11 \quad 9 \quad 5 \quad 12 \quad 6 \quad 7 \quad 9 \quad 11 \quad 11,$$

obtaining

$$5 \quad 6 \quad 7 \quad 9 \quad 9 \quad 11 \quad 11 \quad 11 \quad 12 \quad 12. \tag{A}$$

As on p. 1051 you obtain a stem-and-leaf plot of these very simple data in the form

$$
\begin{array}{rr|l}
1 & 0 & 5 \\
2 & 0 & 6 \\
3 & 0 & 7 \\
5 & 0 & 9\,9 \\
8 & 1 & 1\,1\,1 \\
10 & 1 & 2\,2
\end{array}
$$

The first column gives the cumulative absolute frequency. Division by 10 (the number of data values) gives the cumulative relative frequency. The absolute frequencies of the values are 1, 1, 1, 2, 3, 2, respectively. Division of these values by 10 gives the relative frequencies. You can summarize these numbers in the following table.

Value	5	6	7	9	11	12
Absolute frequency	1	1	1	2	3	2
Relative frequency	0.1	0.1	0.1	0.2	0.3	0.2
Cumulative absolute frequency	1	2	3	5	8	10
Cumulative relative frequency	0.1	0.2	0.3	0.5	0.8	1.0

An example of a histogram is shown on p. 1051. For drawing it in the present problem you need the relative frequencies in your table. The figure shows the result. For drawing the boxplot of the data you need the quartiles q_L (the lower quartile), q_M (the median or middle quartile), and q_U (the upper quartile). If the number of data values is odd, then the median is one of the values, for instance, if 5 values are given, the median is the third value. If the number of data values is even, it is customary to take the average of the two values in the middle. In the present problem you have 10 values and take the sum of the fifth and the sixth, divided by 2, that is, by (A),

$$q_M = \frac{1}{2}(9 + 11) = 10.$$

There are 5 values 5, 6, 7, 9, 9 below the median; hence the third of them is the lower quartile

$$q_L = 7.$$

Similarly, there are 5 values 11, 11, 11, 12, 12 above the median, and the third of these values is the upper quartile

$$q_U = 11.$$

Observing that the smallest data value is 5 and the largest is 12, you can now draw the boxplot as shown in the figure on the next page.

11. **Mean and variance**. These two quantities (and the standard deviation, which is the square root of the variance) will be much more important in our further work than the quartiles just considered. The mean measures the average size of the data values. It is the arithmetic mean of the values, as defined in (5). In the present problem you have $n = 10$ values and obtain

$$\bar{x} = \frac{1}{10}(5 + 6 + \dots + 12 + 12) = 9.3.$$

This is a sum of 10 values. More simply, you can take the numerically different values in the table in Prob. 1 just considered, multiply each by its absolute frequency, and take the sum. This gives

$$\bar{x} = \frac{1}{10}(5 + 6 + 7 + 2 \cdot 9 + 3 \cdot 11 + 2 \cdot 12) = 9.3.$$

This differs from the median 10 (see Prob. 1 above). It is more useful for measuring the average size of the data values because it takes into account the size of each value, not merely its position in the ordered data.

Using $\bar{x} = 9.3$, you calculate from (6) the variance

$$s^2 = \frac{1}{9}[(5 - 9.3)^2 + (6 - 9.3)^2 + \dots + (12 - 9.3)^2 + (12 - 9.3)^2] = 6.455556.$$

More simply, by taking equal data values together as before, you obtain

$$s^2 = \frac{1}{9}[(5 - 9.3)^2 + (6 - 9.3)^2 + (7 - 9.3)^2 + 2 \cdot (8 - 9.3)^2 + 3 \cdot (11 - 9.3)^2 + 2 \cdot (12 - 9.3)^2] = 6.455556.$$

By taking the square root you obtain the standard deviation

$$s = 2.540779.$$

This differs considerably from the interquartile range IQR= $11 - 7 = 4$ (see Prob. 1 above).

The variance and standard deviation are better for measuring the spread (the variation) of the data because, as in the case of $\bar{x}$, these quantities take into account the size of each data value, not merely its position in the data ordered in ascending order (as in (A) in Prob. 1).

17. **Reduction of data by deleting outliers**. In data such as

$$4, \quad 1, \quad 3, \quad 10, \quad 2$$

the value 10 lies so far away from the other values that you can suspect an error in measuring or recording the data. If you are reasonably sure that some error has occurred, human or by the apparatuses used, you can reduce the data by omitting the obviously erroneous value (or values), called an *outlier*. This will generally change both the mean as well as the variance and hence the standard deviation.

Suppose that those data come from a problem such that you must assume that 10 is the outlier. Its omission will decreases the mean

$$\bar{x} = \frac{1}{5}(4 + 1 + 3 + 10 + 2) = 4$$

to

$$\bar{x}_{red} = \frac{1}{4}(4 + 1 + 3 + 2) = 2.5.$$

Similarly, the variance

$$s^2 = \frac{1}{4}[(4 - 4)^2 + (1 - 4)^2 + (3 - 4)^2 + (10 - 4)^2 + (2 - 4)^2] = 12.5$$

is reduced to

$$s_{red}^2 = \frac{1}{3}[(4 - 2.5)^2 + (1 - 2.5)^2 + (3 - 2.5)^2 + (2 - 2.5)^2] = 1.666667.$$

Now take square roots to see that the standard deviation

$$s = \sqrt{12.5} = 3.535534$$

is reduced to

$$s_{red} = \sqrt{\frac{5}{3}} = 1.290994.$$

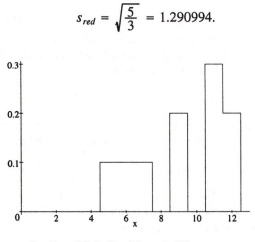

Section 22.1. Problem 1. Histogram

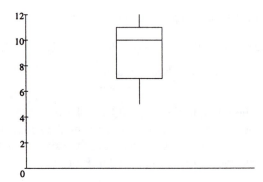

Section 22.1. Problem 1. Boxplot

Sec. 22.2   Experiments, Outcomes, Events

Problem Set 22.2. Page 1057

5. **Sample space**. This sample space S consists of 10 points (outcomes), which you can list as follows (D = Defective, N = Nondefective).

Number of trials	Outcome
1	*D*
2	*ND*
3	*NND*
...	...
10	*NNNNNNNNND*

This is the end of the possibilities because you have only 10 bolts in the lot. If you drew with replacement, your list of outcomes would go on and your sample space would be infinite.

9. Subsets. The empty set, three one-point subsets, three two-point subsets, and S itself.

13. De Morgan's laws. Without referring to any diagrams and just using the definitions you can obtain the first law by starting on the left:

$A \cup B$ is the set of all points in A or B (or both).

Hence its complement

$(A \cup B)^c$ is the set of points neither in A nor in B.

Now on the right,

A^c is the set of points not in A,

B^c is the set of points not in B.

Hence the intersection

$A^c \cap B^c$ is the set of points simultaneously not in A and not in B,

that is, the set of points neither in A nor in B. This proves De Morgan's first law.

In the second law on the left,

$A \cap B$ is the set of points simultaneously in both A and B.

Call this intersection C. Hence the complement

$(A \cap B)^c$ is the set of points not in C.

On the right of the second law,

A^c is the set of points not in A,

B^c is the set of points not in B.

Hence the union

$D = A^c \cup B^c$ is the set of points either not in A or not in B.

Now if a point is in A but not in B, it is in B^c, hence it is in D. Similarly, if a point is in B but not in A, it is in D. This shows that D consists of the points that are not simultaneously in both A and B, that is, D consists of the points not in the intersection $C = A \cap B$. This proves De Morgan's second law.

Sec. 22.3　Probability

Problem Set 22.3. Page 1063

1. Rolling two fair dice. Problems like this amount to counting cases that favor the event in question, namely, the event of obtaining 4, 5, or 6. You can arrange the $6 \cdot 6$ possibilities in a square of 6 rows and 6 columns,

$$
\begin{array}{cccccc}
(1,1) & (1,2) & (1,3) & (1,4) & (1,5) & (1,6) \\
(2,1) & (2,2) & (2,3) & (2,4) & (2,5) & (2,6) \\
(3,1) & (3,2) & (3,3) & (3,4) & (3,5) & (3,6) \\
(4,1) & (4,2) & (4,3) & (4,4) & (4,5) & (4,6) \\
(5,1) & (5,2) & (5,3) & (5,4) & (5,5) & (5,6) \\
(6,1) & (6,2) & (6,3) & (6,4) & (6,5) & (6,6)
\end{array}
$$

where the first number is the number the first die turns up and the second number is the number the second die turns up. The sum 4 corresponds to 3 pairs of values $(1, 3)$, $(2, 2)$, $(3, 1)$ in a sloping line. The sum 5 is obtained from the 4 pairs of values $(1, 4)$, $(2, 3)$. ... in the neighboring sloping line, and the sum 6 from the 5 pairs of values $(1, 5)$, $(2, 4)$, ... in the next sloping line. Together,

$$3 + 4 + 5 = 12$$

of the 36 equally likely cases favor the event in question, which thus has the probability $12/36 = 1/3$.

Of course, you need not write down all these cases; you can find the answer simply by noting that

$$4 = 1 + 3 = 2 + 2 = 3 + 1 \qquad\qquad \text{3 cases}$$
$$5 = 1 + 4 = 2 + 3 = 3 + 2 = 4 + 1 \qquad\qquad \text{4 cases}$$
$$6 = 1 + 5 = 2 + 4 = 3 + 3 = 4 + 2 = 5 + 1 \qquad \text{5 cases}$$

which adds up to 12 cases, as before.

3. Sampling with replacement. "At least one" often suggests considering the complementary event "None". Now you have 30 screws, 10 of them left-handed. Hence the probability of obtaining a left-handed screw is $10/30 = 1/3$. You draw with replacement. Hence before drawing the second screw, you put the screw drawn (which may be left- or right-handed) into the box and mix the content thoroughly, so that the probability of drawing another left-handed screw is the same as before, namely, $10/30 = 1/3$. The point of drawing "with replacement" is that the previous drawings do not influence the further ones. This is the feature of independent events, and you now get from (14) the probability

$$(1/3) \cdot (1/3) = 1/9 = 11\%.$$

This is the probability of the complement "No right-handed screw" of the event "At least one right-handed screw", whose probability you are supposed to determine. Hence the Complementation Rule (Theorem 1) gives the answer $8/9 = 89\%$, approximately.

15. Shooting. Here assume independence in case (b), which is more or less realistic. "At least once" suggests the use of the complementation rule and gives the answer in the form $1 - (3/4)^2 = 0.4375$ shown on p. A47 of Appendix 2 of the book, which is substantially less than in case (a). You can check the result by calculating and summing the probabilities of the three possible outcomes (H = hit, M = miss)

$$HH \qquad HM \qquad MH.$$

Because of independence you obtain

$$\frac{1}{4} \cdot \frac{1}{4} + \frac{1}{4} \cdot \frac{3}{4} + \frac{3}{4} \cdot \frac{1}{4} = \frac{7}{16} = 0.4375.$$

In this and similar problems you may do well to check the results by calculating the probability of the outcomes not yet considered, so that you have available all the probabilities and you can see whether they add up to 1. In the present problem this is quite simple. The probability of MM in case (b) is $(3/4)^2 = 9/16$, which, together with 7/16 above adds up to 1.

19. Extension of Theorem 4 to three events. Theorem 4 states that under the given conditions,

$$P(A \cap B) = P(A)P(B|A). \tag{13}$$

(You will not need the other formula in (13).) The idea of the proof is introducing suitable notations so that you can reduce the present case to that in the theorem. Write D and E instead of A and B in (13). Then you have

$$P(D \cap E) = P(D)P(E|D). \tag{13*}$$

Now comes the trick that will do it. Put

$$D = A \cap B.$$

Since forming intersections is associative, that is,

$$D \cap E = (A \cap B) \cap E = A \cap B \cap E,$$

formula (13*) takes the form

$$P(A \cap B \cap E) = P(A \cap B)P(E|(A \cap B)). \tag{13**}$$

You are almost done. Set $E = C$ on both sides of (13**). (So here you made a little detour because you could have retained C and get away without introducing E; this is typical of many proofs that one first gets enough elbowroom, perhaps more than one eventually needs in completing a proof, but this often not clear at the beginning.) Furthermore, on the right side of (13**) insert $P(A \cap B)$ from (13). Then you obtain precisely the formula to be proved.

Go over the proof before you leave the problem, so that you fully understand the idea of this proof.

Sec. 22.4   Permutations and Combinations

Problem Set 22.4. Page 1068

3. Number of different samples. The order in which you obtain the four objects of a certain sample does not matter. Hence you are concerned with *combinations*, not permutations. A certain object may appear in a sample only once, not twice or more times. Hence you are dealing with combinations *without repetition*. Theorem 3 now gives the answer

$$\binom{50}{4} = \frac{50 \cdot 49 \cdot 48 \cdot 47}{1 \cdot 2 \cdot 3 \cdot 4} = \frac{5\,527\,200}{24} = 230\,300.$$

From the answer you see that counting cases would be hopeless, even in the case of smaller populations.

You could check the result as follows. You have 50 possibilities for picking a first object, then 49 possibilities of picking a second, then 48 for picking a third, and finally 47 for picking a fourth. Hence you have

$$50 \cdot 49 \cdot 48 \cdot 47 = 5527200 \qquad\qquad (A)$$

possibilities of picking 4 objects from the given 50 objects in a given order. Now the order does not matter in sampling; what matters is just which objects you got in your sample. Hence you have to divide (A) by the number of permutations of 4 objects, which is 4! This is the number of different orders in which you can arrange the 4 objects. You thus obtain the same answer as before.

13. Defectives. (a) The lot consists of 6 nondefectives and 2 defectives. The number of samples of size 3 is

$$\binom{8}{3} = \frac{8 \cdot 7 \cdot 6}{1 \cdot 2 \cdot 3} = 56.$$

This follows as in Prob. 3.

(b) If you want no defectives, put them aside and conclude that you now sample from a population of the remaining 6 objects. There are

$$\binom{6}{3} = \frac{6 \cdot 5 \cdot 4}{1 \cdot 2 \cdot 3} = 20$$

different samples of size 3.

(c) 1 defective can be selected from 2 in 2 ways. The other 2 objects in a sample of size 3 can be selected from the 6 nondefective objects in

$$\binom{6}{2} = \frac{6 \cdot 5}{1 \cdot 2} = 15$$

ways. Hence the total number of possibilities is $2 \cdot 15 = 30$ (combine each of the two former with each of the 15 latter possibilities).

(d) 2 defectives leaves you with 1 nondefective in a sample of size 3, that is, with 6 choices.

You can now check your result. (b), (c), (d) exhaust all possibilities; hence the sum of the three values should equal the total number of possibilities, which is 56. Indeed,

$$20 + 30 + 6 = 56.$$

Keep this way of checking in mind; it is useful in other perhaps more involved applications.

16e. Binomial coefficients satisfy a very large number of relationships, a small selection of which is included in Chap. 24 of Ref. [1] in Appendix 1. Formula (14) is one of the most useful ones of them. To prove it, start from

$$(1 + x)^p (1 + x)^q = (1 + x)^{p+q}$$

and develop $(1 + x)^p$, $(1 + x)^q$ as well as the right side in powers of x by means of the binomial theorem. Then equate the coefficients of the power x^r on both sides. On the right you have just one term, namely,

$$\left(\begin{matrix} p+q \\ r \end{matrix} \right) x^r.$$

One the left you have

$$\left[\left(\begin{matrix} p \\ 0 \end{matrix} \right) 1 + \left(\begin{matrix} p \\ 1 \end{matrix} \right) x + \left(\begin{matrix} p \\ 2 \end{matrix} \right) x^2 + \dots + \left(\begin{matrix} p \\ p \end{matrix} \right) x^p \right] \left[\left(\begin{matrix} q \\ 0 \end{matrix} \right) 1 + \left(\begin{matrix} q \\ 1 \end{matrix} \right) x + \left(\begin{matrix} q \\ 2 \end{matrix} \right) x^2 + \dots + \left(\begin{matrix} q \\ q \end{matrix} \right) x^q \right].$$

Now you obtain x^r by multiplying $x^0 \cdot x^r$, then $x \cdot x^{r-1}$, then $x^2 \cdot x^{r-2}$, ..., finally, $x^r \cdot x^0$. These are $r+1$ products. The corresponding coefficients are

$$\left(\begin{matrix} p \\ 0 \end{matrix} \right) \left(\begin{matrix} q \\ r \end{matrix} \right), \quad \left(\begin{matrix} p \\ 1 \end{matrix} \right) \left(\begin{matrix} q \\ r-1 \end{matrix} \right), \quad \left(\begin{matrix} p \\ 2 \end{matrix} \right) \left(\begin{matrix} q \\ r-2 \end{matrix} \right), \quad \left(\begin{matrix} p \\ 3 \end{matrix} \right) \left(\begin{matrix} q \\ r-3 \end{matrix} \right), \dots, \left(\begin{matrix} p \\ r-1 \end{matrix} \right) \left(\begin{matrix} q \\ 1 \end{matrix} \right), \quad \left(\begin{matrix} p \\ r \end{matrix} \right) \left(\begin{matrix} q \\ 0 \end{matrix} \right).$$

The sum of these terms is exactly the left side of (14).

Sec. 22.5   Random Variables, Probability Distributions

Problem Set 22.5. Page 1074

1. **Discrete distribution.** This distribution is discrete. It has the possible values 1, 2, 3. Its probability function $f(x)$ has the values

$$f(1) = 1/14, \quad f(2) = 4/14, \quad f(3) = 9/14;$$

see the figure. These are the probabilities with which the values 1, 2, 3 are assumed. $f(x)$ satisfies (6); otherwise it could not serve as a probability function.

From $f(x)$ you can obtain the distribution function $F(x)$ by summing values of $f(x)$, as indicated in (4), by which the distribution function is defined. Since $x = 1$ is the smallest possible value, you see that $F(x) = 0$ for $x < 1$. At $x = 1$ it jumps up to $1/14 = f(1)$; that is, $F(1) = 1/14$. Then it remains constant and equal to $1/14$ until it reaches $x = 2$, where the next jump occurs, this time of height $4/14 = f(2)$. In formulas,

$$F(x) = f(1) = 1/14 \quad \text{if} \quad 1 \leq x < 2$$

and

$$F(2) = f(1) + f(2) = 1/14 + 4/14 = 5/14.$$

This value is retained until the next (and last) jump occurs at $x = 3$, this time of height $f(3) = 9/14$. Thus

$$F(x) = 5/14 \quad \text{if} \quad 2 \leq x < 3$$

and

$$F(3) = f(1) + f(2) + f(3) = 1/14 + 4/14 + 9/14 = 1.$$

This last value 1 is typical. It is always reached at the last jump and is then retained for all larger x; thus

$$F(x) = 1 \quad \text{if} \quad x \geq 3.$$

If a distribution has infinitely many possible values, then $F(x) = 1$ may not be reached but appear as the sum of the infinite series of the probabilities of the infitely many possible values. An illustration is given in Example 4. Waiting problems are of interest in connection with ticket counters, telephone services, and so on.

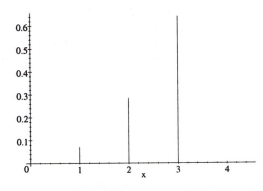

Section 22.5. Problem 1. Probability function

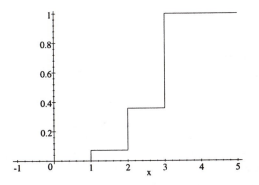

Section 22.5. Problem 1. Distribution function

7. **Continuous distribution. Percentage points. Uniform distribution.** The figures show the density and the distribution function

$$F(x) = \begin{cases} 0 & \text{if } x < 2 \\ \int_2^x 0.25\,dv = 0.25x - 0.5 & \text{if } 2 \leqq x \leqq 6 \\ 1 & \text{if } x > 6. \end{cases}$$

This distribution is called the *uniform distribution* (on the interval from 2 to 6). Note that the area under the curve of $f(x)$ equals 1, as it should be because of (10) on p. 1073. The solutions of (a), (b), and (c) follow from (1), that is,

$$P(X \leqq x) = F(x).$$

(a) $F(c) = 0.25c - 0.5 = 0.9$, hence $c = 4(0.9 + 0.5) = 5.6$. Indeed, $c = 5.6$ is the point such that 90% of the area under the curve of $f(x)$ lies to the left and 10% to the right of this point.

(b) Here you have to use that the complement of $X \geqq x$ is $X < x$. Hence the probability of $X \geqq x$ is 1 minus the probability of $X < x$. Now the given distribution is *continuous*. Hence probabilities are given by integrals, so that $X < x$ has the same probability as $X \leqq x$. But the latter is given by $F(x)$, as follows from (1). Setting $x = c$, you thus have

$$P(X \geqq c) = 1 - F(c) = 0.5,$$

hence

$$F(c) = 0.25c - 0.5 = 0.5, \quad c = 4.$$

This can also be seen by inspection because c is the point such that 50% of the area under the curve of

$f(x)$ lies to the left and 50% to the right of this point.

(c) $F(c) = 0.25c - 0.5 = 0.05$, $c = (0.50 + 0.05)/0.25 = 2.2$. Indeed, 0.2 are 5% of the length 4 of the interval on which $f(x)$ is not zero. The determination of percentage points as in the present problem will occur quite frequently in Chap. 23 on statistical methods, so that you may do well to make sure that you understand the details of this problem, which is particularly simple, due to the fact that $f(x)$ is piecewise constant.

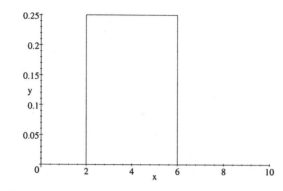

Section 22.5. Problem 7. Density of the uniform distribution on the interval from 2 to 6

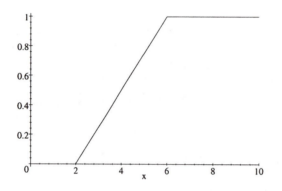

Section 22.5. Problem 7. Distribution function of the uniform distribution on the interval from 2 to 6

19. **Complements of events**, such as those in the present problem, must be considered quite frequently in applications. Now the complement of a set S is the set of all points not in S; see Sec. 22.2. With this and the real line in mind you will obtain the following answers. The complement of $X \leq b$ is $X > b$. Note that $X = b$ is included in the given event, hence it must not occur in the complement. The complement of $X < b$ is $X \geq b$. Here, b belongs to the complement. Similarly in the next two cases. The complement of $X \geq c$ is $X < c$, and the complement of $X > c$ is $X \leq c$. If you have an interval, such as $b \leq X \leq c$, the complement consists of all numbers that lie outside this interval, so that they are either smaller than b or larger than c. The given interval is *closed*; this means that the points b and c are regarded as points of the interval; hence they do not belong to the complement, which therefore is called *open*. The complement consist of the two open infinite intervals $X < b$, stretching to $-\infty$, and $X > c$, stretching to ∞. These are called *infinite intervals* because they extend to $-\infty$ and ∞, respectively. An important point in this problem is the careful distinction between strict inequalities, such as $X < b$, and those that include equality, such as $X \leq b$. In the case of *continuous* distributions it does not matter, as just explained in Prob. 7(b), but in the case of *discrete* distributions it becomes essential.

Sec. 22.6   Mean and Variance of a Distribution

Problem Set 22.6. Page 1078

1. Discrete distribution. The sum of the probabilities of all four possible values is

$$k\left[\binom{3}{0} + \binom{3}{1} + \binom{3}{2} + \binom{3}{3}\right] = k[1 + 3 + 3 + 1] = 8k.$$

According to (6) in Sec. 22.5, this sum must be equal to 1. Hence $k = 1/8 = 0.125$. This gives the probabilities

$$f(0) = 0.125, \quad f(1) = 0.375, \quad f(2) = 0.375, \quad f(3) = 0.125.$$

To obtain the mean, multiply each probability $f(x)$ by the corresponding x and take the sum over all possible values $x = 0, 1, 2, 3$; see (1a). This gives

$$\mu = 0 \cdot 0.125 + 1 \cdot 0.375 + 2 \cdot 0.375 + 3 \cdot 0.175 = 1.5.$$

You can now obtain the variance from (2a) in the form

$$\sigma^2 = (0 - 1.5)^2 \cdot 0.125 + (1 - 1.5)^2 \cdot 0.375 + (2 - 1.5)^2 \cdot 0.375 + (3 - 1.5)^2 \cdot 0.125 = 0.75.$$

Note that the values $x = 0$ and 3 contribute 75% to the variance, three times as much as the other two values, because they lie farther away from the mean than the latter. This is typical.

3. Continuous distribution. The density is

$$f(x) = 2x \quad \text{if} \quad 0 \leq x \leq 1$$

and 0 otherwise. (This cannot be the probability function of a discrete distribution because the possible values of the latter are discrete.) The mean μ is obtained from (1b), where you have to integrate from 0 to 1 only, because $f(x) = 0$ for other values. Check that the area under the curve of $f(x)$ equals 1. This is easy. The integral of $2x$ is x^2, which gives 1 when evaluated at the limits of integration. Hence (10) in Sec. 22.5 is satisfied. By integration you now obtain from (1b)

$$\mu = \int_0^1 x\, 2x\, dx = \frac{2}{3}x^3 \bigg|_0^1 = \frac{2}{3}.$$

You can now obtain the variance from (2b) by integration, again from 0 to 1 only, for the same reason as before.

$$\sigma^2 = \int_0^1 \left(x - \frac{2}{3}\right)^2 (2x)\, dx = \int_0^1 \left(2x^3 - \frac{8x^2}{3} + \frac{8x}{9}\right) dx = \frac{1}{18}.$$

15. Expected value. The expected value (expected gain in a game, expected profit in a business) is the mean value of the corresponding random variable. Actually, in the present problem the random variable is

$$X = \text{Number of turkeys sold},$$

not the profit directly. Notice that 5, 6, 7, 8 are possible values. Now the sum of the corresponding probabilities is 1. Hence 5, 6, 7, 8 are *all* the possible values of X. Equation (1a) thus gives the expected number of turkeys sold on any given day, namely,

$$\mu = 5 \cdot 0.1 + 6 \cdot 0.3 + 7 \cdot 0.4 + 8 \cdot 0.2 = 6.7.$$

Hence the expected sale per day is 6 or 7 turkeys. Multiplication by 3.50 gives the expected profit $6.7 \cdot 3.50 = \$23.45$.

Sec. 22.7   Binomial, Poisson, and Hypergeometric Distributions

Problem Set 22.7. Page 1083

1. Coin tossing, binomial distribution with $p = 1/2$. This is a typical application of the binomial distribution with equal probabilities of success and failure (equal probabilities of heads and tails), provided the coins are fair, so that $p = q = 1/2$, and tossing is done in an orderly way, so that independence of trials is guaranteed. The probability function of the binomial distribution is given by (2), If $p = q = 1/2$, you see that in (2) you have

$$p^x q^{n-x} = \left(\frac{1}{2}\right)^{x+n-x} = \left(\frac{1}{2}\right)^n.$$

Hence in this case, (2) takes the simple form (2*).

In the problem, you toss 5 coins. Hence $n = 5$, and (2*) becomes

$$f(x) = \binom{5}{x}\left(\frac{1}{2}\right)^5, \qquad x = 0, 1, ..., 5. \tag{A}$$

This is the probability of obtaining precisely x heads in a trial, which consists of simultaneously tossing the 5 coins. This is the answer to the first question in the problem.

Since you are supposed to calculate several probabilities, it may be wise to calculate all the 6 probabilities from (A). For the binomial coefficients you obtain

$$1 \quad 5 \quad 10 \quad 10 \quad 5 \quad 1. \tag{B}$$

Multiplication by $(1/2)^5 = 1/32$ gives the corresponding probabilities. The sum of the values in (B) equals 32. Hence the sum of all the probabilities equals 1, as it should be.

From (B) you can readily obtain the answers to the further questions, as follows. You obtain the event *"No heads"* with the probability $f(0) = 1/32$. The complement of this event is *"At least one head."* By the complementation rule (Sec. 22.3) it has the probability $1 - f(0) = 31/32$.

Similarly, the event *"More than 4 heads"* is the same as *"5 heads"* and has the probability $f(5) = 1/32$. The complement is *"Not more than 4 heads"*. By the complementation rule, the corresponding probability is 31/32.

9. Poisson distribution. The information that the resistors have 60 ohms and the characterization of defective ones are irrelevant. Essential are the probability of defectives, 0.1% or $p = 0.001$, and the lot size $n = 200$. Since p is small and n is large, you can expect good approximations from the Poisson distribution. Example 2 will give you some help in doing the present problem, which is even simpler than Example 2.

Obviously, the guarantee will be violated if a lot contains 1 or several defectives. This suggests considering the event

$$\text{"At least one defective in a lot."} \tag{C}$$

"At least one" often is the signal for switching to the complementary event

$$\text{"No defectives in a lot"} \tag{D}$$

and then using the complementation rule (Sec. 22.3). Since $p = 0.001$ and $n = 200$, you will be dealing with the probability function (5) with the mean $\mu = np = 0.2$, that is,

$$f(x) = \frac{0.2^x}{x!}e^{-0.2} = 0.818731\frac{0.2^x}{x!}. \tag{E}$$

Hence the probability of obtaining no defectives in a given lot is $f(0) = 0.818731$. You thus obtain the answer that the probability of at least 1 defective in a lot is $1 - f(0) = 0.181269$; practically, 18%. This is the probability that a given lot will violate the guarantee.

This would be considered as much too large in practice. What you could do is to give the guarantee that there will be at most 1 defective in a lot. This would reduce the violation rate from 18% to about 1.8%. (Can you calculate this?)

In the present problem the exact distribution is the binomial distribution with $p = 0.001$ and $n = 200$, hence $\mu = 0.2$, that is,

$$f(x) = \binom{200}{x}0.001^x(1 - 0.001)^{200-x}.$$

It follows that the probability of obtaining no defectives in a lot is

$$f(0) = (1 - 0.001)^{200} = 0.818649.$$

Hence your approximate value obtained from the Poisson distribution is practically the same.

15. **Multinomial distribution.** You should first note that for $k = 2$ the multinomial distribution reduces to the binomial distribution. Indeed, you first obtain, writing $x_1 = x, x_2 = n - x$ (which follows from $x_1 + ... + x_k = n$ with $n = 2$), $p_1 = p, p_2 = q = 1 - p$ (which follows from $p_1 + ... + p_k = 1$ with $k = 2$)

$$f(x_1, x_2) = \frac{n!}{x_1! x_2!} p_1^{x_1} p_2^{x_2} = \frac{n!}{x! (n-x)!} p^x q^{n-x} = \binom{n}{x} p^x q^{n-x}.$$

From this you see that the idea of the derivation of the formula in the problem is the same as that in the text given for the binomial distribution. Namely, the product of the powers of $p_1, ..., p_k$ is the probability of obtaining in n trials

$$x_1 \text{ times the event } A_1,$$

$$x_2 \text{ times the event } A_2,$$

and so on, in one particular order, for instance, in each of the first x_1 trials you obtain A_1, then in each of the next x_2 trials you obtain $A_2,...$, finally, in each of the last x_k trials you obtain A_k. This is one order of obtaining those events. This probability must now be multiplied by the number of permutations of n things divided into k classes of alike things differing from class to class, because this is the number of permutations of the n outcomes obtained, as given in Theorem 1(b) in Sec. 22.4 (which was also used in the derivation of the binomial distribution, for precisely the same purpose). This number is given by the quotient of the factorials in the formula of the problem. This solves the problem.

Sec. 22.8   Normal Distribution

Problem Set 22.8. Page 1090

1. **Use of normal tables.** If x-values are given and probabilities wanted, use Table A7 in Appendix 5. If probabilities are given and x-values wanted, use Table A8. First make sure that you do understand the difference between these two tasks. Furthermore, those tables are given for the normal distribution with mean 0 and variance 1. This is called the *standardized normal distribution*. And your task in most cases is to apply formula (4), which expresses the distribution function F of the given normal distribution in terms of the distribution function Φ of the standardized normal distribution. Before you do the problem, take another look at Example 2, which is of the same type. In the problem, the given mean is $\mu = 10$, the variance is $\sigma^2 = 4$, hence the standard deviation is $\sigma = 2$. Can you visualize this distribution? Its density is as in Fig. 487, but shifted 10 units to the right (why?) and its shape is even flatter than the flattest curve in that figure (why?). With those values for μ and σ, formula (4) takes the form

$$F(x) = \Phi\left(\frac{x - 10}{2}\right).$$

For x you now have to insert the given value 12 and note that $P(X > 12)$, not $P(X \le 12)$ is wanted. This calls for the use of the complementation rule. From Table 7 you thus obtain

$$P(X > 12) = 1 - P(X \le 12) = 1 - F(12) = 1 - \Phi\left(\frac{12 - 10}{2}\right) = 1 - \Phi(1) = 1 - 0.8413 = 0.1587.$$

This is about 16%.

Is there a connection to the 16% in Fig. 489? There is. You have $\mu + \sigma = 10 + 2 = 12$. Hence you were asking for the probability corresponding to the right "tail" in the left part of Fig. 489.

$P(X < 10)$ asks for the probability that the normal random variable X assume any value between $-\infty$ and the mean. This probability is always 50%, regardless of the size of the mean and the variance, as follows from the fact that the bell-shaped curve of the normal density is symmetric with respect to the mean.

In the next case you calculate $P(X < 11) = P(X \le 11) = F(11) = \Phi(0.5)$ and then use Table A7, obtaining 0.6915.

Similarly in the last case, where you have to find the probability corresponding to an interval and must convert two expressions from F to Φ. Intervals occur often. For this reason the text gives the corresponding formula (5). With the given $a = 9$ and $b = 13$, and with μ and σ as before you obtain

$$F(13) - F(9) = \Phi(1.5) - \Phi(-0.5).$$

Table A7 has no values for negative z, for the good reason that because of the symmetry of the normal density,

$$\Phi(-z) = 1 - \Phi(z).$$

Hence you obtain

$$F(13) - F(9) = \Phi(1.5) - (1 - \Phi(0.5)) = 0.9332 - 1 + 0.6915 = 0.6247.$$

Can you find out whether this last result seems reasonable? Well, the interval has length $4 = 2\sigma$. In Fig. 489(a) (not (b)!) the probability corresponding to the interval of length 2σ equals 68%, as indicated. Your present interval is not symmetrically located with respect to the mean, and this causes a slight loss in probability, as seems obvious from the density curve (the bell-shaped curve). Hence 0.6247 seems reasonable.

13. **Unknown value for a given probability.** This is a problem of the second kind, where you need Table A8. Read first the beginning of Prob. 1, where the distinction is again explained. You will be concerned with a normal distribution whose mean is 1000 and whose standard deviation is 100. Hence its distribution function $F(x)$ is expressed by that of the stadardized normal distribution as shown in (4). With those values you have

$$F(x) = \Phi\left(\frac{x - 1000}{100}\right). \tag{A}$$

Now the problem asks for an x such that the probability of observing any value of X (which is the sick-leave time) greater than x is 20% = 0.2. Also, since the distribution function $F(x)$ gives the probability $P(X \leq x)$, not $P(X > x)$, you have to look for the probability of the event of *not* exceding that "critical" x to be determined. Thus the equation for determining that x is

$$P(X \leq x) = F(x) = 80\% = 0.8.$$

From this and (A) you have

$$\Phi\left(\frac{x - 1000}{100}\right) = 0.8. \tag{B}$$

From this and Table A8 in Appendix 5 you find, corresponding to 80%,

$$z(\Phi) = 0.842.$$

Because of (B) this means that $(x - 1000)/100 = 0.842$. Solving algebraically for x, you finally obtain the answer $x = 1000 + 0.842 \cdot 100 = 1084$. This means that the company should budget about 1084 hours (practically: 1100 hours) of monthly sick-leave time. Note that this rounded figure corresponds to the point $x = \mu + \sigma$, for which the normal probability to the right of it is 16%; see Fig. 489 on p. 1088. This gives you an opportunity of checking whether your result is reasonable.

Sec. 22.9 Distributions of Several Random Variables

Problem Set 22.9. Page 1099

1. **Probabilities for a two-dimensional random variable.** This is a continuous random variable and distribution. Ordinarily you would have to perform integration because in the general case such probabilities are given by (7) or other formulas involving double integrals. In the present problem, however, the density $f(x,y)$ is constant in the rectangle $4 \leq x \leq 10$, $0 \leq y \leq 5$, whose sides are 6 and 5, respectively. The volume under the surface of $f(x,y)$ (which is a portion of a plane over the rectangle and

coincides with the xy-plane outside the rectangle) equals the area $5 \cdot 6 = 30$ of the rectangle times the height k. This volume must equal 1; this is the two-dimensional analog of (10) in Sec. 22.5. Hence $k = 1/30$. The two probabilities can also be obtained by considering rectangles. $X \leq 8$ and $3 \leq Y \leq 4$ corresponds to the rectangle $4 \leq X \leq 8$ and $3 \leq Y \leq 4$, which has width 4 and height 1, hence area 4, so that the first answer is

$$P(X \leq 8,\ 3 \leq Y \leq 4) = \frac{1}{30} \cdot 4 = \frac{2}{15}.$$

Similarly, the portion of the rectangle $9 \leq X \leq 13$ and $0 \leq Y \leq 1$ inside the large rectangle in which $f(x,y)$ is not 0 is $9 \leq X \leq 10$ and $0 \leq Y \leq 1$ and has the area 1. This gives the probability

$$P(9 \leq X \leq 13,\ Y \leq 1) = \frac{1}{30} \cdot 1 = \frac{1}{30}.$$

5. **Marginal density.** The given two-dimensional distribution (called the *uniform distribution in a rectangle*) has the density

$$f(x, y) = \frac{1}{(\beta_1 - \alpha_1)(\beta_2 - \alpha_2)}. \tag{A}$$

for (x,y) in the rectangle, and $f(x,y) = 0$ outside the rectangle. The marginal distributions of a continuous distribution are continuous. Hence they have a density. The density $f_2(y)$ of the marginal distribution of Y is obtained from (16) by integration over x. Since $f(x,y)$ is different from 0 only from $x = \alpha_1$ to $x = \beta_1$ (see Fig. 492), you have to integrate the constant function $f(x,y)$ over x between these limits. This gives $\beta_1 - \alpha_1$ times $f(x,y)$ in (A). But the factor $\beta_1 - \alpha_1$ cancels $\beta_1 - \alpha_1$ in the denominator of $f(x,y)$, so that you obtain

$$f_2(y) = \frac{1}{\beta_2 - \alpha_2} \quad \text{if} \quad \alpha_2 < y < \beta_2$$

and $f_2(y) = 0$ outside this y-interval. Hence the two-dimensional uniform distribution has as its marginal distribution with respect to y a one-dimensional uniform distribution (see Example 2 in Sec. 22.6, except for the notations). Similarly for the marginal distribution with respect to the other random variable, X.

7. **Addition of means and variances.** For adding means use Theorem 1, which is valid regardless of whether the random variables concerned are independent or dependent. You thus obtain for the mean thickness of a core

$$50 \cdot 0.5 + 49 \cdot 0.05 = 27.45 \ [\text{mm}],$$

which equals a little over 1 in. The first term results from the metal layers and the second from the paper layers. To obtain the standard deviation of the cores you have to use the addition theorem for variances (Theorem 3). Hence from the given standard deviation you must first find the variance of a single metal sheet, which is $0.05^2 = 0.0025 \ [\text{mm}^2]$, and of a single paper layer, which is $0.02^2 = 0.0004 \ [\text{mm}^2]$. Theorem 3 requires independence of the random variables concerned. It seems reasonable to assume that this requirement is (practically) satisfied in the present problem. Hence you obtain the variance of a core $50 \cdot 0.0025 + 49 \cdot 0.0004 = 0.1446 \ [\text{mm}^2]$. Note that the paper layers contribute to this value much less than the metal layers do. From this by taking the square root you obtain the standard deviation of the thickness of a core $\sqrt{0.1446} = 0.3803 \ [\text{mm}]$.

CHAPTER 23. Mathematical Statistics

Sec. 23.2 Estimation of Parameters

Problem Set 23.2. Page 1108

1. **Normal distribution.** This problem is similar to Example 1 in the text, but simpler since $\mu = 0$ is given. The likelihood function remains the same, with $\mu = 0$ in the auxiliary function h. Similarly, its logarithm remains

$$\ln l = -n \ln (\sqrt{2\pi}) - n \ln \sigma - h.$$

In Example 1 you had two parameters μ and σ and two equations for determining estimates of them. You now have only one parameter to be estimated, namely, σ, and you use the second equation only. This equation looks the same as before, with $\mu = 0$. By solving it algebraically, as in Example 1, you obtain the estimate

$$\tilde{\sigma}^2 = \frac{1}{n}(x_1 + x_2^2 + \dots + x_n^2).$$

5. **Exponential distribution.** By definition the density of the exponential distribution is

$$f(x) = \theta e^{-\theta x}$$

if $x \geq 0$ and 0 for negative x. It involves the parameter θ, for which you are supposed to find a maximum likelihood estimate. You begin with the likelihood function, which for a given sample $x_1, \dots, x_n$ is

$$l = \theta e^{-\theta x_1} \theta e^{-\theta x_2} \dots \theta e^{-\theta x_n} = \theta^n \exp[-\theta(x_1 + \dots + x_n)].$$

Next you calculate the logarithm of this, obtaining

$$\ln l = n \ln \theta - \theta(x_1 + \dots + x_n).$$

You now take the partial derivative of this with respect to θ and equate it to zero. This gives

$$\frac{n}{\theta} - (x_1 + \dots + x_n) = 0.$$

Solving algebraically for θ and denoting the solution (the desired maximum likelihood estimate) by $\hat{\theta}$ gives

$$\theta = \hat{\theta} = \frac{n}{x_1 + \dots + x_n} = \frac{1}{\bar{x}}.$$

Thus the estimate is the reciprocal of the sample mean. This looks strange, but finds its plausible explanation if you calculate the mean of the distribution, using the general definition (1b) of the mean of a continuous distribution in Sec. 22.6. You then obtain, by integrating by parts and noting that the integral-free expression obtained in this process is zero at both limits of integration

$$\mu = \int_{-\infty}^{\infty} x f(x)\, dx = \int_{0}^{\infty} x \theta e^{-\theta x}\, dx = \frac{1}{\theta}.$$

Hence the maximum likelihood estimate of $\theta = 1/\mu$ is $1/\bar{x}$. This makes good sense. The figure shows the density of the exponential distribution for three values of the parameter θ, namely, $\theta = 1, 2, 3$. You see that for large θ the mean $1/\theta$ is small. This agrees with the fact that then the curve is steep, with the area under the curve concentrated near small values of $x (> 0)$.

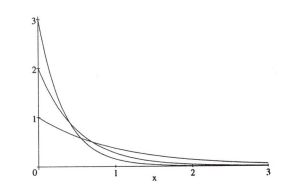

Section 23.2. Problem 5. Densities for $\theta = 1, 2, 3$

Sec. 23.3   Confidence Intervals

Problem Set 23.3. Page 1117

3. **Confidence interval for the mean when the variance is known.** You must sharply distinguish between the two cases of a known σ and an unknown σ.

   Case 1. When σ is known, your calculation (for a normal distribution) will involve only the normal distribution itself.

   Case 2. When σ is unknown, you need another distribution, namely, the t-distribution in Table A9 in Appendix 5.

   The present problem concerns Case 1 because σ is known, $\sigma = 2.5$. Hence you need no further distribution besides the normal distribution. The reason is simple. You can regard the sample mean

$$\bar{x} = \frac{1}{n}(x_1 + \ldots + x_n)$$

as an observed value of the random variable

$$\overline{X} = \frac{1}{n}(X_1 + \ldots + X_n),$$

and $\overline{X}$ is normal with mean μ and variance σ^2/n. Here $X_1, \ldots, X_n$ are independent random variables all having the same distribution, namely, the normal distribution of the population from which the sample is taken, with mean μ and variance σ^2. This is stated in Theorem 1(b). On p. 1111 it is shown that from this theorem there follows formula (7), that is,

$$P(\overline{X} - k \leqq \mu \leqq \overline{X} + k) = \gamma. \tag{A}$$

Here the confidence level γ must be chosen. In our problem it is required that $\gamma = 0.99$. The letter k in (A) is just a short notation for $c\sigma/\sqrt{n}$. In our problem, $k = c \cdot 2.5/\sqrt{6}$. Here c depends on the choice of γ and can be calculated from (6). To make things easier, you find c (as well as all your steps) in Table 23.1, which has been obtained from (6) and (7). There you see that for $\gamma = 0.99$ the critical value c is $c = 2.576$. With this you obtain $k = 2.576 \cdot 2.5/\sqrt{6} = 2.629119$.

   The next step is Step 3, in which you need the sample mean

$$\bar{x} = \frac{1}{6}(30.8 + 30.0 + 29.9 + 30.1 + 31.7 + 34.0) = 31.083\,333.$$

You can now obtain the confidence interval from (3) in the form

$$\text{CONF}_{0.99}(31.083 - 2.629 \leqq \mu \leqq 31.083 + 2.629), \tag{B}$$

that is,

$$\text{CONF}_{0.99}(28.454 \leqq \mu \leqq 33.713). \tag{C}$$

You see that the sample mean $\bar{x}$ is the midpoint of the interval, and its length is $2k$. It is rather long, but this should not surprise you because the sample size 6 is small. If you want a shorter interval, you have to take a larger sample (if you can); see Example 2 and Fig. 494. Or you can take a smaller γ, e.g. 0.95 [if you can allow a larger chance of being wrong, that is, of obtaining an interval that does not contain the unknown population mean μ (1 in about 20 cases if you choose $\gamma = 0.95$)].

Note that in (B) and (C) three digits have been dropped because it would not make too much sense to give 6D-values for the endpoints of an interval that is over 5 units long and is obtained from sample values that are rounded to 1D. Rather, you could drop even more digits if you wish.

9. **Confidence interval for the mean when the variance is unknown.** This is Case 2 (see the beginning of the previous problem), in which the normal distribution is no longer sufficient, but you need the t-distribution. Values needed for the present purpose are given in Table A9 in Appendix 5.

What is the difference compared to Case 1? We had $k = \sigma c/\sqrt{n}$. This can no longer be used because σ is unknown. Instead you have to use $k = sc/\sqrt{n}$, as shown in Table 23.2 (Sec. 23.3), with c determined from equation (9), that is, since $\gamma = 0.99$ is required,

$$F(c) = (1/2)(1 + \gamma) = (1/2)(1 + 0.99) = 0.995.$$

The size of the given sample is $n = 20$. Hence the number of degrees of freedom of the t-distribution to be used is $n - 1 = 19$. For $F(z) = 0.995$ you find in the column for 19 degrees of freedom the value $c = 2.86$. This is your value of c in Step 4 of Table 23.2, p. 1112. The variance of a sample of size $n = 20$ is given as $s^2 = 0.09$ cm^2. Hence the sample standard deviation is $s = 0.3$. With these values you can now calculate in Step 4 of Table 23.2 (p. 1112)

$$k = s c/\sqrt{n} = 0.3 \cdot 2.86/\sqrt{20} = 0.191855.$$

The given sample mean is $\bar{x} = 15.50$. This is the midpoint of the confidence interval, whose endpoints are $\bar{x} - k$ and $\bar{x} + k$. See (10) in Table 23.2 on p. 1112. Numerically, by inserting $\bar{x}$ and k,

$$\text{CONF}_{0.99}(15.308 \leq \mu \leq 15.692).$$

17. **Functions of random variables.** If X is normal with mean 40 and variance 4, then $3X$ has 3 times the mean of X, which is 120, and $3^2 = 9$ times the variance of X, which equals 36. This follows from Team Project 14(g) in Sec. 22.8, which concerns transformations of the present kind.

Similarly, $5X - 2$ has the mean $5 \cdot 40 - 2 = 198$ and the variance $5^2 \cdot 4 = 100$. Note that the variance is translation invariant, that is, -2 in $5X - 2$ has no effect on the variance of $5X - 2$.

Sec. 23.4   Testing of Hypotheses, Decisions

Problem Set 23.4. Page 1127

1. **Test for the mean when the variance is unknown.** The sample is

$$1, \quad -1, \quad 1, \quad 3, \quad -8, \quad 6, \quad 0.$$

Normality of the corresponding population is assumed. You are requested to test the hypothesis

$$\mu = \mu_0 = 0 \tag{A}$$

against the alternative

$$\mu > 0. \tag{B}$$

As in the previous section you must sharply distinguish between two cases.
Case 1. The variance σ^2 of the population is known (pp. 1122-1124),
Case 2. The variance of the population is unknown (pp.1124-1125).

The present problem belongs to Case 2 because the variance σ^2 of the population is not given. Hence you proceed as in Example 3 on p. 1124. That is, you use the t-distribution (Table A9 in Appendix 5), as follows. From the sample you calculate an observed value

$$t = \frac{\bar{x} - \mu_0}{s/\sqrt{n}} = \frac{x}{s/\sqrt{n}}. \tag{C}$$

of the random variable

$$T = \frac{\bar{X} - \mu_0}{S/\sqrt{n}} = \frac{\bar{X}}{S/\sqrt{n}} \tag{D}$$

used in Example 3. Here S^2 is a random variable such that the sample variance s^2 is an observed value of S^2. This variable S^2 is given in (12), p. 1114, in the section on confidence intervals. This is not just by chance, but the determination of confidence intervals and the testing of hypotheses are based on the same theory; they are related tasks resulting by looking at situations from two different angles.

From Table A9 you calculate a critical value c, which you then compare with t in (C). From (A) and (B) you see that the test is right-sided. This is illustrated in the upper part of Fig. 497 (p. 1120). Since $\alpha = 5\%$ is required, your c will be the 95%-point of the t-distribution with $n - 1 = 7 - 1 = 6$ degrees of freedom ($n = 7$ is the sample size). In the row of Table A9 for $F(z) = 0.95$ and the column for 6 degrees of freedom you find the value

$$c = 1.94. \tag{E}$$

You now calculate t from (C) and compare it with c. If $t \leq c$, the hypothesis is accepted. If $t > c$, the hypothesis is rejected.

The calculation will give you $\bar{x} = 0.286$ (actually, 2/7, but it would not make sense to carry along more digits), and $s = 4.309$ (recall that we use (2), p. 1106, with $n - 1 = 6$ in the denominator, which is better than n, perhaps used by your CAS, Maple, for instance). From this and (C) you obtain

$$t = \frac{0.286 - 0}{4.309/\sqrt{7}} = 0.18 < c = 1.94.$$

Hence the hypothesis is accepted.

17. **Comparison of means.** The answer on p. A49 in Appendix 2 gives you the idea of how to proceed. Case A (paired comparison) would not be appropriate because there is no indication that the automobiles used to obtain the two samples were the same, and even if they were, you would have to know the two corresponding values of the mileage, so that you could not pair. Since the samples sizes are the same ($n_1 = n_2 = 16$), you can use (12), which is obtained from (11) by obvious algebraic simplifications. Since the means and the standard deviations are given, the solution of the problem amounts to inserting these values into (12). This gives $t_0 = 3.328$.

The hypothesis is that brand B is not better than brand A. The alternative is that B is better than A. This is a right-sided test. You obtain c from Table A9 in Appendix 5 with $n_1 + n_2 - 2 = 32 - 2 = 30$ degrees of freedom, as is mentioned without proof in Example 5. No α is given. $\alpha = 5\%$ or $\alpha = 1\%$ are the usual choices. In the present application the table gives 1.70 as the 95%-point and 2.46 as the 99%-point. Hence for both choices of α the hypothesis is rejected because 3.328 is larger than both of these values. Hence on the basis of the given samples you can assert that brand B is significantly better than brand A; the higher values are not just due to chance effects.

Sec. 23.5   Quality Control

Problem Set 23.5. Page 1132

1. **Control of mean and standard deviation.** The sample mean $\bar{x}$ is an observed value of $\bar{X}$ (defined by (4) on p. 1110), which is normal with mean 1 (if the hypothesis is true) and standard deviation $\sigma/\sqrt{4} = 0.01$. The solution of the problem follows from (1) on p. 1129. You should understand that this is a two-sided test with $\alpha = 1\%$, the values -2.58 and 2.58 in (1) corresponding to the 0.5% and 99.5% points of the standardized normal distribution, and $\sigma/\sqrt{n}$ being the standard deviation of $\bar{X}$. The upper part of Fig. 501 indicates that the area under the density curve of $\bar{X}$ is divided into three parts, namely, 0.5% lies above

UCL, 0.5% below LCL, and 99% between these two points. Since $n = 4$, your simple calculation gives

$$\text{LCL} = 1 - 2.58 \cdot \frac{0.02}{\sqrt{n}} = 0.9742, \quad \text{UCL} = 1.0258 .$$

9. **Number of defectives.** If in a production process, $p\%$ (for instance, 3%, thus $p = 0.03$) of the items are defective, then the probability of obtaining x defectives in n independent trials is given by the binomial distribution, whose probability function is (2) in Sec. 22.7. This distribution has the mean np and the variance $npq = np(1 - np)$. Now since you are required to use the three-sigma limits as LCL and UCL on a control chart for the mean, you obtain the formulas in the answer, which are

$$\text{LCL} = \mu - 3\sigma = np - 3\sqrt{npq}, \quad \text{UCL} = \mu + 3\sigma = np + 3\sqrt{npq} .$$

This explains these formulas. Note that in the case of the normal distribution the three-sigma limits would correspond to a choice of $\alpha = 0.3\%$ (see (6c) in Sec. 22.8).

Sec. 23.6   Acceptance Sampling

Problem Set 23.6. Page 1136

1. **Sampling plan with $c = 1$.** This means that a lot is accepted if it contains one or no dull knive. Conclude from this that in (3) you have to take the sum of the first two terms (corresponding to $x = 0$ and $x = 1$). The figure shows the corresponding OC curve, given by

$$P(A, \theta) = e^{-\mu}(1 + \mu) = e^{-20\theta}(1 + 20\theta).$$

From this formula you obtain the three specific values given in the answer, namely, 0.9825 ($\theta = 1\% = 0.01$), 0.9384 ($\theta = 2\%$), and 0.4060 ($\theta = 10\%$).

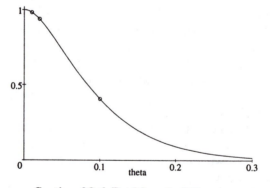

Section 23.6. Problem 1. OC curve

9. **Influence of the sample size n.** The formula in the answer is

$$P(A; \theta) = (1 - \theta)^n + n\theta(1 - \theta)^{n-1}.$$

This is the sum of the first two terms of the binomial distribution with probability of success $p = \theta$ (the probability of drawing a defective in a single trial), giving the probability of obtaining none or 1 defective in n independent trials; practically speaking, in drawing a small sample from a very large lot. The curves are the steeper the larger the sample size is.

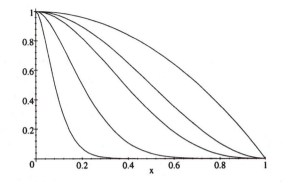

Section 23.6. Problem 9. Sampling plans with $c = 1$ and $n = 2$ (upper curve), 3, 4, 8, 20

Sec. 23.7   Goodness of Fit. Chi-Square Test

Problem Set 23.7. Page 1140

1. **Coin flipping.** By definition, a coin is fair if heads and tails have the same probability, namely, 1/2. Hence in 100 trials you should expect 50 heads and 50 tails, except for random deviations. The problem thus amounts to deciding whether a result of 40 heads and 60 tails could still be possible if the coin is fair, that is, the result could still be due to randomness, or whether the deviation is significant, that is, whether one must assume that the coin is not fair. Accordingly, you have to test the hypothesis that the coin is fair, against the alternative that it is not fair. To perform a chi-square test, calculate χ_0^2 from (1) in Table 23.7 (Sec. 23.7). For this you need

$$b_1 = 40 \quad \text{Number of heads observed}$$

$$e_1 = 50 \quad \text{Number of heads expected if the hypothesis is true}$$

and similarly

$$b_2 = 60 \quad \text{Number of tails observed}$$

$$e_2 = 50 \quad \text{Number of tails expected if the hypothesis is true.}$$

Thus $K = 2$ in Table 23.7, and you obtain a sum of two terms, namely,

$$\chi_0^2 = \frac{(b_1 - e_1)^2}{e_1} + \frac{(b_2 - e_2)^2}{e_2} = \frac{(40 - 50)^2}{50} + \frac{(60 - 50)^2}{50} = 4.$$

(If you want to proceed literally as in Table 23.7, assign, for instance, $x = 1$ to 'head' and $x = 0$ to 'no head' (that is, 'tail') and then divide the x-axis into 2 intervals, one containing $x = 0$ and the other $x = 1$; this gives the formula for χ_0^2.) The significance level $\alpha = 5\%$ is required in the problem. You now obtain the critical c from the table of the χ^2-distribution (Table A10 in Appendix 5). For $K - 1 = 1$ degree of freedom you find that the equation $P(\chi^2 \leq c) = 1 - \alpha = 0.95$ has the solution $c = 3.84$. Now the above $\chi_0^2 = 4$, which measures the deviation, is greater than this critical value. This means that if the hypothesis is true, the probability that an observed value of χ_0^2 falls anywhere in the interval from 3.84 to ∞ is only 5%. Since in the present case this has happened, you reject the hypothesis and assert that the coin used in obtaining that sample is not fair. 4 is not much greater than 3.84, and Table A10 shows you (in the column for 1 degree of freedom) that $\alpha = 2.5\%$ (thus 97.5%, the next value given in the table) would lead to the acceptance of the hypothesis. This is a common situation that should not surprise you.

Sec. 23.8 Nonparametric Tests

Problem Set 23.8. Page 1143

3. **Sign test**. From the results of the 11 trials drop the three that do not contribute to the decision; this is as in Example 1 of the text. This leaves you with 8 results. 7 of them (when A turned out to be better than B) you can regard as positive and the remaining 1 (when B was better than A) as negative. The hypothesis is that A is not better than B, The alternative is that A is better than B. This you can infer from the wording of the problem. Under the hypothesis, positive and negative values should have the same probability $p = q = 1/2$. Consider the random variable $X = $ *Number of positive values among* 8 *values*. If the hypothesis is true, its possible values have the probabilities

$$f(x) = P(X = x) = \binom{8}{x}\left(\frac{1}{2}\right)^8 = 0.00391 \binom{8}{x}.$$

This is a binomial distribution with $p = 1/2$; see Sec. 22.7. Conclude from this that under the hypothesis the probability of observing so many positive values, 7 or even more (namely, 8) is given by

$$f(7) + f(8) = 0.00391(8 + 1) = 0.0352.$$

This is less than the usual 5%, so you reject the hypothesis and assert that filters of type A are better than filters of type B. Note that in this test the hypothesis would be accepted under a significance level $\alpha = 1\% = 0.01$.

11. **Test for trend**. The sample is

22 19 21 20 25 18 27 30 26 24.

Proceed as in Example 2 of the text by listing the transpositions in the sample, beginning with the first sample value, 22, then considering the next, 19, and so on, in the given order.

22	precedes	19 21 20 18	4	transpositions
19	"	18	1	"
21	"	20 18	2	"
20	"	18	1	"
25	"	18 24	2	"
27	"	26 24	2	"
30	"	26 24	2	"
26	"	24	1	"

Hence the sample contains 15 transpositions. Its size is $n = 10$. The hypothesis is that there is no trend, that is, it makes no difference whether the animals receive different amounts of food (of course, within reasonable limits, which are not indicated in the problem). The problem requires that you test this against the alternative of positive trend.

Let $T = $ *Number of transpositions*, as in the Example 2 of the text. You need the probability

$$P(T \le 15) \tag{A}$$

of *f* obtaining 15 or fewer transpositions in a sample of $n = 10$ values. For this, Table A12 in Appendix 5 gives the probability 0.108, almost 11%, certainly large enough for accepting the hypothesis that there is no trend. If in an experiment a somewhat unexpected result turns up, as in the present case, one should indicate precisely under what conditions the result was obtained (for instance, state the kind of food and the range of the amounts given) and one should conduct further experiments.

Sec. 23.9   Regression Analysis. Fitting Straight Lines

Problem Set 23.9. Page 1150

1. Linear regression (straight line). The regression line (2) is

$$y = k_0 + k_1 x.$$

Its coefficients k_0 and k_1 are obtained by solving the normal equations (10). For setting up these equations, calculate

$$n = 4 \qquad \text{the number of sample values (4 pairs)}$$

$$\sum x_j = 30 \qquad \text{the sum of the } x\text{-values in the sample}$$

$$\sum x_j^2 = 306 \quad \text{the sum of their squares}$$

$$\sum y_j = 119 \quad \text{the sum of the } y\text{-values}$$

$$\sum x_j y_j = 1141 \quad \text{the sum of the products of the } x\text{-and } y\text{-values}$$

Hence the linear system of equations (10) in the unknowns k_0 and k_1 takes the form

$$4k_0 + 30k_1 = 119$$
$$30k_0 + 306k_1 = 1141.$$

You can solve it by elimination or by Cramer's rule (p. 342), obtaining

$$k_0 = 6.74074, \qquad k_1 = 3.0679.$$

Hence the regression line is

$$y = 6.74074 + 3.0679x.$$

The figure shows this line as well as the given data as points in the xy-plane. They lie close to the line, which justifies the use of a straight line in this regression analysis. (Note that the scale on the y-axis is different from that on the x-axis.)

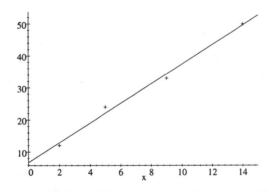

Section 23.9. Problem 1. Data and regression line

7. Confidence interval for the slope κ_1 of the regression line. So far, probability has not been involved in connection with the least squares principle, which is a *geometrical* principle. If you want to know to what extent you can "trust" the sample regression line, in particular its slope k_1, you may determine a confidence interval for the slope κ_1 of the regression line of the population, but this requires that you make assumptions involving probability, for instance, that the random variable Y for which y is an observed value, is normal for each fixed x (see Assumption (A2) on p. 1148) and that you have independence in sampling (see Assumption (A3)). For solving the problem, proceed according to Table 23.12 on p. 1149, as follows.

1st Step. $\gamma = 95\%$ is required.

2nd Step. Determine c from (13), that is,

$$F(c) = \frac{1}{2}(1 + \gamma) = 0.975,$$

and the t-table (Table A9) in Appendix 5. Since the given sample consists of $n = 4$ pairs of values, you have to look in the column for $n - 2 = 2$ degrees of freedom, where you will find the value 4.30. This is c.

3rd Step. In the necessary calculations you can use values from Prob. 1. Calculate $3s_x^2$ from (9), using $\sum x_j = 30$ and $\sum x_j^2 = 306$. This gives (multiply by $n - 1 = 3$ on both sides)

$$3s_x^2 = 306 - \frac{1}{4} \cdot 30^2 = 81.$$

Next use $\sum y_j = 119$ and $\sum x_j y_j = 1141$ for obtaining from (8)

$$3s_{xy} = 1141 - \frac{1}{4} \cdot 30 \cdot 119 = 248.5.$$

In (14) you will need

$$\sum y_j^2 = 4309,$$

as calculated from the sample. With this you obtain from (14)

$$3s_y^2 = 4309 - \frac{1}{4} \cdot 119^2 = 768.75.$$

In (15) you will need $k_1^2 = 3.0679^2 = 9.4120$. With this, (15) gives

$$q_0 = 768.75 - 9.4120 \cdot 81 = 6.377.$$

4th Step. Now $c = 4.30$ (see before) is needed. You finally calculate

$$K = 4.30\sqrt{\frac{6.378}{281}} = 0.8531.$$

This is half the length of the confidence interval. The midpoint is the slope $k_1 = 3.0679$ of the sample regression line (see before). Hence from (16) in Table 23.12 in Sec. 23.9 you obtain the answer

$$\text{CONF}_{0.95}(3.0679 - 0.8531 \leqq \kappa_1 \leqq 3.0679 + 0.8531),$$

that is,

$$\text{CONF}_{0.95}(2.2148 \leqq \kappa_1 \leqq 3.9210).$$

It may surprise you that the points in the figure to Prob. 1 lie close to the regression line, but, nevertheless, the confidence interval for the slope of the regression line is relatively large, perhaps larger than you had expected. However, this is due to the fact that the sample used is very small ($n = 4$).